D1201994

Applied Probability and Stochastic Processes

Second Edition

Richard M. Feldman · Ciriaco Valdez-Flores

Applied Probability and Stochastic Processes

Second Edition

Richard M. Feldman
Texas A&M University
Industrial and Systems
Engineering Department
College Station
Texas 77843-3131
USA
richf@tamu.edu

Ciriaco Valdez-Flores
Sielken & Associates
Consulting, Inc.
3833 Texas Avenue
Bryan
Texas 77802
USA
ciriaco@sielkenassociates.com

ISBN 978-3-642-05155-5 e-ISBN 978-3-642-05158-6
DOI 10.1007/978-3-642-05158-6
Springer Heidelberg Dordrecht London New York

Library of Congress Control Number: 2009940615

Originally published by PWS Publishing/Thomson Publishing, USA, 1995

Cover design: eStudio Calamar S.L.

Printed on acid-free paper

Springer is part of Springer Science+Business Media (www.springer.com)

This book is dedicated to our wives, Alice Feldman and Nancy Vivas-Valdez, whose patience, love, and support keep us going and make life worthwhile.

Preface

This book is a result of teaching stochastic processes to junior and senior undergraduates and beginning graduate students over many years. In teaching such a course, we have realized a need to furnish students with material that gives a mathematical presentation while at the same time providing proper foundations to allow students to build an intuitive feel for probabilistic reasoning. We have tried to maintain a balance in presenting advanced but understandable material that sparks an interest and challenges students, without the discouragement that often comes as a consequence of not understanding the material. Our intent in this text is to develop stochastic processes in an elementary but mathematically precise style and to provide sufficient examples and homework exercises that will permit students to understand the range of application areas for stochastic processes.

We also practice active learning in the classroom. In other words, we believe that the traditional practice of lecturing continuously for 50 to 75 minutes is not a very effective method for teaching. Students should somehow engage in the subject matter during the teaching session. One effective method for active learning is, after at most 20 minutes of lecture, to assign a small example problem for the students to work and one important tool that the instructor can utilize is the computer. Sometimes we are fortunate to lecture students in a classroom containing computers with a spreadsheet program, usually Microsoft's Excel. For a course dealing with random variables, Excel is an ideal tool that can be used both within the classroom and for homework to demonstrate probabilistic concepts. In order to take full advantage of this, we have moved the chapter on simulation to the second chapter in the book. It is not necessary to cover all the sections of the simulation chapter, but we suggest covering at least Sects. 2.1 and 2.3 so that simulation can then be easily used with Excel throughout the course to demonstrate random processes and used during the lecture to actively engage the students in the lecture material.

The only prerequisites for an undergraduate course using this textbook is a previous course covering calculus-based probability and statistics and familiarity with basic matrix operations. Included at the end of most chapters is an appendix covering the use of Excel for the problems of the chapter; thus, a familiarity with Excel would be helpful but not necessary. For students needing a review of matrices, some of the basic operations are given in Appendix A at the end of the book.

This book could also be used for an introductory course to stochastic processes at the graduate level, in which case an additional prerequisite of linear programming should be required if the chapter on Markov decision theory is to be covered. It would also be helpful to expect graduate students to be competent programmers in some scientific programming language. There are two chapters covering advanced topics that would be skipped in an undergraduate course: Chap. 12 — Markov Decision Theory — and Chap. 13 — Advanced Queueing Theory. Knowledge of linear programming is necessary for Chap. 12, and a programming language or VBA would be very helpful in implementing the concepts in Chap. 13.

The book is organized as follows: The first three chapters are background material to be used throughout the book. The first chapter is a review of probability. It is intended simply as a review; the material is too terse if students have not previously been exposed to probability. However, our experience is that most students do not learn probability until two or three exposures to it, so this chapter should serve as an excellent summary and review for most students. The second chapter is an introduction to simulation so that it can be used to demonstrate concepts covered in future chapters. Included in this chapter is material covering random number generation (Sect. 2.2) which we sometimes skip when teaching our undergraduate course in stochastic processes since Excel has its own generator. The third chapter is a review of statistics which is only presented because some statistical concepts will be covered in later chapters, but this in not central to the text. We expect students to have already been exposed to this material and we generally skip this chapter for our undergraduate course and refer to it as needed.

The fourth chapter begins the introduction to random processes and covers the basic concepts of Poisson processes. The fifth chapter covers Markov chains in some detail. The approach in this chapter and in Chap. 6 is similar to the approach taken by Çinlar (*Introduction to Stochastic Processes*, Prentice-Hall, 1975). The homework problems cover a wide variety of modeling situations as an attempt is made to begin the development of "modelers". Chapter 6 is an introduction to continuous time Markov processes. The major purpose of the chapter is to give the tools necessary for the development of queueing models; therefore, the emphasis in the chapter is on steady-state analyses. The final section of Chapter 6 is a brief treatment of the time-dependent probabilities for Markov processes. This final section can be skipped for most undergraduate classes. Queueing theory is covered in Chaps. 7 and 8, where Chap. 7 deals with the basics of single queues and Chap. 8 introduces queueing networks. As in the Markov chain chapter, an attempt has been made to develop a wide variety of modeling situations through the homework problems.

Chapter 9 has two sections: the first deals with the specifics of event-driven simulations while the second introduces some of the statistical issues for output analysis. If the mechanical details of simulation (like future events lists) are not of interest to the instructor, the first section of Chap. 9 can be skipped with no loss of continuity. Chapter 2 together with the second section of Chap. 9 should yield an excellent introduction to simulation. No programming language is assumed since our purpose is not to produce experts in simulation, but simply to introduce the concepts and develop student interest in simulation. If simulation is covered adequately by other courses, Chap. 9 can be easily skipped.

Chapters 10 and 11 introduce a change in tactics and present two chapters dealing with specific problem domains: the first is inventory and the second is replacement. Applied probability can be taught as a collection of techniques useful for a wide variety of applications, or it can be taught as various application areas for which randomness plays an important role. The first nine chapters focus on particular techniques with some applications being emphasized through examples and the homework problems. The next two chapters focus on two problem domains that have been historically important in applied probability and stochastic processes. It was difficult to decide on the proper location for these two chapters. There is some Markov chain references in the last section of the inventory chapter; therefore, it is best to start with Chaps. 1, 2, 4, and 5 for most courses. After covering Markov chains, it would be appropriate and easily understood if the next chapter taught was either Markov processes (Chap. 6), inventory theory (Chap. 10), or replacement theory (Chap. 11). It simply depends on the inclination of the instructor.

Chapters 12 and 13 are only included for advanced students. Chapter 12 covers Markov decision processes, and Chap. 13 is a presentation of phase-type distributions and the matrix geometric approach to queueing systems adopted from the work of Neuts (*Matrix-Geometric Solutions in Stochastic Models*, Johns Hopkins University Press, 1981).

We are indebted to many of our colleagues for their invaluable assistance and professional support. For this second edition, we especially thank Guy L. Curry and Don T. Phillips for their contributions and encouragement. We are also grateful to Brett Peters, the department head of Industrial and Systems Engineering at Texas A&M University, and to Robert L. Sielken of Sielken & Associates Consulting for their continuous support. Section 1.6 and parts of Chap. 8 come from Guy Curry and can also be found in the text by Curry and Feldman, *Manufacturing Systems Modeling and Analysis*, Springer-Verlag, 2009. Finally, we acknowledge our thanks through the words of the psalmist, "Give thanks to the Lord, for He is good; His love endures forever." (Psalms 107:1, NIV)

College Station, Texas
December 2009

Richard M. Feldman
Ciriaco Valdez-Flores

Contents

Chapter 1
Basic Probability Review

The background material needed for this textbook is a general understanding of probability and the properties of various distributions; thus, before discussing the modeling of the various manufacturing and production systems, it is important to review the fundamental concepts of basic probability. This material is not intended to teach probability theory, but it is used for review and to establish a common ground for the notation and definitions used throughout the book. For those already familiar with probability, this chapter can easily be skipped.

1.1 Basic Definitions

To understand probability , it is best to envision an experiment for which the outcome (result) is unknown. All possible outcomes must be defined and the collection of these outcomes is called the sample space. Probabilities are assigned to subsets of the sample space, called events. We shall give the rigorous definition for probability. However, the reader should not be discouraged if an intuitive understanding is not immediately acquired. This takes time and the best way to understand probability is by working problems.

Definition 1.1. An element of a *sample space* is an *outcome*. A set of outcomes, or equivalently a subset of the sample space, is called an *event*. □

Definition 1.2. A *probability space* is a three-tuple $(\Omega, \mathscr{F}, \Pr)$ where Ω is a sample space, $\mathscr{F}$ is a collection of events from the sample space, and Pr is a probability measure that assigns a number to each event contained in $\mathscr{F}$. Furthermore, Pr must satisfy the following conditions, for each event A, B within $\mathscr{F}$:

- $\Pr(\Omega) = 1$,
- $\Pr(A) \geq 0$,
- $\Pr(A \cup B) = \Pr(A) + \Pr(B)$ if $A \cap B = \phi$, where ϕ denotes the empty set,
- $\Pr(A^c) = 1 - \Pr(A)$, where A^c is the complement of A. □

R.M. Feldman, C. Valdez-Flores, *Applied Probability and Stochastic Processes*, 2nd ed., DOI 10.1007/978-3-642-05158-6_1,

It should be noted that the collection of events, $\mathscr{F}$, in the definition of a probability space must satisfy some technical mathematical conditions that are not discussed in this text. If the sample space contains a finite number of elements, then $\mathscr{F}$ usually consists of all the possible subsets of the sample space. The four conditions on the probability measure Pr should appeal to one's intuitive concept of probability. The first condition indicates that something from the sample space must happen, the second condition indicates that negative probabilities are illegal, the third condition indicates that the probability of the union of two disjoint (or mutually exclusive) events is the sum of their individual probabilities and the fourth condition indicates that the probability of an event is equal to one minus the probability of its complement (all other events). The fourth condition is actually redundant but it is listed in the definitions because of its usefulness.

A probability space is the full description of an experiment; however, it is not always necessary to work with the entire space. One possible reason for working within a restricted space is because certain facts about the experiment are already known. For example, suppose a dispatcher at a refinery has just sent a barge containing jet fuel to a terminal 800 miles down river. Personnel at the terminal would like a prediction on when the fuel will arrive. The experiment consists of all possible weather, river, and barge conditions that would affect the travel time down river. However, when the dispatcher looks outside it is raining. Thus, the original probability space can be restricted to include only rainy conditions. Probabilities thus restricted are called conditional probabilities according to the following definition.

Definition 1.3. Let $(\Omega, \mathscr{F}, \Pr)$ be a probability space where A and B are events in $\mathscr{F}$ with $\Pr(B) > 0$. The *conditional probability* of A given B, denoted $\Pr(A|B)$, is

$$\Pr(A|B) = \frac{\Pr(A \cap B)}{\Pr(B)} .$$

□

Venn diagrams are sometimes used to illustrate relationships among sets. In the diagram of Fig. 1.1, assume that the probability of a set is proportional to its area. Then the value of $\Pr(A|B)$ is the proportion of the area of set B that is occupied by the set $A \cap B$.

Example 1.1. A telephone manufacturing company makes radio phones and plain phones and ships them in boxes of two (same type in a box). Periodically, a quality control technician randomly selects a shipping box, records the type of phone in the box (radio or plain), and then tests the phones and records the number that were defective. The sample space is

$$\Omega = \{(r,0),(r,1),(r,2),(p,0),(p,1),(p,2)\} ,$$

where each outcome is an ordered pair; the first component indicates whether the phones in the box are the radio type or plain type and the second component gives the number of defective phones. The set $\mathscr{F}$ is the set of all subsets, namely,

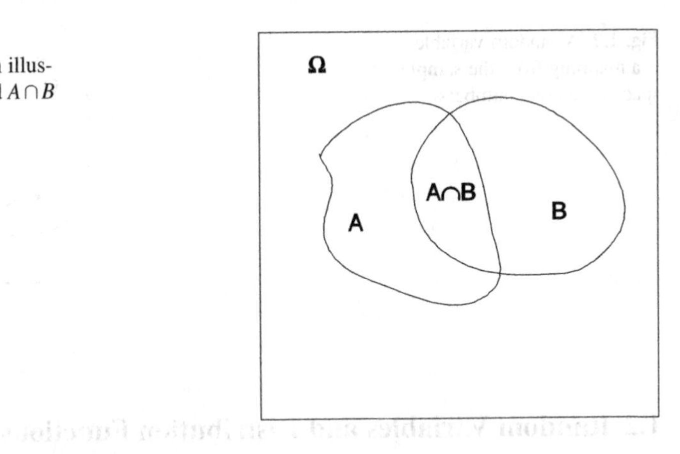

Fig. 1.1 Venn diagram illustrating events A, B, and $A \cap B$

$$\mathscr{F} = \{\phi, \{(r,0)\}, \{(r,1)\}, \{(r,0),(r,1)\}, \cdots, \Omega\} .$$

There are many legitimate probability laws that could be associated with this space. One possibility is

$$\Pr\{(r,0)\} = 0.45 , \ \Pr\{(p,0)\} = 0.37 ,$$
$$\Pr\{(r,1)\} = 0.07 , \ \Pr\{(p,1)\} = 0.08 ,$$
$$\Pr\{(r,2)\} = 0.01 , \ \Pr\{(p,2)\} = 0.02 .$$

By using the third property in Definition 1.2, the probability measure can be extended to all events; for example, the probability that a box is selected that contains radio phones and at most one phone is defective is given by

$$\Pr\{(r,0),(r,1)\} = 0.52 .$$

Now let us assume that a box has been selected and opened. We observe that the two phones within the box are radio phones, but no test has yet been made on whether or not the phones are defective. To determine the probability that at most one phone is defective in the box containing radio phones, define the event A to be the set $\{(r,0),(r,1),(p,0),(p,1)\}$ and the event B to be $\{(r,0),(r,1),(r,2)\}$. In other words, A is the event of having at most one defective phone, and B is the event of having a box of radio phones. The probability statement can now be written as

$$\Pr\{A|B\} = \frac{\Pr(A \cap B)}{\Pr(B)} = \frac{\Pr\{(r,0),(r,1)\}}{\Pr\{(r,0),(r,1),(r,2)\}} = \frac{0.52}{0.53} = 0.991 .$$

□

• *Suggestion: Do Problems 1.1–1.2 and 1.19.*

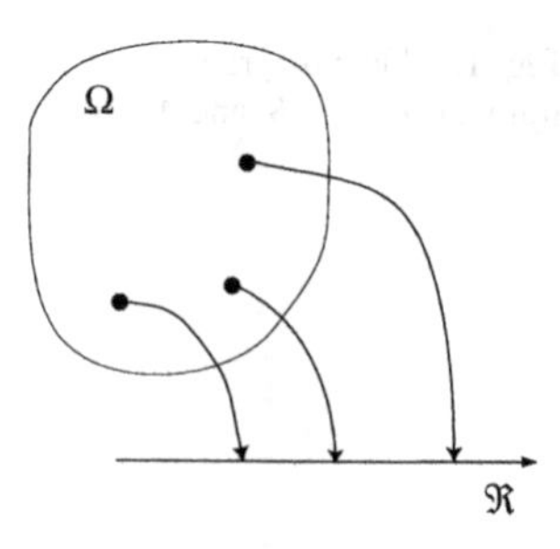

Fig. 1.2 A random variable is a mapping from the sample space to the real numbers

1.2 Random Variables and Distribution Functions

It is often cumbersome to work with the outcomes directly in mathematical terms. Random variables are defined to facilitate the use of mathematical expressions and to focus only on the outcomes of interest.

Definition 1.4. A *random variable* is a function that assigns a real number to each outcome in the sample space. □

Figure 1.2 presents a schematic illustrating a random variable. The name "random variable" is actually a misnomer, since it is not random and is not a variable. As illustrated in the figure, the random variable simply maps each point (outcome) in the sample space to a number on the real line[1].

Revisiting Example 1.1, let us assume that management is primarily interested in whether or not at least one defective phone is in a shipping box. In such a case a random variable D might be defined such that it is equal to zero if all the phones within a box are good and equal to 1 otherwise; that is,

$$\begin{aligned} D(r,0) &= 0\,, \quad D(p,0) = 0\,, \\ D(r,1) &= 1\,, \quad D(p,1) = 1\,, \\ D(r,2) &= 1\,, \quad D(p,2) = 1\,. \end{aligned}$$

The set $\{D = 0\}$ refers to the set of all outcomes for which $D = 0$ and a legitimate probability statement would be

$$\Pr\{D = 0\} = \Pr\{(r,0),(p,0)\} = 0.82\,.$$

To aid in the recognition of random variables, the notational convention of using only capital Roman letters (or possibly Greek letters) for random variables is followed. Thus, if you see a lower case Roman letter, you know immediately that it can not be a random variable.

[1] Technically, the space into which the random variable maps the sample space may be more general than the real number line, but for our purposes, the real numbers will be sufficient.

Random variables are either discrete or continuous depending on their possible values. If the possible values can be counted, the random variable is called discrete; otherwise, it is called continuous. The random variable D defined in the previous example is discrete. To give an example of a continuous random variable, define T to be a random variable that represents the length of time that it takes to test the phones within a shipping box. The range of possible values for T is the set of all positive real numbers, and thus T is a continuous random variable.

A cumulative distribution function (CDF) is often used to describe the probability measure underlying the random variable. The cumulative distribution function (usually denoted by a capital Roman letter or a Greek letter) gives the probability accumulated up to and including the point at which it is evaluated.

Definition 1.5. The function F is the *cumulative distribution function* for the random variable X if

$$F(a) = \Pr\{X \le a\}$$

for all real numbers a. □

The CDF for the random variable D defined above is

$$F(a) = \begin{cases} 0 & \text{for } a < 0 \\ 0.82 & \text{for } 0 \le a < 1 \\ 1.0 & \text{for } a \ge 1\,. \end{cases} \tag{1.1}$$

Figure 1.3 gives the graphical representation for F. The random variable T defined to represent the testing time for phones within a randomly chosen box is continuous and there are many possibilities for its probability measure since we have not yet defined its probability space. As an example, the function G (see Fig. 1.10) is the cumulative distribution function describing the randomness that might be associated with T:

$$G(a) = \begin{cases} 0 & \text{for } a < 0 \\ 1 - \mathrm{e}^{-2a} & \text{for } a \ge 0\,. \end{cases} \tag{1.2}$$

Property 1.1. *A cumulative distribution function F has the following properties:*

- $\lim_{a \to -\infty} F(a) = 0$,
- $\lim_{a \to +\infty} F(a) = 1$,
- $F(a) \le F(b)$ *if* $a < b$,
- $\lim_{a \to b^+} F(a) = F(b)$.

The first and second properties indicate that the graph of the cumulative distribution function always begins on the left at zero and limits to one on the right. The third property indicates that the function is nondecreasing. The fourth property indicates that the cumulative distribution function is right-continuous. Since the distribution function is monotone increasing, at each discontinuity the function value is defined

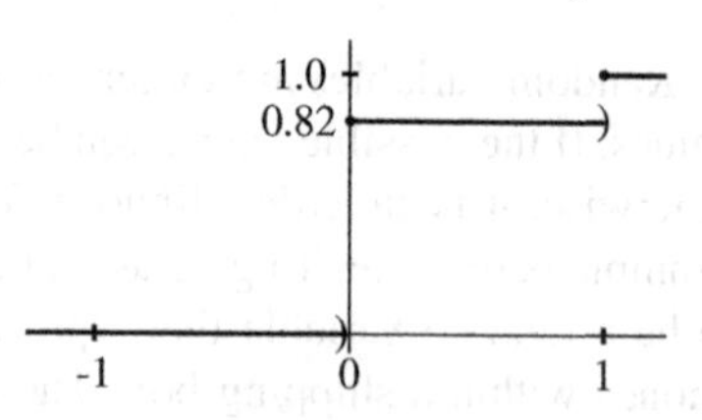

Fig. 1.3 Cumulative distribution function for Eq. (1.1) for the discrete random variable D

by the larger of two limits: the limit value approaching the point from the left and the limit value approaching the point from the right.

It is possible to describe the random nature of a discrete random variable by indicating the size of jumps in its cumulative distribution function. Such a function is called a probability mass function (denoted by a lower case letter) and gives the probability of a particular value occurring.

Definition 1.6. The function f is the *probability mass function* (pmf) of the discrete random variable X if

$$f(k) = \Pr\{X = k\}$$

for every k in the range of X. □

If the pmf is known, then the cumulative distribution function is easily found by

$$\Pr\{X \leq a\} = F(a) = \sum_{k \leq a} f(k) \,. \tag{1.3}$$

The situation for a continuous random variable is not quite as easy because the probability that any single given point occurs must be zero. Thus, we talk about the probability of an interval occurring. With this in mind, it is clear that a mass function is inappropriate for continuous random variables; instead, a probability density function (denoted by a lower case letter) is used.

Definition 1.7. The function g is called the *probability density function* (pdf) of the continuous random variable Y if

$$\int_a^b g(u)\mathrm{d}u = \Pr\{a \leq Y \leq b\}$$

for all a, b in the range of Y. □

From Definition 1.7 it should be seen that the pdf is the derivative of the cumulative distribution function and

$$G(a) = \int_{-\infty}^{a} g(u)\mathrm{d}u \,. \tag{1.4}$$

The cumulative distribution functions for the example random variables D and T are defined in Eqs. (1.1 and 1.2). We complete that example by giving the pmf for D and the pdf for T as follows:

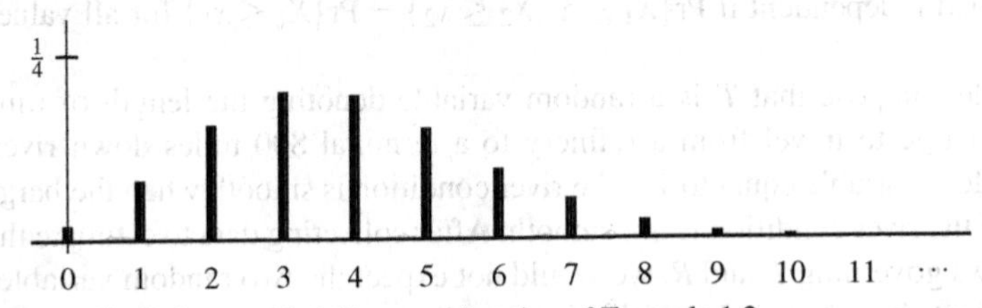

Fig. 1.4 The Poisson probability mass function of Example 1.2

$$f(k) = \begin{cases} 0.82 & \text{for } k = 0 \\ 0.18 & \text{for } k = 1 \,. \end{cases} \tag{1.5}$$

and

$$g(a) = \begin{cases} 2\mathrm{e}^{-2a} & \text{for } a \geq 0 \\ 0 & \text{otherwise} \,. \end{cases} \tag{1.6}$$

Example 1.2. Discrete random variables need not have finite ranges. A classical example of a discrete random variable with an infinite range is due to Rutherford, Chadwick, and Ellis from 1920 [10, pp. 209–210]. An experiment was performed to determine the number of α-particles emitted by a radioactive substance in 7.5 seconds. The radioactive substance was chosen to have a long half-life so that the emission rate would be constant. After 2608 experiments, it was found that the number of emissions in 7.5 seconds was a random variable, N, whose pmf could be described by

$$\Pr\{N = k\} = \frac{(3.87)^k \mathrm{e}^{-3.87}}{k!} \quad \text{for } k = 0, 1, \cdots \,.$$

It is seen that the discrete random variable N has a countably infinite range and the infinite sum of its pmf equals one. In fact, this distribution is fairly important and will be discussed later under the heading of the Poisson distribution. Figure 1.4 shows its pmf graphically. □

The notion of independence is very important when dealing with more than one random variable. Although we shall postpone the discussion on multivariate distribution functions until Sect. 1.5, we introduce the concept of independence at this point.

Definition 1.8. The random variables $X_1, \cdots, X_n$ are *independent* if

$$\Pr\{X_1 \leq x_1, \cdots, X_n \leq x_n\} = \Pr\{X_1 \leq x_1\} \times \cdots \times \Pr\{X_n \leq x_n\}$$

for all possible values of $x_1, \cdots, x_n$. □

Conceptually, random variables are independent if knowledge of one (or more) random variable does not "help" in making probability statements about the other random variables. Thus, an alternative definition of independence could be made using conditional probabilities (see Definition 1.3) where the random variables X_1

and X_2 are called independent if $\Pr\{X_1 \leq x_1 | X_2 \leq x_2\} = \Pr\{X_1 \leq x_1\}$ for all values of x_1 and x_2.

For example, suppose that T is a random variable denoting the length of time it takes for a barge to travel from a refinery to a terminal 800 miles down river, and R is a random variable equal to 1 if the river condition is smooth when the barge leaves and 0 if the river condition is not smooth. After collecting data to estimate the probability laws governing T and R, we would not expect the two random variables to be independent since knowledge of the river conditions would help in determining the length of travel time.

One advantage of independence is that it is easier to obtain the distribution for sums of random variables when they are independent than when they are not independent. When the random variables are continuous, the pdf of the sum involves an integral called a *convolution*.

Property 1.2. *Let X_1 and X_2 be independent continuous random variables with pdf's given by $f_1(\cdot)$ and $f_2(\cdot)$. Let $Y = X_1 + X_2$, and let $h(\cdot)$ be the pdf for Y. The pdf for Y can be written, for all y, as*

$$h(y) = \int_{-\infty}^{\infty} f_1(y-x) f_2(x) \mathrm{d}x \,.$$

Furthermore, if X_1 and X_2 are both nonnegative random variables, then

$$h(y) = \int_{0}^{y} f_1(y-x) f_2(x) \mathrm{d}x \,.$$

Example 1.3. Our electronic equipment is highly sensitive to voltage fluctuations in the power supply so we have collected data to estimate when these fluctuations occur. After much study, it has been determined that the time between voltage spikes is a random variable with pdf given by (1.6), where the unit of time is hours. Furthermore, it has been determined that the random variables describing the time between two successive voltage spikes are independent. We have just turned the equipment on and would like to know the probability that within the next 30 minutes at least two spikes will occur.

Let X_1 denote the time interval from when the equipment is turned on until the first voltage spike occurs, and let X_2 denote the time interval from when the first spike occurs until the second occurs. The question of interest is to find $\Pr\{Y \leq 0.5\}$, where $Y = X_1 + X_2$. Let the pdf for Y be denoted by $h(\cdot)$. Property 1.2 yields

$$\begin{aligned} h(y) &= \int_0^y 4\mathrm{e}^{-2(y-x)} \mathrm{e}^{-2x} \mathrm{d}x \\ &= 4\mathrm{e}^{-2y} \int_0^y \mathrm{d}x = 4y\mathrm{e}^{-2y} \,, \end{aligned}$$

for $y \geq 0$. The pdf of Y is now used to answer our question, namely,

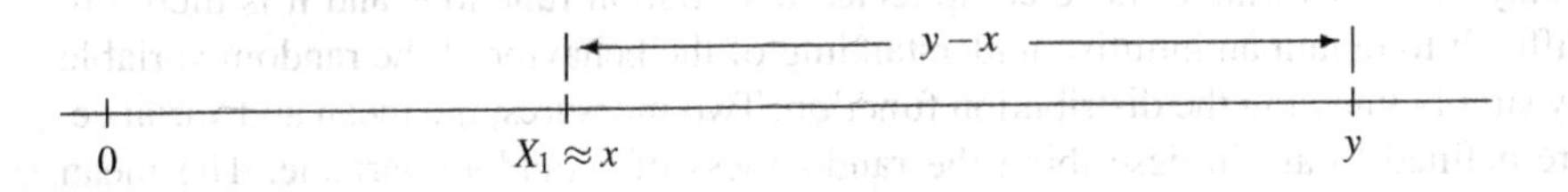

Fig. 1.5 Time line illustrating the convolution

$$\Pr\{Y \le 0.5\} = \int_0^{0.5} h(y)\mathrm{d}y = \int_0^{0.5} 4y\mathrm{e}^{-2y}\mathrm{d}y = 0.264\,.$$

□

It is also interesting to note that the convolution can be used to give the cumulative distribution function if the first pdf in the above property is replaced by the CDF; in other words, for *nonnegative* random variables we have

$$H(y) = \int_0^{y} F_1(y-x) f_2(x)\mathrm{d}x\,. \tag{1.7}$$

Applying (1.7) to our voltage fluctuation question yields

$$\Pr\{Y \le 0.5\} \equiv H(0.5) = \int_0^{0.5} (1-\mathrm{e}^{-2(0.5-x)})2\mathrm{e}^{-2x}\mathrm{d}x = 0.264\,.$$

We rewrite the convolution of Eq. (1.7) slightly to help in obtaining an intuitive understanding of why the convolution is used for sums. Again, assume that X_1 and X_2 are independent, nonnegative random variables with pdf's f_1 and f_2, then

$$\Pr\{X_1 + X_2 \le y\} = \int_0^{y} F_2(y-x) f_1(x)\mathrm{d}x\,.$$

The interpretation of $f_1(x)\mathrm{d}x$ is that it represents the probability that the random variable X_1 falls in the interval $(x, x+\mathrm{d}x)$ or, equivalently, that X_1 is approximately x. Now consider the time line in Fig. 1.5. For the sum to be less than y, two events must occur: first, X_1 must be some value (call it x) that is less than y; second, X_2 must be less than the remaining time that is $y-x$. The probability of the first event is approximately $f_1(x)\mathrm{d}x$, and the probability of the second event is $F_2(y-x)$. Since the two events are independent, they are multiplied together; and since the value of x can be any number between 0 and y, the integral is from 0 to y.

- *Suggestion: Do Problems 1.3–1.6.*

1.3 Mean and Variance

Many random variables have complicated distribution functions and it is therefore difficult to obtain an intuitive understanding of the behavior of the random variable by simply knowing the distribution function. Two measures, the mean and variance, are defined to aid in describing the randomness of a random variable. The mean equals the arithmetic average of infinitely many observations of the random variable and the variance is an indication of the variability of the random variable. To illustrate this concept we use the square root of the variance which is called the *standard deviation*. In the 19*th* century, the Russian mathematician P. L. Chebyshev showed that for any given distribution, *at least* 75% of the time the observed value of a random variable will be within two standard deviations of its mean and *at least* 93.75% of the time the observed value will be within four standard deviations of the mean. These are general statements, and specific distributions will give much tighter bounds. (For example, a commonly used distribution is the normal "bell shaped" distribution. With the normal distribution, there is a 95.44% probability of being within two standard deviations of the mean.) Both the mean and variance are defined in terms of the expected value operator, that we now define.

Definition 1.9. Let h be a function defined on the real numbers and let X be a random variable. The *expected value* of $h(X)$ is given, for X discrete, by

$$E[h(X)] = \sum_k h(k) f(k)$$

where f is its pmf, and for X continuous, by

$$E[h(X)] = \int_{-\infty}^{\infty} h(s) f(s) \mathrm{d}s$$

where f is its pdf. □

Example 1.4. A supplier sells eggs by the carton containing 144 eggs. There is a small probability that some eggs will be broken and he refunds money based on broken eggs. We let B be a random variable indicating the number of broken eggs per carton with a pmf given by

k	$f(k)$
0	0.779
1	0.195
2	0.024
3	0.002

A carton sells for \$4.00, but a refund of 5 cents is made for each broken egg. To determine the expected income per carton, we define the function h as follows

k	$h(k)$
0	4.00
1	3.95
2	3.90
3	3.85

Thus, $h(k)$ is the net revenue obtained when a carton is sold containing k broken eggs. Since it is not known ahead of time how many eggs are broken, we are interested in determining the *expected* net revenue for a carton of eggs. Definition 1.9 yields

$$E[h(B)] = 4.00 \times 0.779 + 3.95 \times 0.195 + 3.90 \times 0.024 + 3.85 \times 0.002 = 3.98755 .$$

□

The expected value operator is a linear operator, and it is not difficult to show the following property.

Property 1.3. *Let X and Y be two random variables with c being a constant, then*

- $E[c] = c$,
- $E[cX] = cE[X]$,
- $E[X+Y] = E[X] + E[Y]$.

In the egg example since the cost per broken egg is a constant ($c = 0.05$), the expected revenue per carton could be computed as

$$\begin{aligned} E[4.0 - 0.05B] &= 4.0 - 0.05E[B] \\ &= 4.0 - 0.05\,(\,0 \times 0.779 + 1 \times 0.195 + 2 \times 0.024 + 3 \times 0.002\,) \\ &= 3.98755 . \end{aligned}$$

The expected value operator provides us with the procedure to determine the mean and variance.

Definition 1.10. The *mean*, μ or $E[X]$, and *variance*, σ^2 or $V[X]$, of a random variable X are defined as

$$\mu = E[X], \quad \sigma^2 = E[(X-\mu)^2] ,$$

respectively. The *standard deviation*, σ, is the square root of the variance. □

Property 1.4. *The following are often helpful as computational aids:*

- $V[X] = \sigma^2 = E[X^2] - \mu^2$
- $V[cX] = c^2 V[X]$
- *If* $X \geq 0$, $E[X] = \int_0^\infty [1 - F(s)]\mathrm{d}s$ *where* $F(x) = \Pr\{X \leq x\}$
- *If* $X \geq 0$, *then* $E[X^2] = 2\int_0^\infty s[1 - F(s)]\mathrm{d}s$ *where* $F(x) = \Pr\{X \leq x\}$.

Example 1.5. The mean and variance calculations for a discrete random variable can be easily illustrated by defining the random variable N to be the number of defective phones within a randomly chosen box from Example 1.1. In other words, N has the pmf given by

$$\Pr\{N = k\} = \begin{cases} 0.82 & \text{for } k = 0 \\ 0.15 & \text{for } k = 1 \\ 0.03 & \text{for } k = 2\,. \end{cases}$$

The mean and variance is, therefore, given by

$$\begin{aligned} E[N] &= 0 \times 0.82 + 1 \times 0.15 + 2 \times 0.03 \\ &= 0.21, \end{aligned}$$

$$\begin{aligned} V[N] &= (0 - 0.21)^2 \times 0.82 + (1 - 0.21)^2 \times 0.15 + (2 - 0.21)^2 \times 0.03 \\ &= 0.2259\,. \end{aligned}$$

Or, an easier calculation for the variance (Property 1.4) is

$$\begin{aligned} E[N^2] &= 0^2 \times 0.82 + 1^2 \times 0.15 + 2^2 \times 0.03 \\ &= 0.27 \end{aligned}$$

$$\begin{aligned} V[N] &= 0.27 - 0.21^2 \\ &= 0.2259\,. \end{aligned}$$

□

Example 1.6. The mean and variance calculations for a continuous random variable can be illustrated with the random variable T whose pdf was given by Eq. 1.6. The mean and variance is therefore given by

$$E[T] = \int_0^\infty 2s\mathrm{e}^{-2s}\mathrm{d}s = 0.5\,,$$

$$V[T] = \int_0^\infty 2(s - 0.5)^2 \mathrm{e}^{-2s}\mathrm{d}s = 0.25\,.$$

Or, an easier calculation for the variance (Property 1.4) is

$$E[T^2] = \int_0^\infty 2s^2 e^{-2s} ds = 0.5 ,$$

$$V[T] = 0.5 - 0.5^2 = 0.25 .$$

□

The final definition in this section is used often as a descriptive statistic to give an intuitive feel for the variability of processes.

Definition 1.11. The *squared coefficient of variation*, C^2, of a nonnegative random variable T is the ratio of the the variance to the mean squared; that is,

$$C^2[T] = \frac{V[T]}{E[T]^2} .$$

□

- *Suggestion: Do Problems 1.7–1.14.*

1.4 Important Distributions

There are many distribution functions that are used so frequently that they have become known by special names. In this section, some of the major distribution functions are given. The student will find it helpful in years to come if these distributions are committed to memory. There are several textbooks (for example, see [9, chap. 6] for an excellent summary and [6], [7], [8] for in-depth descriptions) that give more complete descriptions of distributions, and we recommend gaining a familiarity with a variety of distribution functions before any serious modeling is attempted.

Uniform-Discrete: The random variable N has a discrete uniform distribution if there are two integers a and b such that the pmf of N can be written as

$$f(k) = \frac{1}{b-a+1} \quad \text{for } k = a, a+1, \cdots, b . \tag{1.8}$$

Then,

$$E[N] = \frac{a+b}{2}; \quad V[N] = \frac{(b-a+1)^2 - 1}{12} .$$

Example 1.7. Consider rolling a fair die. Figure 1.6 shows the uniform pmf for the "number of dots" random variable. Notice in the figure that, as the name "uniform" implies, all the probabilities are the same. □

Bernoulli: (Named after Jacob Bernoulli, 1654-1705.) The random variable N has a Bernoulli distribution if there is a number $0 < p < 1$ such that the pmf of N can be written as

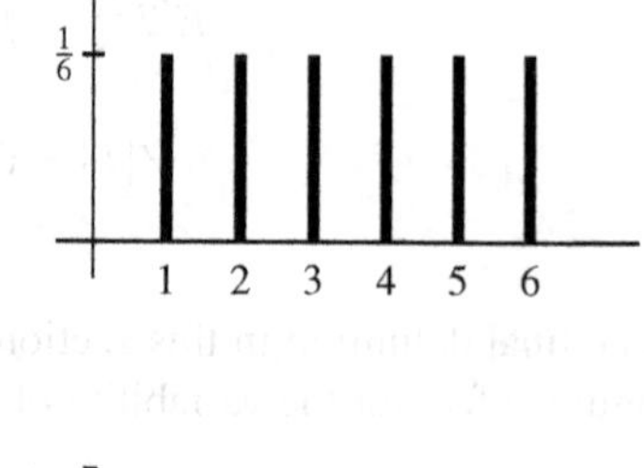

Fig. 1.6 A discrete uniform probability mass function

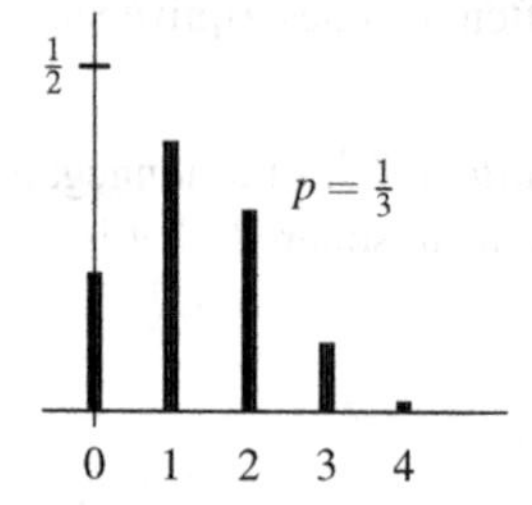

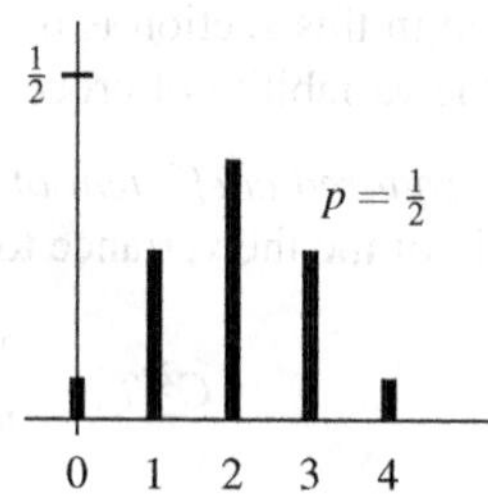

Fig. 1.7 Two binomial probability mass functions

$$f(k) = \begin{cases} 1-p & \text{for } k = 0 \\ p & \text{for } k = 1 . \end{cases} \tag{1.9}$$

Then,

$$E[N] = p; \;\; V[N] = p(1-p); \;\; C^2[N] = \frac{1-p}{p} .$$

Binomial: (From Jacob Bernoulli and published posthumously in 1713 by his nephew.) The random variable N has a binomial distribution if there is a number $0 < p < 1$ and a positive integer n such that the pmf of N can be written as

$$f(k) = \frac{n!}{k!(n-k)!} p^k (1-p)^{n-k} \text{ for } k = 0, 1, \cdots, n . \tag{1.10}$$

Then,

$$E[N] = np; \;\; V[N] = np(1-p); \;\; C^2[N] = \frac{1-p}{np} .$$

The number p is often though of as the probability of a success. The binomial pmf evaluated at k thus gives the probability of k successes occurring out of n trials. The binomial random variable with parameters p and n is the sum of n (independent) Bernoulli random variables each with parameter p.

Example 1.8. We are monitoring calls at a switchboard in a large manufacturing firm and have determined that one third of the calls are long distance and two thirds of the calls are local. We have decided to pick four calls at random and would like to know how many calls in the group of four are long distance. In other words, let N be a random variable indicating the number of long distance calls in the group of four. Thus, N is binomial with $n = 4$ and $p = 1/3$. It also happens that in this company, half of the individuals placing calls are women and half are men. We would also like to know how many of the group of four were calls placed by men. Let M denote the number of men placing calls; thus, M is binomial with $n = 4$ and $p = 1/2$. The

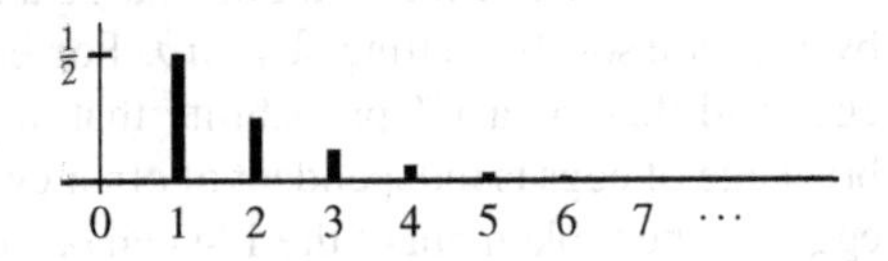

Fig. 1.8 A geometric probability mass function

pmf's for these two random variables are shown in Fig. 1.7. Notice that for $p = 0.5$, the pmf is symmetric, and as p varies from 0.5, the graph becomes skewed. □

Geometric: The random variable N has a geometric distribution if there is a number $0 < p < 1$ such that the pmf of N can be written as

$$f(k) = p(1-p)^{k-1} \text{ for } k = 1,2,\cdots . \tag{1.11}$$

Then,

$$E[N] = \frac{1}{p}; \quad V[N] = \frac{1-p}{p^2}; \quad C^2[N] = 1-p.$$

The idea behind the geometric random variable is that it represents the number of trials until the first success occurs. In other words, p is thought of as the probability of success for a single trial, and we continually perform the trials until a success occurs. The random variable N is then set equal to the number of trial that we had to perform. Note that although the geometric random variable is discrete, its range is infinite.

Example 1.9. A car saleswoman has made a statistical analysis of her previous sales history and determined that each day there is a 50% chance that she will sell a luxury car. After careful further analysis, it is also clear that a luxury car sale on one day is independent of the sale (or lack of it) on another day. On New Year's Day (a holiday in which the dealership was closed) the saleswoman is contemplating when she will sell her first luxury car of the year. If N is the random variable indicating the day of the first luxury car sale ($N = 1$ implies the sale was on January 2), then N is distributed according to the geometric distribution with $p = 0.5$, and its pmf is shown in Fig. 1.8. Notice that theoretically the random variable has an infinite range, but for all practical purposes the probability of the random variable being larger than seven is negligible. □

Poisson: (By Simeon Denis Poisson, 1781-1840; published in 1837.) The random variable N has a Poisson distribution if there is a number $\lambda > 0$ such that the pmf of N can be written as

$$f(k) = \frac{\lambda^k e^{-\lambda}}{k!} \text{ for } k = 0,1,\cdots . \tag{1.12}$$

Then,

$$E[N] = \lambda; \quad V[N] = \lambda; \quad C^2[N] = 1/\lambda .$$

The Poisson distribution is the most important discrete distribution in stochastic modeling. It arises in many different circumstances. One use is as an approximation to the binomial distribution. For n large and p small, the binomial is approximated by the Poisson by setting $\lambda = np$. For example, suppose we have a box of 144 eggs and there is a 1% probability that any one egg will break. Assuming that the breakage of eggs is independent of other eggs breaking, the probability that exactly 3 eggs will be broken out of the 144 can be determined using the binomial distribution with $n = 144$, $p = 0.01$, and $k = 3$; thus

$$\frac{144!}{141!3!}(0.01)^3(0.99)^{141} = 0.1181 ,$$

or by the Poisson approximation with $\lambda = 1.44$ that yields

$$\frac{(1.44)^3 \mathrm{e}^{-1.44}}{3!} = 0.1179 .$$

In 1898, L. V. Bortkiewicz [10, p. 206] reported that the number of deaths due to horse-kicks in the Prussian army was a Poisson random variable. Although this seems like a silly example, it is very instructive. The reason that the Poisson distribution holds in this case is due to the binomial approximation feature of the Poisson. Consider the situation: there would be a small chance of death by horse-kick for any one person (i.e., p small) but a large number of individuals in the army (i.e., n large). There are many analogous situations in modeling that deal with large populations and a small chance of occurrence for any one individual within the population. In particular, arrival processes (like arrivals to a bus station in a large city) can often be viewed in this fashion and thus described by a Poisson distribution. Another common use of the Poisson distribution is in population studies. The population size of a randomly growing organism often can be described by a Poisson random variable. W. S. Gosset (1876–1937), using the pseudonym of Student, showed in 1907 that the number of yeast cells in 400 squares of haemocytometer followed a Poisson distribution. The number of radioactive emissions during a fixed interval of time is also Poisson as indicated in Example 1.2. (Fig. 1.4 also shows the Poisson pmf.)

Uniform-Continuous: The random variable X has a continuous uniform distribution if there are two numbers a and b with $a < b$ such that the pdf of X can be written as

$$f(s) = \begin{cases} \frac{1}{b-a} & \text{for } a \le s \le b \\ 0 & \text{otherwise .} \end{cases} \tag{1.13}$$

Then its cumulative probability distribution is given by

$$F(s) = \begin{cases} 0 & \text{for } s < a \\ \frac{s-a}{b-a} & \text{for } a \le s < b \\ 1 & \text{for } s \ge b , \end{cases}$$

and

$$E[X] = \frac{a+b}{2}; \quad V[X] = \frac{(b-a)^2}{12}; \quad C^2[X] = \frac{(b-a)^2}{3(a+b)^2} .$$

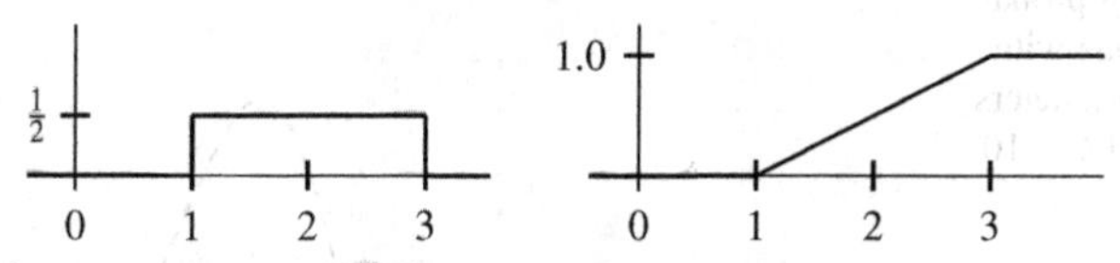

Fig. 1.9 The probability density function and cumulative distribution function for a continuous uniform distribution between 1 and 3

Fig. 1.10 Exponential CDF (solid line) and pdf (dashed line) with a mean of 1/2

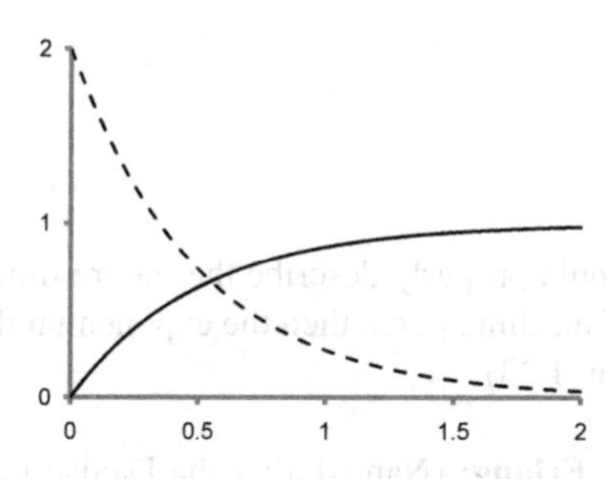

The graphs for the pdf and CDF of the continuous uniform random variables are the simplest of the continuous distributions. As shown in Fig. 1.9, the pdf is a rectangle and the CDF is a "ramp" function.

Exponential: The random variable X has an exponential distribution if there is a number $\lambda > 0$ such that the pdf of X can be written as

$$f(s) = \begin{cases} \lambda e^{-\lambda s} & \text{for } s \geq 0 \\ 0 & \text{otherwise} \end{cases}. \tag{1.14}$$

Then its cumulative probability distribution is given by

$$F(s) = \begin{cases} 0 & \text{for } s < 0, \\ 1 - e^{-\lambda s} & \text{for } s \geq 0; \end{cases}$$

and

$$E[X] = \frac{1}{\lambda}; \quad V[X] = \frac{1}{\lambda^2}; \quad C^2[X] = 1 .$$

The exponential distribution is an extremely common distribution in probabilistic modeling. One very important feature is that the exponential distribution is the only continuous distribution that contains no memory. Specifically, an exponential random variable X is said to be memoryless if

$$\Pr\{X > t + s | X > t\} = \Pr\{X > s\} . \tag{1.15}$$

That is if, for example, a machine's failure time is due to purely random events (like voltage surges through a power line), then the exponential random variable

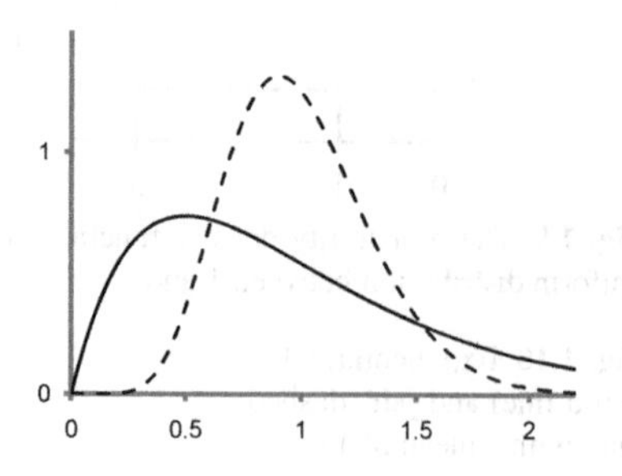

Fig. 1.11 Two Erlang probability density functions with mean 1 and shape parameters $k = 2$ (solid line) and $k = 10$ (dashed line)

would properly describe the failure time. However, if failure is due to the wear out of machine parts, then the exponential distribution would not be suitable (see Problem 1.23).

Erlang: (Named after the Danish mathematician A. K. Erlang, 1878–1929, for his extensive use of it and his pioneering work in queueing theory in the early 1900's while working for the Copenhagen Telephone Company.) The nonnegative random variable X has an Erlang distribution if there is a positive integer k and a positive number β such that the pdf of X can be written as

$$f(s) = \frac{k(ks)^{k-1} e^{-(k/\beta)s}}{\beta^k (k-1)!} \quad \text{for } s \geq 0 . \tag{1.16}$$

Then,

$$E[X] = \beta; \; V[X] = \frac{\beta^2}{k}; \; C^2[X] = \frac{1}{k} .$$

Note that the constant β is often called the scale factor because changing its value is equivalent to either stretching or compressing the x-axis, and the constant k is often called the shape parameter because changing its value changes the shape of the pdf.

The usefulness of the Erlang is due to the fact that an Erlang random variable with parameters k and β is the sum of k (independent) exponential random variables each with mean β/k. In modeling process times, the exponential distribution is often inappropriate because the standard deviation is as large as the mean. Engineers usually try to design systems that yield a standard deviation of process times significantly smaller than their mean. Notice that for the Erlang distribution, the standard deviation decreases as the square root of the parameter k increases so that processing times with a small standard deviation can often be approximated by an Erlang random variable.

Figure 1.11 illustrates the effect of the parameter k by graphing the pdf for a type-2 Erlang and a type-10 Erlang. (The parameter k establishes the "type" for the Erlang distribution.) Notice that a type-1 Erlang is an exponential random variable so its pdf would have the form shown in Fig. 1.10.

Gamma: The Erlang distribution is part of a larger class of nonnegative random variables called gamma random variables. It is a common distribution used to describe process times and has two parameters: a shape parameter, α, and a scale parameter, β. A shape parameter is so named because varying its value results in different shapes for the pdf. Varying the scale parameter does not change the shape of the distribution, but it tends to "stretch" or "compress" the x-axis. Before giving the density function for the gamma, we must define the *gamma function* because it is used in the definition of the gamma distribution. The gamma function is defined, for $x > 0$, as

$$\Gamma(x) = \int_0^\infty s^{x-1} e^{-s} ds\,. \tag{1.17}$$

One useful property of the gamma function is the relationship $\Gamma(x+1) = x\Gamma(x)$, for $x \geq 1$. Thus, if x is a positive integer, $\Gamma(x) = (x-1)!$. (For some computational issues, see the appendix to this chapter.) The density function for a gamma random variable is given by

$$f(s) = \frac{s^{\alpha-1} e^{-s/\beta}}{\beta^\alpha \Gamma(\alpha)} \quad \text{for } s \geq 0\,. \tag{1.18}$$

Then,

$$E[X] = \beta\alpha;\ \ V[X] = \beta^2\alpha;\ \ C^2[X] = \frac{1}{\alpha}\,.$$

Notice that if it is desired to determine the shape and scale parameters for a gamma distribution with a known mean and variance, the inverse relationships are

$$\alpha = \frac{E[X]^2}{V[X]} \quad \text{and} \quad \beta = \frac{E[X]}{\alpha}\,.$$

Weibull: In 1939, W. Weibull [2, p. 73] (a Swedish engineer who lived from 1887 to 1979) developed a distribution for describing the breaking strength of various materials. Since that time, many statisticians have shown that the Weibull distribution can often be used to describe failure times for many different types of systems. The Weibull distribution has two parameters: a scale parameter, β, and a shape parameter, α. Its cumulative distribution function is given by

$$F(s) = \begin{cases} 0 & \text{for } s < 0 \\ 1 - e^{-(s/\beta)^\alpha} & \text{for } s \geq 0\,. \end{cases} \tag{1.19}$$

Both scale and shape parameters can be any positive number. As with the gamma distribution, the shape parameter determines the general shape of the pdf (see Fig. 1.12) and the scale parameter either expands or contracts the pdf. The moments of the Weibull are a little difficult to express because they involve the gamma function (1.17). Specifically, the moments for the Weibull distribution are

$$E[X] = \beta\Gamma(1+\frac{1}{\alpha});\quad E[X^2] = \beta^2\Gamma(1+\frac{2}{\alpha});\quad E[X^3] = \beta^3\Gamma(1+\frac{3}{\alpha})\,. \tag{1.20}$$

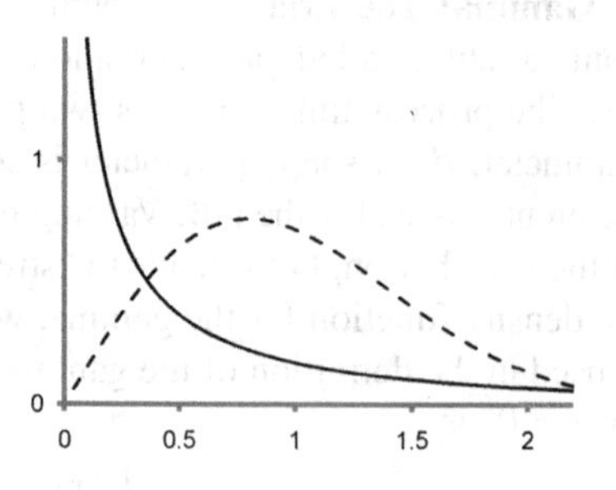

Fig. 1.12 Two Weibull probability density functions with mean 1 and shape parameters $\alpha = 0.5$ (solid line) and $\alpha = 2$ (dashed line)

It is more difficult to determine the shape and scale parameters for a Weibull distribution with a known mean and variance, than it is for the gamma distribution because the gamma function must be evaluated to determine the moments of a Weibull. Some computational issues for obtaining the shape and scale parameters of a Weibull are discussed in the appendix to this chapter.

When the shape parameter is greater than 1, the shape of the Weibull pdf is unimodal similar to the Erlang with its type parameter greater than 1. When the shape parameter equals 1, the Weibull pdf is an exponential pdf. When the shape parameter is less than 1, the pdf is similar to the exponential except that the graph is asymptotic to the y-axis instead of hitting the y-axis. Figure 1.12 provides an illustration of the effect that the shape parameter has on the Weibull distribution. Because the mean values were held constant for the two pdf's shown in the figure, the value for β varied. The pdf plotted with a solid line in the figure has $\beta = 0.5$ that, together with $\alpha = 0.5$, yields a mean of 1 and a standard deviation of 2.236; the dashed line is pdf that has $\beta = 1.128$ that, together with $\alpha = 2$, yields a mean of 1 and a standard deviation 0.523.

Normal: (Discovered by A. de Moivre, 1667-1754, but usually attributed to Karl Gauss, 1777-1855.) The random variable X has a normal distribution if there are two numbers μ and σ with $\sigma > 0$ such that the pdf of X can be written as

$$f(s) = \frac{1}{\sigma\sqrt{2\pi}} \mathrm{e}^{-(s-\mu)^2/(2\sigma^2)} \text{ for } -\infty < s < \infty . \tag{1.21}$$

Then,

$$E[X] = \mu; \;\; V[X] = \sigma^2; \;\; C^2[X] = \frac{\sigma^2}{\mu^2} .$$

The normal distribution is the most common distribution recognized by most people by its "bell shaped" curve. Its pdf and CDF are shown below in Fig. 1.13 for a normally distributed random variable with mean zero and standard deviation one.

Although the normal distribution is not widely used in stochastic modeling, it is, without question, the most important distribution in statistics. The normal distribution can be used to approximate both the binomial and Poisson distributions.

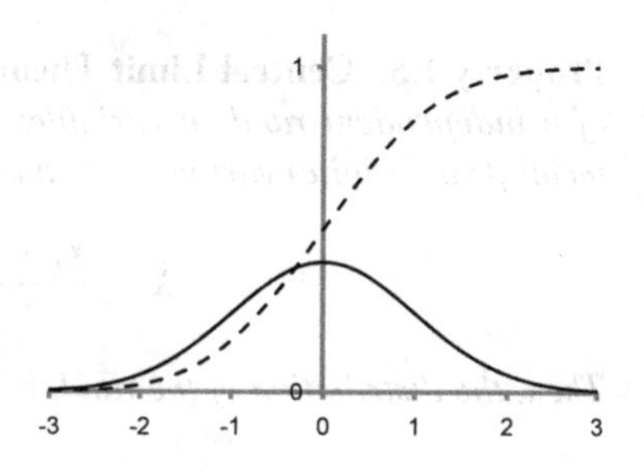

Fig. 1.13 Standard normal pdf (solid line) and CDF (dashed line)

A common rule-of-thumb is to approximate the binomial whenever n (the number of trials) is larger than 30. If $np < 5$, then use the Poisson for the approximation with $\lambda = np$. If $np \geq 5$, then use the normal for the approximation with $\mu = np$ and $\sigma^2 = np(1-p)$. Furthermore, the normal can be used to approximate the Poisson whenever $\lambda > 30$. When using a continuous distribution (like the normal) to approximate a discrete distribution (like the Poisson or binomial), the interval between the discrete values is usually split halfway. For example, if we desire to approximate the probability that a Poisson random variable will take on the values 29, 30, or 31 with a continuous distribution, then we would determine the probability that the continuous random variable is between 28.5 and 31.5.

Example 1.10. A software company has received complaints regarding their responsiveness for customer service. They have decided to analyze the arrival pattern of phone calls to customer service and have determined that the arrival of complaints in any one hour is a Poisson random variable with a mean of 120. The company would like to know the probability that in any one hour 140 or more calls arrive. To determine that probability, let N be a Poisson random variable with $\lambda = 120$, let X be a random variable with $\mu = \sigma^2 = 120$ and let Z be a standard normal random variable (i.e., Z is normal with mean 0 and variance 1). The above question is answered as follows:

$$\begin{aligned}
\Pr\{N \geq 140\} &\approx \Pr\{X > 139.5\} \\
&= \Pr\{Z > (139.5 - 120)/10.95\} \\
&= \Pr\{Z > 1.78\} = 1 - 0.9625 = 0.0375\,.
\end{aligned}$$

□

The importance of the normal distribution is due to its property that sample means from almost any practical distribution will limit to the normal; this property is called the *Central Limit Theorem.* We state this property now even though it needs the concept of statistical independence that is not yet defined. However, because the idea should be somewhat intuitive, we state the property at this point since it is so central to the use of the normal distribution.

Property 1.5. Central Limit Theorem. *Let $\{X_1, X_2, \cdots, X_n\}$ be a sequence of n independent random variables each having the same distribution with mean μ and (finite) variance σ^2, and define*

$$\overline{X} = \frac{X_1 + X_2 + \cdots + X_n}{n} .$$

Then, the distribution of the random variable Z defined by

$$Z = \frac{\overline{X} - \mu}{\sigma/\sqrt{n}}$$

approaches a normal distribution with zero mean and standard deviation of one as n gets large.

Log Normal: The final distribution that we briefly mention is based on the normal distribution. Specifically, if X is a normal random variable with mean μ_N and variance σ_N^2, the random variable $Y = e^X$ is called a log-normal random variable with mean μ_L and variance σ_L^2. (Notice that the name arrises because the random variable defined by the natural log of Y; namely $\ln(Y)$, is normally distributed.) This distribution is always non-negative and can have a relatively large right-hand tail. It is often used for modeling repair times and also for modeling many biological characteristics. It is not difficult to obtain the mean and variance of the log-normal distribution from the characteristics of the normal:

$$\mu_L = e^{\mu_N + \frac{1}{2}\sigma_N^2} , \quad \text{and } \sigma_L^2 = \mu_L^2 \times (e^{\sigma_N^2} - 1) . \tag{1.22}$$

Because the distribution is skewed to the right (long right-hand tail), the mean is to the right of the mode which is given by $e^{\mu_N - \sigma_N^2}$. If the mean and variance of the log-normal distribution is known, it is straight forward to obtain the characteristics of the normal random variable that generates the log-normal, specifically

$$\sigma_N^2 = \ln(c_L^2 + 1) , \quad \text{and } \mu_N = \ln(\mu_L) - \frac{1}{2}\sigma_N^2 , \tag{1.23}$$

where the squared coefficient of variation is given by $c_L^2 = \sigma_L^2/\mu_L^2$.

Example 1.11. The height of trees within a stand of Loblolly pine trees is highly variable and is distributed according to a log-normal distribution with mean 21.2 m and standard deviation of 52.1 m. Notice that the squared coefficient of variation is 5.983. (A squared coefficient of variation greater 1.0 implies a significant amount of variability.) Let H denote a random variable representing the height of the trees. Then there must be another random variable, call it X, such that $H = eX$ where $E[X] = 2.087$ and its standard deviation is 1.394. □

Skewness: Before moving to the discussion of more than one random variable, we mention an additional descriptor of distributions. The first moment gives the central tendency for random variables, and the second moment is used to measure variability. The third moment, that was not discussed previously, is useful as a measure of skewness (i.e., non-symmetry). Specifically, the coefficient of skewness, ν, for a random variable T with mean μ and standard deviation σ is defined by

$$\nu = \frac{E[(T-\mu)^3]}{\sigma^3}, \tag{1.24}$$

and the relation to the other moments is

$$E[(T-\mu)^3] = E[T^3] - 3\mu E[T^2] + 2\mu^3 .$$

A symmetric distribution has $\nu = 0$; if the mean is to the left of the mode, $\nu < 0$ and the left-hand side of the distribution will have the longer tail; if the mean is to the right of the mode, $\nu > 0$ and the right-hand side of the distribution will have the longer tail. For example, $\nu = 0$ for the normal distribution, $\nu = 2$ for the exponential distribution, $\nu = 2/\sqrt{k}$ for a type-k Erlang distribution, and for a gamma distribution, we have $\nu = 2/\sqrt{\alpha}$. The log normal distribution can have a relatively large positive skew and its skewness coefficient is given as $\nu = (e^{\sigma_N^2}+2)\sqrt{e^{\sigma_N^2}-1}$. The Weibull distribution can have both positive and negative skewness coefficients. The pdf's shown in Fig. 1.12 have skewness coefficients of 3.9 and 0.63, respectively, for the solid line figure and dashed line graphs. Thus, the value of ν can help complete the intuitive understanding of a particular distribution.

- *Suggestion: Do Problems 1.15–1.18.*

1.5 Multivariate Distributions

The analysis of physical phenomena usually involves many distinct random variables. In this section we discuss the concepts involved when two random variables are defined. The extension to more than two is left to the imagination of the reader and the numerous textbooks that have been written on the subject.

Definition 1.12. The function F is called the *joint cumulative distribution function* for X_1 and X_2 if

$$F(a,b) = \Pr\{X_1 \le a, X_2 \le b\}$$

for a and b any two real numbers. □

In a probability statement as in the right-hand-side of the above equation, the comma means intersection of events and is read as "The probability that X_1 is less than or equal to *a and* X_2 is less than or equal to b". The initial understanding of joint probabilities is easiest with discrete random variables.

Definition 1.13. The function f is a *joint pmf* for the discrete random variables X_1 and X_2 if

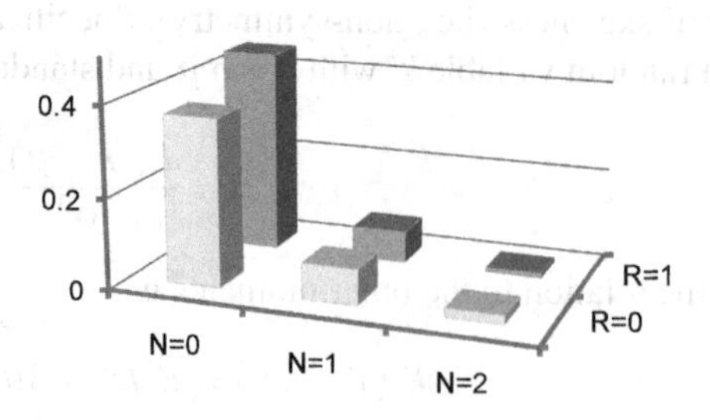

Fig. 1.14 Probability mass function for the two discrete random variables from Example 1.12

$$f(a,b) = \Pr\{X_1 = a, X_2 = b\}$$

for each (a,b) in the range of (X_1, X_2). □

For the single-variable pmf, the height of the pmf at a specific value gives the probability that the random variable will equal that value. It is the same for the joint pmf except that the graph is in three-dimensions. Thus, the height of the pmf evaluated at a specified ordered pair gives the probability that the random variables will equal those specified values (Fig. 1.14).

It is sometimes necessary to obtain from the joint pmf the probability of one random variable without regard to the value of the second random variable.

Definition 1.14. The *marginal pmf* for X_1 and X_2, denoted by f_1 and f_2, respectively, are

$$f_1(a) = \Pr\{X_1 = a\} = \sum_k f(a,k)$$

for a in the range of X_1, and

$$f_2(b) = \Pr\{X_2 = b\} = \sum_k f(k,b)$$

for b in the range of X_2. □

Example 1.12. We return again to Example 1.1 to illustrate these concepts. The random variable R will indicate whether a randomly chosen box contains radio phones or plain phones; namely, if the box contains radio phones then we set $R = 1$ and if plain phones then $R = 0$. Also the random variable N will denote the number of defective phones in the box. Thus, according to the probabilities defined in Example 1.1, the joint pmf,

$$f(a,b) = \Pr\{R = a, N = b\},$$

has the probabilities as listed in in the main portion of Table 1.1. By summing in the "margins", we obtain the marginal pmf for R and N separately as shown in the

Table 1.1 Joint and marginal probability mass functions of Example 1.12

	$N=0$	$N=1$	$N=2$	$f_1(\cdot)$
$R=0$	0.37	0.08	0.02	0.47
$R=1$	0.45	0.07	0.01	0.53
$f_2(\cdot)$	0.82	0.15	0.03	

margins of Table 1.1. Thus we see, for example, that the probability of choosing a box with radio phones (i.e., $\Pr\{R=1\}$) is 53%, the probability of choosing a box of radio phones that has one defective phone (i.e., $\Pr\{R=1, N=1\}$) is 7%, and the probability that both phones in a randomly chosen box (i.e., $\Pr\{N=2\}$) are defective is 3%. □

Continuous random variables are treated in an analogous manner to the discrete case. The major difference in moving from one continuous random variable to two is that probabilities are given in terms of a volume under a surface instead of an area under a curve (see Fig. 1.15 for representation of a joint pdf).

Definition 1.15. The functions g, g_1, and g_2 are the *joint pdf* for X_1 and X_2, the *marginal pdf* for X_1, and the *marginal pdf* for X_2, respectively, as the following hold:

$$\Pr\{a_1 \le X_1 \le b_1, a_2 \le X_2 \le b_2\} = \int_{a_2}^{b_2}\int_{a_1}^{b_1} g(s_1,s_2)\mathrm{d}s_1\mathrm{d}s_2$$

$$g_1(a) = \int_{-\infty}^{\infty} g(a,s)\mathrm{d}s$$

$$g_2(b) = \int_{-\infty}^{\infty} g(s,b)\mathrm{d}s\,,$$

where

$$\Pr\{a \le X_1 \le b\} = \int_a^b g_1(s)\mathrm{d}s$$

$$\Pr\{a \le X_2 \le b\} = \int_a^b g_2(s)\mathrm{d}s\,.$$

□

We return now to the concept of conditional probabilities (Definition 1.3). The situation often arises in which the experimentalist has knowledge regarding one random variable and would like to use that knowledge in predicting the value of the other (unknown) random variable. Such predictions are possible through conditional probability functions

Definition 1.16. Let f be a joint pmf for the discrete random variables X_1 and X_2 with f_2 the marginal pmf for X_2. Then the *conditional pmf* for X_1 given that $X_2 = b$ is defined, if $\Pr\{X_2 = b\} \ne 0$, to be

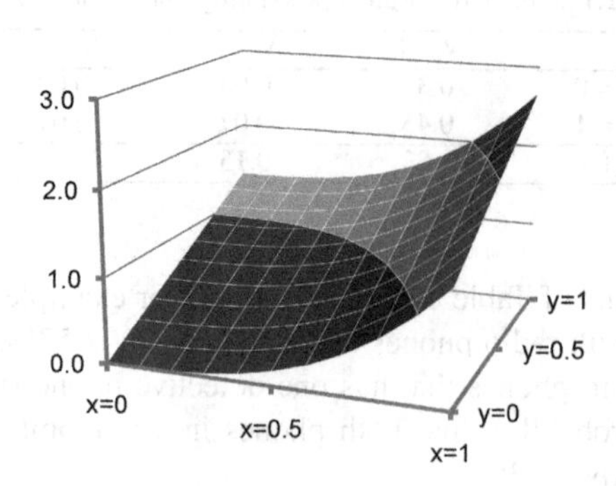

Fig. 1.15 Probability density function for the two continuous random variables from Example 1.13

$$f_{1|b}(a) = \frac{f(a,b)}{f_2(b)} ,$$

where

$$\Pr\{X_1 = a|X_2 = b\} = f_{1|b}(a) .$$

□

Definition 1.17. Let g be a joint pdf for continuous random variables X_1 and X_2 with g_2 the marginal pdf for X_2. Then the *conditional pdf* for X_1 given that $X_2 = b$ is defined to be

$$g_{1|b}(a) = \frac{g(a,b)}{g_2(b)} ,$$

where

$$\Pr\{a_1 \le X_1 \le a_2|X_2 = b\} = \int_{a_1}^{a_2} g_{1|b}(s)\mathrm{d}s .$$

□

The conditional statements for X_2 given a value for X_1 are made similarly to Definitions 1.16 and 1.17 with the subscripts reversed. These conditional statements can be illustrated by using Example 1.12. It has already been determined that the probability of having a box full of defective phones is 3%; however, let us assume that it is already known that we have picked a box of radio phones. Now, given a box of radio phones, the probability of both phones being defective is

$$f_{2|a=1}(2) = \frac{f(1,2)}{f_1(1)} = \frac{0.01}{0.53} = 0.0189 ;$$

thus, knowledge that the box consisted of radio phones enabled a more accurate prediction of the probabilities that both phones were defective. Or to consider a different situation, assume that we know the box has both phones defective. The probability that the box contains plain phones is

$$f_{1|b=2}(0) = \frac{f(0,2)}{f_2(2)} = \frac{0.02}{0.03} = 0.6667\,.$$

Example 1.13. Let X and Y be two continuous random variables with joint pdf given by

$$f(x,y) = \frac{4}{3}(x^3 + y) \text{ for } 0 \le x \le 1, 0 \le y \le 1\,.$$

Utilizing Definition 1.15, we obtain

$$f_1(x) = \frac{4}{3}(x^3 + 0.5) \text{ for } 0 \le x \le 1$$

$$f_2(y) = \frac{4}{3}(y + 0.25) \text{ for } 0 \le y \le 1\,.$$

To find the probability that Y is less than or equal to 0.5, we perform the following steps:

$$\begin{aligned} \Pr\{Y \le 0.5\} &= \int_0^{0.5} f_2(y)\mathrm{d}y \\ &= \frac{4}{3}\int_0^{0.5} (y + 0.25)\mathrm{d}y = \frac{1}{3}\,. \end{aligned}$$

To find the probability that Y is less than or equal to 0.5 given we know that $X = 0.1$, we perform

$$\begin{aligned} \Pr\{Y \le 0.5 | X = 0.1\} &= \int_0^{0.5} f_{2|0.1}(y)\mathrm{d}y \\ &= \int_0^{0.5} \frac{0.1^3 + y}{0.1^3 + 0.5}\mathrm{d}y \\ &= \frac{0.1255}{0.501} \approx \frac{1}{4}\,. \end{aligned}$$

□

Example 1.14. Let U and V be two continuous random variables with joint pdf given by

$$g(u,v) = 8u^3 v \text{ for } 0 \le u \le 1, 0 \le v \le 1\,.$$

The marginal pdf's are

$$\begin{aligned} g_1(u) &= 4u^3 \quad \text{for } 0 \le u \le 1 \\ g_2(v) &= 2v \quad \text{for } 0 \le v \le 1\,. \end{aligned}$$

The following two statements are easily verified.

$$\Pr\{0.1 \le V \le 0.5\} = \int_{0.1}^{0.5} 2v\mathrm{d}v = 0.24$$

$$\Pr\{0.1 \leq V \leq 0.5 | U = 0.1\} = 0.24 .$$

□

The above example illustrates independence. Notice in the example that knowledge of the value of U did not change the probabilities regarding the probability statement of V.

Definition 1.18. Let f be the joint probability distribution (pmf if discrete and pdf if continuous) of two random variables X_1 and X_2. Furthermore, let f_1 and f_2 be the marginals for X_1 and X_2, respectively. If

$$f(a,b) = f_1(a) f_2(b)$$

for all a and b, then X_1 and X_2 are called *independent*. □

Independent random variables are much easier to work with because of their separability. However, in the use of the above definition, it is important to test the property for *all* values of a and b. It would be easy to make a mistake by stopping after the equality was shown to hold for only one particular pair of a, b values. Once independence has been shown, the following property is very useful.

Property 1.6. *Let X_1 and X_2 be independent random variables. Then*

$$E[X_1 X_2] = E[X_1] E[X_2]$$

and

$$V[X_1 + X_2] = V[X_1] + V[X_2] .$$

Example 1.15. Consider again the random variables R and N defined in Example 1.12. We see from the marginal pmf's given in that example that $E[R] = 0.53$ and $E[N] = 0.21$. We also have

$$\begin{aligned} E[R \cdot N] &= 0 \times 0 \times 0.37 + 0 \times 1 \times 0.08 + 0 \times 2 \times 0.02 \\ &\quad + 1 \times 0 \times 0.45 + 1 \times 1 \times 0.07 + 1 \times 2 \times 0.01 = 0.09 . \end{aligned}$$

Thus, it is possible to say that the random variables R and N are not independent since $0.53 \times 0.21 \neq 0.09$. If, however, the expected value of the product of two random variables equals the product of the two individual expected values, the claim of independence is *not* proven. □

We close this section by giving two final measures that are used to express the relationship between two dependent random variables. The first measure is called the *covariance* and the second measure is called the *correlation coefficient*.

Definition 1.19. The *covariance* of two random variables, X_1 and X_2, is defined by

$$cov(X_1,X_2) = E[(X_1 - E[X_1])(X_2 - E[X_2])].$$

□

Property 1.7. *The following is often helpful as a computational aid:*

$$cov(X_1,X_2) = E[X_1X_2] - \mu_1\mu_2 ,$$

where μ_1 and μ_2 are the means for X_1 and X_2, respectively.

Comparing Property 1.6 to Property 1.7, it should be clear that random variables that are independent have zero covariance. However, it is possible to obtain random variables with zero covariance that are not independent. (See Example 1.17 below.) The covariance is also important when determining the variance of the sum of two random variables that are not independent so that Property 1.6 cannot be used.

Property 1.8. *Let X_1 and X_2 be two random variables, then*

$$V[X_1 + X_2] = V[X_1] + V[X_2] - cov(X_1,X_2) .$$

A principle use of the covariance is in the definition of the correlation coefficient, that is a measure of the linear relationship between two random variables.

Definition 1.20. Let X_1 be a random variable with mean μ_1 and variance σ_1^2. Let X_2 be a random variable with mean μ_2 and variance σ_2^2. The *correlation coefficient* , denoted by ρ, of X_1 and X_2 is defined by

$$\rho = \frac{cov(X_1,X_2)}{\sqrt{V(X_1)V(X_1)}} = \frac{E[X_1X_2] - \mu_1\mu_2}{\sigma_1\sigma_2} .$$

□

The correlation coefficient is always between negative one and positive one. A negative correlation coefficient indicates that if one random variable happens to be large, the other random variable is likely to be small. A positive correlation coefficient indicates that if one random variable happens to be large, the other random variable is also likely to be large. The following examples illustrate this concept.

Example 1.16. Let X_1 and X_2 denote two discrete random variables, where X_1 ranges from 1 to 3 and X_2 ranges from 10 to 30. Their joint and marginal pmf's are given in Table 1.2.

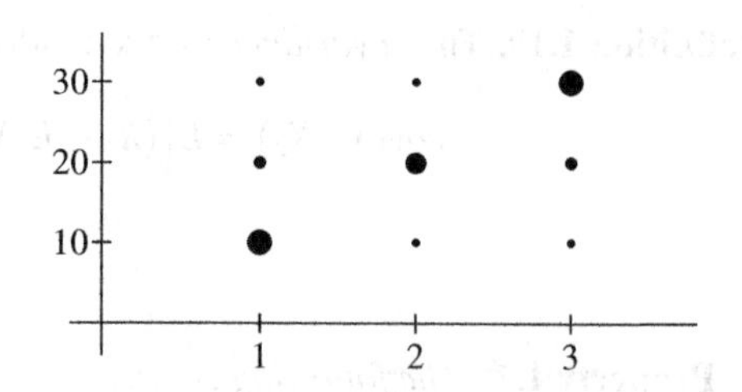

Fig. 1.16 Graphical representation for conditional probabilities of X_2 given X_1 from Example 1.16, where the correlation coefficient is 0.632

Table 1.2 Marginal probability mass functions of Example 1.16

	$X_2 = 10$	$X_2 = 20$	$X_2 = 30$	$f_1(\cdot)$
$X_1 = 1$	0.28	0.08	0.04	0.4
$X_1 = 2$	0.04	0.12	0.04	0.2
$X_1 = 3$	0.04	0.08	0.28	0.4
$f_2(\cdot)$	0.36	0.28	0.36	

The following facts should not be difficult to verify: $\mu_1 = 2.0$, $\sigma_1^2 = 0.8$, $\mu_2 = 20.0$, $\sigma_2^2 = 72.0$, and $E[X_1 X_2] = 44.8$. Therefore the correlation coefficient of X_1 and X_2 is given by

$$\rho = \frac{44.8 - 2 \times 20}{\sqrt{0.8 \times 72}} = 0.632 \,.$$

The conditional probabilities will help verify the intuitive concept of a positive correlation coefficient. Figure 1.16 contains a graph illustrating the conditional probabilities of X_2 given various values of X_1; the area of each circle in the figure is proportional to the conditional probability. Thus, the figure gives a visual representation that as X_1 increases, it is likely (but *not* necessary) that X_2 will increase. For example, the top right-hand circle represents $\Pr\{X_2 = 30 | X_1 = 3\} = 0.7$, and the middle right-hand circle represents $\Pr\{X_2 = 20 | X_1 = 3\} = 0.2$.

As a final example, we switch the top and middle right-hand circles in Fig. 1.16 so that the appearance is not so clearly linear. (That is, let $\Pr\{X_1 = 3, X_2 = 20\} = 0.28$, $\Pr\{X_1 = 3, X_2 = 30\} = 0.08$, and all other probabilities the same.) With this change, μ_1 and σ_1^2 remain unchanged, $\mu_2 = 18$, $\sigma_2^2 = 48.0$, $\mathrm{cov}(X_1, X_2) = 2.8$ and the correlation coefficient is $\rho = 0.452$. Thus, as the linear relationship between X_1 and X_2 weakens, the value of ρ becomes smaller. □

If the random variables X and Y have a linear relationship (however "fuzzy"), their correlation coefficient will be non-zero. Intuitively, the square of the correlation coefficient, ρ^2, indicates that amount of variability that is due to that linear relationship. For example, suppose that the correlation between X and Y is 0.8 so that $\rho^2 = 0.64$. Then 64% of the variability in Y is due the variability of X through their linear relationship.

Example 1.17. Let X_1 and X_2 denote two discrete random variables, where X_1 ranges from 1 to 3 and X_2 ranges from 10 to 30. Their joint and marginal pmf's are given in Table 1.3.

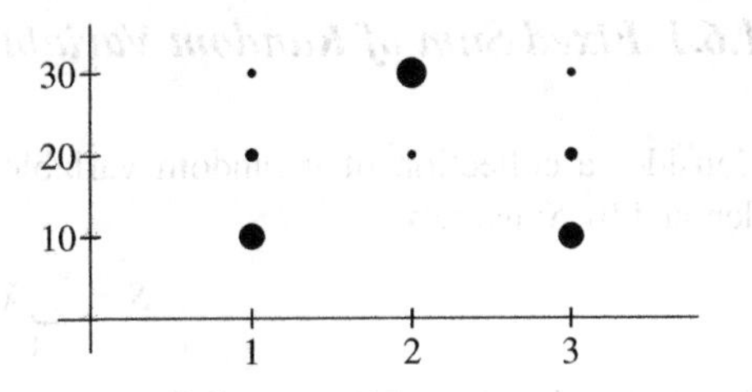

Fig. 1.17 Graphical representation for conditional probabilities of X_2 given X_1 from Example 1.17, where the correlation coefficient is zero

Table 1.3 Marginal probability mass functions of Example 1.17

	$X_2 = 10$	$X_2 = 20$	$X_2 = 30$	$f_1(\cdot)$
$X_1 = 1$	0.28	0.08	0.04	0.4
$X_1 = 2$	0.00	0.02	0.18	0.2
$X_1 = 3$	0.28	0.08	0.04	0.4
$f_2(\cdot)$	0.56	0.18	0.26	

Again, we give the various measures and allow the reader to verify their accuracy: $\mu_1 = 2$, $\mu_2 = 17$, and $E[X_1X_2] = 34$. Therefore the correlation coefficient of X_1 and X_2 is zero so there is no *linear* relation between X_1 and X_2; however, the two random variables are clearly dependent. If X_1 is either one or three, then the most likely value of X_2 is 10; whereas, if X_1 is 2, then it is impossible for X_2 to have the value of 10; thus, the random variables must be dependent. If you observe the representation of the conditional probabilities in Fig. 1.17, then the lack of a linear relationship is obvious. □

- *Suggestion: Do Problems 1.20–1.25.*

1.6 Combinations of Random Variables

This probability review is concluded with a discussion of a problem type that will be frequently encountered in the next several chapters; namely, combinations of random variables. The properties of the sum of a fixed number of random variables is a straightforward generalization of previous material; however when the sum has a random number of terms, an additional variability factor must be taken into account. The final combination discussed in this section is called a mixture of random variables. An example of a mixture is the situation where the random processing time at a machine will be from different probability distributions based on the (random) product type being processed. Each of these three combinations of random variables are considered in turn. The material in this section is taken from [3], and is especially useful in the queueing chapters.

1.6.1 Fixed Sum of Random Variables

Consider a collection of n random variables, $X_1, X_2, \cdots, X_n$ and let their sum be denoted by S; namely,

$$S = \sum_{i=1}^{n} X_i . \tag{1.25}$$

By a generalization of Property 1.3, we have

$$\begin{aligned} E[S] &= E[X_1 + X_2 + \cdots + X_n] \\ &= E[X_1] + E[X_2] + \cdots + E[X_n] . \end{aligned} \tag{1.26}$$

Note that (1.26) is valid even if the random variables are not independent.

The variance of the random variable S is obtained in a similar manner to the expected value

$$\begin{aligned} V[S] &= E[(S - E[S])^2] \\ &= E[S^2] - E[S]^2 \\ &= E[(X_1 + X_2 + \cdots + X_n)^2] - (E[X_1] + E[X_2] + \cdots + E[X_n])^2 \\ &= \sum_{i=1}^{n} E[X_i^2] + 2\sum_{i=1}^{n}\sum_{j>i}^{n} E[X_i X_j] - (E[X_1] + E[X_2] + \cdots + E[X_n])^2 \\ &= \sum_{i=1}^{n} \left(E[X_i^2] - E[X_i]^2\right) + 2\sum_{i=1}^{n}\sum_{j>i}^{n} \left(E[X_i X_j] - E[X_i]E[X_j]\right) \\ &= \sum_{i=1}^{n} V[X_i] + 2\sum_{i=1}^{n}\sum_{j>i}^{n} cov[X_i, X_j] . \end{aligned} \tag{1.27}$$

Notice that when the random variables are pair-wise independent, i.e., X_i and X_j are independent for all i and j, then $E[X_i X_j] = E[X_i]E[X_j]$ and Property 1.6 is generalized indicating that the variance of the sum of n independent random variables is the sum of the individual variances. In addition, when $X_1, \cdots, X_n$ are independent *and* identically distributed (called *i.i.d.*), we have that

$$\begin{aligned} E[S] &= nE[X_1] \\ V[S] &= nV[X_1] . \end{aligned} \tag{1.28}$$

1.6.2 Random Sum of Random Variables

Before discussing the random sum of random variables, we need a property of conditional expectations. For this discussion we follow the development in [4] in which these properties are developed assuming discrete random variables because the discrete case is more intuitive than the continuous case. (Although the development below only considers the discrete case, our main result — given as Property 1.9 — is true for both discrete and continuous random variables.)

Let Y and X be two random variables. The conditional probability that the random variable Y takes on a value b given that the random variable X takes the value a is written as

$$\Pr\{Y=b|X=a\} = \frac{\Pr\{Y=b,X=a\}}{\Pr\{X=a\}}, \quad \text{if } \Pr\{X=a\} \neq 0$$

(see Definition 1.16). Thus, the conditional expectation of Y given that $X=a$ changes as the value a changes so it is a function, call it g, of a; namely,

$$E[Y|X=a] = \sum_b b \Pr\{Y=b|X=a\} = g(a) .$$

Hence, the conditional expectation of Y given X is a random variable since it depends on the value of X, expressed as

$$E[Y|X] = g(X) . \tag{1.29}$$

Taking the expectation on both sides of (1.29), yields the (unconditional) expectation of Y and gives the following important property.

Property 1.9. *Let Y and X be any two random variables with finite expectation. The conditional expectation of Y given X is a random variable with expectation given by*

$$E\left[E[Y|X]\right] = E[Y] .$$

Property 1.9 can now be used to obtain the properties of a random sum of random variables. Let S be defined by

$$S = \sum_{i=1}^{N} X_i ,$$

where $X_1, X_2, \cdots$ is a sequence of *i.i.d.* random variables, and N is a nonnegative discrete random variable independent of each X_i. (When $N=0$, the random sum is interpreted to be zero.) For a fixed n, Eq. (1.28) yields

$$E\left[\sum_{i=1}^{N} X_i|N=n\right] = nE[X_1] , \text{ thus}$$

$$E\left[\sum_{i=1}^{N} X_i|N\right] = NE[X_1] .$$

The expected value of the random sum can be derived from the above result using Property 1.9 regarding conditional expectations as follows:

$$
\begin{aligned}
E[S] &= E\left[E\left[\sum_{i=1}^{N} X_i | N\right]\right] \\
&= E[NE[X_1]] \\
&= E[N]E[X_1] \,.
\end{aligned}
$$

Note that the final equality in the above arises using Property 1.6 regarding independence and the fact that each random variable in an *i.i.d.* sequence has the same mean.

We obtain the variance of the random variable S in a similar fashion, using $V[S] = E[S^2] - E[S]^2$ but we shall leave its derivation for homework with some hints (see Problem 1.28). Thus, we have the following property:

Property 1.10. *Let $X_1, X_2, \cdots$ be a sequence of i.i.d. random variables where for each i, $E[X_i] = \mu$ and $V[X_i] = \sigma^2$. Let N be a nonnegative discrete random variable independent of the i.i.d. sequence, and let $S = \sum_{i=1}^{N} X_i$. Then*

$$
\begin{aligned}
E[S] &= \mu E[N] \\
V[S] &= \sigma^2 E[N] + \mu^2 V[N] \,.
\end{aligned}
$$

Notice that the squared coefficient of variation of the random sum can also be easily written as

$$
C^2[S] = C^2[N] + \frac{C^2[X]}{E[N]} \,, \text{ where } C^2[X] = \frac{\sigma^2}{\mu^2} \,.
$$

1.6.3 Mixtures of Random Variables

The final type of random variable combination that we consider is a mixture of random variables. For example, consider two products processed on the same machine, where the two product types have different processing characteristics. Specifically, let X_1 and X_2 denote the random processing times for types 1 and 2, respectively, and then let T denote the processing time for an arbitrarily chosen part. The processing sequence will be assumed to be random with p_1 and p_2 being the probability that type 1 and type 2, respectively, are to be processed. In other words, T will equal X_1 with probability p_1 and T will equal X_2 with probability p_2. Intuitively, we have the following relationship.

$$
T = \begin{cases} X_1 & \text{with probability } p_1, \\ X_2 & \text{with probability } 1 - p_1 \,. \end{cases}
$$

Thus, T is said to be a mixture of X_1 and X_2. In generalizing this concept, we have the following definition.

Definition 1.21. Let $X_1, \cdots, X_n$ be a sequence of independent random variables and let I be a positive discrete random variable with range $1, \cdots, n$ independent of the $X_1, \cdots, X_n$ sequence. The random variable T is called a *mixture of random variables with index I* if it can be written as

$$T = X_I .$$

□

Making use of Property 1.9, it should not be too difficult to show the following property.

Property 1.11. *Let T be a mixture of $X_1, \cdots, X_n$ where the mean of X_i is μ_i and variance of X_i is σ_i^2. Then*

$$E[T] = \sum_{i=1}^{n} p_i \mu_i$$

$$E[T^2] = \sum_{i=1}^{n} p_i \left(\sigma_i^2 + \mu_i^2\right) ,$$

where $\Pr\{I = i\} = p_i$ are the probabilities associated with the index.

Notice that the above property gives the first and second moment, not the variance directly. If the variance is desired, the equation $V[T] = E[T^2] - E[T]^2$ must be used.

- *Suggestion: Do Problems 1.26–1.28.*

Appendix

In this appendix, two numerical problems are discussed: the computation of the gamma function (Eq. 1.17) and the determination of the shape and scale parameters for the Weibull distribution. We give suggestions for those using Microsoft Excel and those who are interested in doing the computations within a programming environment.

The gamma function: For Microsoft Excel users, the gamma function is evaluated by first obtaining the natural log of the function since Excel provides an automatic function for the log of the gamma instead of the gamma function itself. For example, to obtain the gamma function evaluated at 1.7, use the formula `=EXP(GAMMALN(1.7))`. This yields a value of 0.908639.

For programmers who need the gamma function, there are some good approximations available. A polynomial approximation taken from [5, p. 155] is

$$\Gamma(1+x) \approx 1 + a_1 x + a_2 x^2 + \cdots + a_5 x^5 \quad \text{for } 0 \le x \le 1, \tag{1.30}$$

where the constants are $a_1 = -0.5748646$, $a_2 = 0.9512363$, $a_3 = -0.6998588$, $a_4 = 0.4245549$, and $a_5 = -0.1010678$. (Or if you need additional accuracy, an eight term approximation is also available in [5] or [1, p. 257].) If it is necessary to evaluate $\Gamma(x)$ for $x < 1$ then use the relationship

$$\Gamma(x) = \frac{1}{x}\Gamma(1+x) . \tag{1.31}$$

If it is necessary to evaluate $\Gamma(n+x)$ for $n > 1$ and $0 \leq x \leq 1$, then use the relationship:

$$\Gamma(n+x) = (n-1+x)(n-2+x)\cdots(1+x)\Gamma(1+x) . \tag{1.32}$$

Example 1.18. Suppose we wish to compute $\Gamma(0.7)$. The approximation given by (1.30), yields a result of $\Gamma(1.7) = 0.9086$. Applying (1.31) yields $\Gamma(0.7) = 0.9086/0.7 = 1.298$. Now suppose that we wish to obtain the gamma function evaluated at 5.7. From (1.32), we have $\Gamma(5.7) = 4.7 \times 3.7 \times 2.7 \times 1.7 \times 0.9086 = 72.52$.
□

Weibull parameters: The context for this section is that we know the first two moments of a Weibull distribution (1.20) and would like to determine the shape and scale parameters. Notice that the squared coefficient of variation can be written as $C^2[X] = E[X^2]/(E[X])^2 - 1$; thus, the shape parameter is the value of α that satisfies

$$C^2[X] + 1 = \frac{\Gamma(1+2/\alpha)}{(\Gamma(1+1/\alpha))^2} , \tag{1.33}$$

and the scale parameter is then determined by

$$\beta = \frac{E[X]}{\Gamma(1+1/\alpha)} \tag{1.34}$$

Example 1.19. Suppose we would like to find the parameters of the Weibull random variable with mean 100 and standard deviation 25. We first note that $C^2[X] + 1 = 1.0625$. We then fill in a spreadsheet with the following values and formulas.

	A	B
1	`alpha-guess`	`1`
2	`first moment term`	`=EXP(GAMMALN(1+1/B1))`
3	`second moment term`	`=EXP(GAMMALN(1+2/B1))`
4	`ratio`	`= B3/(B2*B2)`
5	`mean`	`100`
6	`beta-value`	`=B5/B2`

The GoalSeek tool (found under the "Tools" menu in Excel 2003 and under the "What-If" button on the Data Tab for Excel 2007) is ideal for solving (1.33). When GoalSeek is clicked, a dialog box appears with three parameters. For the above spreadsheet, the "Set cell" parameter is set to `B4`, the "To value" parameter is set to `1.0625` (i.e., the value of the $C^2[X] + 1$), and the "By changing cell" parameter is set to `B1`. The results should be that the `B1` cell is changed to 4.519 and the `B6` cell is changed to 109.55. □

Problems

1.1. A manufacturing company ships (by truckload) its product to three different distribution centers on a weekly basis. Demands vary from week to week ranging over 0, 1, and 2 truckloads needed at each distribution center. Conceptualize an experiment where a week is selected and then the number of truckloads demanded at each of the three centers are recorded.
(a) Describe the sample space, i.e., list all outcomes.
(b) How many possible different events are there?
(c) Write the event that represents a total of three truckloads are needed for the week.
(d) If each event containing a single outcome has the same probability, what is the probability that a total demand for three truckloads will occur?

1.2. A library has classified its books into fiction and nonfiction. Furthermore, all books can also be described as hardback and paperback. As an experiment, we shall pick a book at random and record whether it is fiction or nonfiction and whether it is paperback or hardback.
(a) Describe the sample space, i.e., list all outcomes.
(b) Describe the event space, i.e., list all events.
(c) Define a probability measure such that the probability of picking a nonfiction paperback is 0.15, the probability of picking a nonfiction book is 0.30, and the probability of picking a fiction hardback is 0.65.
(d) Using the probabilities from part (c), find the probability of picking a fiction book given that the book chosen is known to be a paperback.

1.3. Let N be a random variable describing the number of defective items in a box from Example 1.1. Draw the graph for the cumulative distribution function of N and give its pmf.

1.4. Let X be a random variable with cumulative distribution function given by

$$G(a) = \begin{cases} 0 & \text{for } a < 0, \\ a^2 & \text{for } 0 \le a < 1, \\ 1 & \text{for } a \ge 1. \end{cases}$$

(a) Give the pdf for X.
(b) Find $\Pr\{X \ge 0.5\}$.
(c) Find $\Pr\{0.5 < X \le 0.75\}$.
(d) Let X_1 and X_2 be independent random variables with their *CDF* given by $G(\cdot)$. Find $\Pr\{X_1 + X_2 \le 1\}$.

1.5. Let T be a random variable with pdf given by

$$f(t) = \begin{cases} 0 & \text{for } t < 0.5, \\ ke^{-2(t-0.5)} & \text{for } t \geq 0.5. \end{cases}$$

(a) Find k.
(b) Find $\Pr\{0.25 \leq T \leq 1\}$.
(c) Find $\Pr\{T \leq 1.5\}$.
(d) Give the cumulative distribution function for T.
(e) Let the independent random variables T_1 and T_2 have their pdf given by $f(\cdot)$. Find $\Pr\{1 \leq T_1 + T_2 \leq 2\}$.
(f) Let $Y = X + T$, where X is independent of T and is defined by the previous problem. Give the pdf for Y.

1.6. Let U be a random variable with pdf given by

$$h(u) = \begin{cases} 0 & \text{for } u < 0, \\ u & \text{for } 0 \leq u < 1, \\ 2-u & \text{for } 1 \leq u < 2, \\ 0 & \text{for } u \geq 2. \end{cases}$$

(a) Find $\Pr\{0.5 < U < 1.5\}$.
(b) Find $\Pr\{0.5 \leq U \leq 1.5\}$.
(c) Find $\Pr\{0 \leq U \leq 1.5,\ 0.5 \leq U \leq 2\}$. (A comma acts as an intersection and is read as an "and".)
(d) Give the cumulative distribution function for U and calculate $\Pr\{U \leq 1.5\} - \Pr\{U \leq 0.5\}$.

1.7. An independent roofing contractor has determined that the number of jobs obtained for the month of September varies. From previous experience, the probabilities of obtaining 0, 1, 2, or 3 jobs have been determined to be 0.1, 0.35, 0.30, and 0.25, respectively. The profit obtained from each job is $300. What is the expected profit for September?

1.8. There are three investment plans for your consideration. Each plan calls for an investment of $25,000 and the return will be one year later. Plan A will return $27,500. Plan B will return $27,000 or $28,000 with probabilities 0.4 and 0.6, respectively. Plan C will return $24,000, $27,000, or $33,000 with probabilities 0.2, 0.5, and 0.3, respectively. If your objective is to maximize the expected return, which plan should you choose? Are there considerations that might be relevant other than simply the expected values?

1.9. Let the random variables A, B, C denote the returns from investment plans A, B, and C, respectively, from the previous problem. What are the mean and standard deviations of the three random variables?

1.10. Let N be a random variable with cumulative distribution function given by

$$F(x) = \begin{cases} 0 & \text{for } x < 1, \\ 0.2 & \text{for } 1 \le x < 2, \\ 0.5 & \text{for } 2 \le x < 3, \\ 0.8 & \text{for } 3 \le x < 4, \\ 1 & \text{for } x \ge 4. \end{cases}$$

Find the mean and standard deviation of N.

1.11. Prove that the $E[(X-\mu)^2] = E[X^2] - \mu^2$ for any random variable X whose mean is μ.

1.12. Find the mean and standard deviation for X as defined in Problem 1.4.

1.13. Show using integration by parts that

$$E[X] = \int_0^b [1 - F(x)]\mathrm{d}x, \quad \text{for } 0 \le a \le x \le b,$$

where F is the CDF of a random variable with support on the interval $[a,b]$ with $a \ge 0$. Note that the lower integration limit is 0 not a. (A random variable is zero outside its interval of support.)

1.14. Find the mean and standard deviation for U as defined in Problem 1.6. Also, find the mean and standard deviation using the last two properties mentioned in Property 1.4.

Use the appropriate distribution from Sect. 1.4 to answer the questions in Problems 1.15–1.18.

1.15. A manufacturing company produces parts, 97% of which are within specifications and 3% are defective (outside specifications). There is apparently no pattern to the production of defective parts; thus, we assume that whether or not a part is defective is independent of other parts.
(a) What is the probability that there will be no defective parts in a box of 5?
(b) What is the probability that there will be exactly 2 defective parts in a box of 5?
(c) What is the probability that there will be 2 or more defective parts in a box of 5?
(d) Use the Poisson distribution to approximate the probability that there will be 4 or more defective parts in a box of 40.
(e) Use the normal distribution to approximate the probability that there will be 20 or more defective parts in a box of 400.

1.16. A store sells two types of tables: plain and deluxe. When an order for a table arrives, there is an 80% chance that the plain table will be desired.
(a) Out of 5 orders, what is the probability that no deluxe tables will be desired?
(b) Assume that each day 5 orders arrive and that today (Monday) an order came for a deluxe table. What is the probability that the first day in which one or more deluxe tables are again ordered will be in three more days (Thursday)? What is the expected number of days until a deluxe table is desired?
(c) Actually, the number of orders each day is a Poisson random variable with a mean of 5. What is the probability that exactly 5 orders will arrive on a given day?

1.17. A vision system is designed to measure the angle at which the arm of a robot deviates from the vertical; however, the vision system is not totally accurate. The results from observations is a continuous random variable with a uniform distribution. If the measurement indicates that the range of the angle is between 9.7 and 10.5 degrees, what is the probability that the actual angle is between 9.9 and 10.1 degrees?

1.18. In an automated soldering operation, the location at which the solder is placed is very important. The deviation from the center of the board is a normally distributed random variable with a mean of 0 inches and a standard deviation of 0.01 inches. (A positive deviation indicates a deviation to the right of the center and a negative deviation indicates a deviation to the left of the center.)
(a) What is the probability that on a given board the actual location of the solder deviated by less than 0.005 inches (in absolute value) from the center?
(b) What is the probability that on a given board the actual location of the solder deviated by more than 0.02 inches (in absolute value) from the center?

1.19. The purpose of this problem is to illustrate the dangers of statistics, especially with respect to categorical data and the use of conditional probabilities. In this example, the data may be used to support contradicting claims, depending on the inclinations of the person doing the reporting! The population in which we are interested is made up of males and females, those who are sick and not sick, and those who received treatment prior to becoming sick and who did not receive prior treatment. (In the questions below, assume that the treatment has no adverse side effects.) The population numbers are as follows.

Males		
	sick	not sick
treated	200	300
not treated	50	50

Females		
	sick	not sick
treated	50	100
not treated	200	370

(a) What is the conditional probability of being sick given that the treatment was received and the patient is a male?
(b) Considering only the population of males, should the treatment be recommended?
(c) Considering only the population of females, should the treatment be recommended?
(d) Considering the entire population, should the treatment be recommended?

1.20. Let X and Y be two discrete random variables where their joint pmf

$$f(a,b) = \Pr\{X = a, Y = b\}$$

is defined by

	0	1	2
10	0.01	0.06	0.03
11	0.02	0.12	0.06
12	0.02	0.18	0.10
13	0.07	0.24	0.09

with the possible values for X being 10 through 13 and the possible values for Y being 0 through 2.
(a) Find the marginal pmf's for X and Y and then find the $\Pr\{X = 11\}$ and $E[X]$.
(b) Find the conditional pmf for X given that $Y = 1$ and then find the $\Pr\{X = 11|Y = 1\}$ and find the $E[X|Y = 1]$.
(c) Are X and Y independent? Why or why not?
(d) Find $\Pr\{X = 13, Y = 2\}$, $\Pr\{X = 13\}$, and $\Pr\{Y = 2\}$. (Now make sure your answer to part (c) was correct.)

1.21. Let S and T be two continuous random variables with joint pdf given by

$$f(s,t) = kst^2 \text{ for } 0 \leq s \leq 1,\ 0 \leq t \leq 1\,,$$

and zero elsewhere.
(a) Find the value of k.
(b) Find the marginal pdf's for S and T and then find the $\Pr\{S \leq 0.5\}$ and $E[S]$.
(c) Find the conditional pdf for S given that $T = 0.1$ and then find the $\Pr\{S \leq 0.5|T = 0.1\}$ and find the $E[S|T = 0.1]$.
(d) Are S and T independent? Why or why not?

1.22. Let U and V be two continuous random variables with joint pdf given by

$$g(u,v) = e^{-u-v} \text{ for } u \geq 0,\ v \geq 0\,,$$

and zero elsewhere.
(a) Find the marginal pdf's for U and V and then find the $\Pr\{U \leq 0.5\}$ and $E[U]$.
(b) Find the conditional pdf for U given that $V = 0.1$ and then find the $\Pr\{U \leq 0.5|V = 0.1\}$ and find the $E[U|V = 0.1]$.
(c) Are U and V independent? Why or why not?

1.23. This problem is to consider the importance of keeping track of history when discussing the reliability of a machine and to emphasize the meaning of Eq. (1.15). Let T be a random variable that indicates the time until failure for the machine. Assume that T has a uniform distribution from zero to two years and answer the question, "What is the probability that the machine will continue to work for at least three more months?"
(a) Assume the machine is new.
(b) Assume the machine is one year old and has not yet failed.
(c) Now assume that T has an exponential distribution with mean one year, and answer parts (a) and (b) again.

(d) Is it important to know how old the machine is in order to answer the question, "What is the probability that the machine will continue to work for at least three more months?"

1.24. Determine the correlation coefficient for the random variables X and Y from Example 1.13.

1.25. A shipment containing 1,000 steel rods has just arrived. Two measurements are of interest: the cross-sectional area and the force that each rod can support. We conceptualize two random variables: A and B. The random variable A is the cross-sectional area, in square centimeters, of the chosen rod, and B is the force, in kilo-Newtons, that causes the rod to break. Both random variables can be approximated by a normal distribution. (A generalization of the normal distribution to two random variables is called a bivariate normal distribution.) The random variable A has a mean of 6.05 cm^2 and a standard deviation of 0.1 cm^2. The random variable B has a mean of 132 kN and a standard deviation of 10 kN. The correlation coefficient for A and B is 0.8.

To answer the questions below use the fact that if X_1 and X_2 are bivariate normal random variables with means μ_1 and μ_2, respectively, variances σ_1 and σ_2, respectively, and a correlation coefficient ρ, the following hold:

- The marginal distribution of X_1 is normal.
- The conditional distribution of X_2 given X_1 is normal.
- The conditional expectation is given by

$$E[X_2|X_1 = x] = \mu_2 + \rho \frac{\sigma_2}{\sigma_1}(x - \mu_1) .$$

- the conditional variance is given by

$$V[X_2|X_1 = x] = \sigma_2^2(1 - \rho^2) .$$

(a) Specifications call for the rods to have a cross-sectional area of between 5.9 cm^2 and 6.1 cm^2. What is the expected number of rods that will have to be discarded because of size problems?
(b) The rods must support a force of 31 kN, and the engineer in charge has decided to use a safety factor of 4; therefore, design specifications call for each rod to support a force of at least 124 kN. What is the expected number of rods that will have to be discarded because of strength problems?
(c) A rod has been selected, and its cross-sectional area measures 5.94 cm^2. What is the probability that it will not support the force required in the specifications?
(d) A rod has been selected, and its cross-sectional area measures 6.08 cm^2. What is the probability that it will not support the force required in the specifications?

1.26. Using Property 1.9, show the following relationship holds for two dependent random variables, X and Y:

$$V[Y] = E[V[Y|X]] + V[E[Y|X]] .$$

1.27. Let X_1 and X_2 be two independent Bernoulli random variables with $E[X_1] = 0.8$ and $E[X_2] = 0.6$. Let $S = X_1 + X_2$.
(a) Give the joint pmf for S and X_1.
(b) Give the marginal pmf for S.
(c) Give the correlation coefficient for S and X_1.
(d) Give the conditional pmf for S given $X_1 = 0$ and $X_1 = 1$.
(e) Demonstrate that Property 1.9 is true where $Y = S$ and $X = X_1$.
(f) Demonstrate that the property given in Problem 1.26 is true where $Y = S$ and $X = X_1$.

1.28. Derive the expression for the variance in Property 1.10. For this proof, you will need to use the following two equations:

$$E[S^2] = E\left[E\left[\left(\sum_{i=1}^{N} X_i\right)^2 |N\right]\right],$$

and

$$E\left[\left(\sum_{i=1}^{n} X_i\right)^2\right] = E\left[\sum_{i=1}^{n} X_i^2 + \sum_{i=1}^{n}\sum_{j\neq i} X_i X_j\right].$$

References

1. Abramowitz, M., and Stegun, I.A., editors (1970). *Handbook of Mathematical Functions*, Dover Publications, Inc., New York.
2. Barlow, R.E., and Proschan, F. (1975). *Statistical Theory of Reliability and Life Testing*, Holt, Rinehart and Winston, Inc., New York.
3. Curry, G.L., and Feldman, R.M. (2009). *Manufacturing Systems Modeling and Analysis*, Springer-Verlag, Berlin.
4. Çinlar, E. (1975). *Introduction to Stochastic Processes*, Prentice-Hall, Inc., Englewood Cliffs, NJ.
5. Hastings, Jr., C. (1955). *Approximations for Digital Computers*, Princeton University Press, Princeton, NJ.
6. Johnson, N.L., Kemp, A.W., and Kotz, S. (2005). *Univariate Discrete Distributions*, 3rd ed., John Wiley Sons, New York.
7. Johnson, N.L., Kotz, S, and Balakrishnan, N. (1994). *Continuous Univariate Distributions, Volume 1*, 2nd ed., John Wiley Sons, New York.
8. Johnson, N.L., Kotz, S, and Balakrishnan, N. (1995). *Continuous Univariate Distributions, Volume 2*, 2nd ed., John Wiley Sons, New York.
9. Law, A.M. (2007). *Simulation Modeling and Analysis*, 4th ed., McGraw Hill, Boston.
10. Rahman, N.A. (1968). *A Course in Theoretical Statistics*, Hafner Publishing Co., New York.

Chapter 2
Basics of Monte Carlo Simulation

Simulation is one of the most widely used probabilistic modeling tools in industry. It is used for the analysis of existing systems and for the selection of hypothetical systems. For example, suppose a bank has been receiving complaints from customers regarding the length of time that customers are spending in line waiting at the drive-in window. Management has decided to add some extra windows; they now need to decide how many to add. Simulation models can be used to help management in determining the number of windows to add. Even though the main focus of this textbook is towards building analytical (as opposed to simulation) models, there will be times when the physical system is too complicated for analytical modeling; in such a case, simulation would be an appropriate tool. The idea behind simulation, applied to this banking problem, is that a computer program would be written to generate randomly arriving customers, and then process each customer through the drive-in facility. In such a manner, the effect of having different numbers of windows could be determined before the expense of building them is incurred. Or, to continue this example, it may be possible (after covering Chap. 7) to build an analytical model of the banking problem; however, the analyst may want to furnish additional evidence for the validity of the analytical model. (When a model is to be used for decision making that involves large capital expenditures, validation efforts are always time consuming and essential.) A simulation could be built to model the same system as the analytical model describes. Then, if the two models agree, the analyst would have confidence in their use.

A final reason for simulation is that it can be used to increase understanding of the process being modeled. In fact, this is the reason that this introduction to simulation is the second chapter. As we introduce various aspects of random processes, simulation will be used to help the intuitive understanding of the dynamics of the process. It is impossible to build a simulation of something not understood, so just the process of developing a simulation of a specific process will force an understanding of that process.

The term "Monte Carlo" is usually applied to a simulation that does not involve the passage of time; thus, this introductory chapter gives the basics of simulation mainly being concerned with the generation of random numbers and random vari-

R.M. Feldman, C. Valdez-Flores, *Applied Probability and Stochastic Processes*, 2nd ed., DOI 10.1007/978-3-642-05158-6_2, © Springer-Verlag Berlin Heidelberg 2010

ates. Because of the prevalence of random number generators within many different types of software, the sections on random numbers can be skipped. The section dealing with generating random variates should not be skipped since it will be used in later chapters to illustrate non-uniform distributions using Excel. A more detailed treatment of simulation is postponed until Chap. 9. In particular, the use of future event lists and the statistical analyses of simulation output are discussed in the second simulation chapter.

One important topic that will not be discussed in this book is the use of a special simulation language. To facilitate the building of simulations, many special purpose simulation languages have been developed. Some of these languages are very easy to learn, and we recommend that students interested in simulation investigate the various simulation languages. For our use, Excel will be sufficient for demonstrating these concepts.

2.1 Simulation by Hand

In this section, we introduce the basic concepts of simulation through two simple examples. The examples use simulation to determine properties of simple probabilistic events. Although the examples can be solved using analytical equations, they should serve to illustrate the concept of simulation.

Example 2.1. Consider a microwave oven salesperson who works in a busy retail store. Often people come into the department merely to browse, while others are interested in purchasing an oven. Of all the people who take up the salesperson's time, 50 percent end up buying one of the three models available and 50 percent do not make a purchase. Of those customers who actually buy an oven, 25 percent purchase the plain model, 50 percent purchase the standard model, and 25 percent purchase the deluxe model. The plain model yields a profit of $30, the standard model yields a profit of $60, and the deluxe model yields a profit of $75.

The salesperson wishes to determine the average profit per customer. (We assume that the salesperson does not have a mathematical background and therefore is unable to calculate the exact expected profit per customer. Hence, an alternative to a mathematical computation for estimating the profit must be used.) One approach for estimating the average profit would be to keep records of all the customers who talk to the salesperson and, based on the data, calculate an estimate for the expected profit per customer. However, there is an easier and less time-consuming method: the process can be simulated.

The basic concept in simulation is to generate random outcomes (for example, we might toss a coin or roll dice to generate outcomes) and then associate appropriate physical behavior with the resultant (random) outcomes. To simulate the microwave oven buying decision, a (fair) coin is tossed with a head representing a customer and a tail representing a browser; thus, there is a 50–50 chance that an individual entering the department is a customer. To simulate the type of microwave that is bought by an interested customer, two fair coins are tossed. Two tails (which occur

25 percent of the time) represent buying the plain model, a head and a tail (which occur 50 percent of the time) represent buying the standard model, and two heads (which occur 25 percent of the time) represent buying the deluxe model. Table 2.1 summarizes the relationship between the random coin tossing results and the physical outcomes being modeled. Table 2.2 shows the results of repeating this process 20 times to simulate 20 customers and their buying decisions.

Table 2.1 Probability laws for Example 2.1

Random value	Simulated outcome
Head	Customer
Tail	Browser
Tail-tail	Buy plain
Tail-head	Buy standard
Head-tail	Buy standard
Head-head	Buy deluxe

Table 2.2 Simulated behavior of 20 customers for Example 2.1

Customer number	Coin toss	Interested?	Two-coin toss	Profit	Cumulative profit	Average profit
1	T	no	–	0	0	0.00
2	H	yes	TH	60	60	30.00
3	H	yes	HH	75	135	45.00
4	H	yes	TH	60	195	48.75
5	T	no	–	0	195	39.00
6	T	no	–	0	195	32.50
7	H	yes	TT	30	225	32.14
8	T	no	–	0	225	28.13
9	H	yes	HT	60	285	31.67
10	H	yes	HH	75	360	36.00
11	H	yes	TT	30	390	35.45
12	H	yes	TT	30	420	35.00
13	T	no	–	0	420	32.31
14	T	no	–	0	420	30.00
15	T	no	–	0	420	28.00
16	H	yes	TH	60	480	30.00
17	H	yes	HT	60	540	31.76
18	T	no	–	0	540	30.00
19	H	yes	HH	75	615	32.37
20	H	yes	HH	75	690	34.50

If we take the final cumulative profit and divide it by the number of customers, the estimate for the expected profit per customer is calculated to be \$34.50. Two facts are immediately obvious. First, the number of interested customers was 12 out of 20, but because there is a 50-50 chance that any given customer will be interested, we expect only 10 out of 20. Furthermore, utilizing some basic probability rules, a person knowledgeable in probability would determine that the theoretical expected profit per customer is only $28.125. Thus, it is seen that the simulation does not provide the exact theoretical values sought but will provide reasonable estimates if enough random numbers are used. □

The simulation is, in fact, just a statistical experiment. This point cannot be overemphasized. *The results of a simulation involving random numbers must be interpreted statistically.* In the next chapter, a summary of the basic statistical concepts needed for analyzing simulations as well as some stochastic processes will be given. Chapter 9 gives the specific application of statistics to the analysis of simulation output. For now it is important to realize that the simulation is simply a statistical experiment performed so that the expense and time needed to perform and/or to observe the actual process can be avoided. For instance, this experimental evaluation might require less than an hour to accomplish, whereas obtaining an estimate of the theoretical profit by observing the actual process for 20 customers might take days.

Example 2.2. Let us consider a game taken from a popular television game show. The contestant is shown three doors, labeled A, B, and C, and told that behind one of the three doors is a pot of gold worth $120,000; there is nothing behind the other two doors. The contestant is to pick a door. After the door is selected, the host will pick from the other two doors one with an empty room. The contestant will then be given the choice of switching doors or keeping the originally selected door. After the contestant decides whether or not to switch, the game is over and the contestant gets whatever is behind the door finally chosen. The question of interest is "should the contestant switch when given the option?" We simulate here the "no-switch" option and refer to the homework Problem 2.2 for the "switch" option. (It is interesting to note that in the column "Ask Marilyn" appearing in *Parade* magazine's September 9, 1990, issue, the columnist, Marilyn vos Savant, indicated that it is always better to switch when given the choice. Her assertion to switch generated an estimated 10,000 responses and 92% of those responding (including many PhDs) wrote to say that Marilyn was wrong! After your simulation, you will be able to make an informed decision as to Marilyn's assertion. See [7, pp. 42–45] for an excellent and entertaining discussion of this game based on TV's "Let's Make a Deal".)

To simulate the "no-switch" policy, it is necessary to simulate the random event of placing the gold behind a door, and simulate the event of selecting a door. Both events will be simulated by rolling a single die and interpreting the outcome according to Table 2.3. Since a switch is never made, it is not necessary to simulate the host selecting a door after the initial selection.

Table 2.3 Random events for the "pot-of-gold" game

Random value of die	Simulated outcome
1 or 2	Choose Door A
3 or 4	Choose Door B
5 or 6	Choose Door C

Table 2.4 Simulated results from the "pot-of-gold" game using a no-switch policy

Game number	Die toss	Door for gold	Die toss	Door chosen	Yield	Cumulative	Average
1	5	*C*	2	*A*	0	0	0
2	3	*B*	5	*C*	0	0	0
3	5	*C*	4	*B*	0	0	0
4	3	*B*	2	*A*	0	0	0
5	2	*A*	1	*A*	$120,000	$120,000	$24,000
6	1	*A*	6	*C*	0	$120,000	$20,000
7	6	*C*	4	*B*	0	$120,000	$17,143
8	4	*B*	4	*B*	$120,000	$240,000	$30,000
9	5	*C*	5	*C*	$120,000	$360,000	$40,000
10	5	*C*	3	*B*	0	$360,000	$36,000
11	4	*B*	2	*A*	0	$360,000	$32,727
12	3	*B*	4	*B*	$120,000	$480,000	$40,000
13	2	*A*	4	*B*	0	$480,000	$36,923
14	1	*A*	3	*B*	0	$480,000	$34,286
15	1	*A*	3	*B*	0	$480,000	$32,000
16	4	*B*	6	*C*	0	$480,000	$30,000
17	5	*C*	6	*C*	$120,000	$600,000	$35,294
18	4	*B*	4	*B*	$120,000	$720,000	$40,000
19	1	*A*	6	*C*	0	$720,000	$37,895
20	2	*A*	4	*B*	0	$720,000	$36,000

The results from twenty simulated games are shown in Table 2.4. Based on these simulated results of the game, the estimate for the expected yield is $36,000 per game. (Of course, it is easy to calculate that the actual expected yield is $40,000 per game, which illustrates that simulations only yield statistical estimates.) □

As can be seen from these two examples, a difficulty in simulation is deciding what information to maintain, and then keeping track of that information. Once the decision is made as to what information is relevant, the simulation itself is straightforward although it can be time consuming to run.

- *Suggestion: Do Problems 2.1 and 2.2.*

2.2 Generation of Random Numbers

System simulation depends heavily on random number generation to model the stochastic nature of the systems being studied. The examples in the previous section simulated random outcomes by having the modeler physically toss a coin or a die. A computer, by contrast, must simulate random outcomes by generating numbers to give the appearance of randomness. This section deals with techniques for generating random numbers which are uniformly distributed between 0 and 1. The next section deals with methods for using random numbers to produce values according to general probability laws (i.e., randomness not restricted to uniform). Future chapters will use simulation to illustrate various random processes by taking advantage of the random number generator of Excel. Since Excel has its own random number generator, this section may be skipped, although the next section dealing with the more general random variates would be worthwhile to read.

Most mathematical table books contain several pages of random numbers. Such random numbers have three properties: (1) the numbers are between zero and one, (2) the probability of selecting a number in the interval (a,b) is equal to $(b-a)$, where $0 \leq a < b \leq 1$, and (3) the numbers are statistically independent. The concept of statistical independence simply means that if several numbers are chosen, knowledge of the value of one number does not provide information that will help in predicting the value of another number. Independence, therefore, rules out the possibility of trends or cyclic patterns occurring within a sequence of random numbers. In terms of probability, these random numbers are thus realizations from a sequence of independent, identically distributed random variables having a continuous uniform distribution between zero and one. If random variable from a distribution other than uniform between 0 and 1 is needed, the numbers are called random variates which are covered in the next section.

Because a computer can follow only very specific steps, random numbers generated by a computer are not truly random but are more properly called "pseudo-random numbers" (although the "pseudo" prefix is usually dropped). Numbers generated by a computer program or subprogram are, therefore, called "random" if statistical tests cannot determine the difference between computer-generated numbers and truly random sequences of numbers (see the next chapter for some such statistical tests).

All methods of random number generation in computers begin with an initial value called the *initial random number seed*. Each time a random number is desired, the random number seed is used to produce the next random number and the seed is transformed into another number that becomes the new seed. Thus, the random number seed is continually changed as the random numbers are generated. A simulation program must always furnish an initial random number seed. If the same initial random number seed is used whenever the simulation program is run, the exact same stream of random numbers will be generated. If a different stream of random numbers is desired, then a different initial random number seed must be furnished. In the following, three methods that have been used for random number generation are presented.

2.2.1 Multiplicative Linear Congruential Generators

One of the most popular methods of generating random numbers until 10 to 15 years ago was the multiplicative linear congruential generator method (abbreviated MLCG). To define this method, a multiplier (call it a) and a modulus (call it m) must be given. The random number seed will always be positive and less than the modulus and the random number associated with the seed is the value of the seed divided by the modulus. To obtain the new seed, the old seed is multiplied by multiplier. The resulting integer might then be larger than the modulus, and if it is too large, it is divided by the modulus and the remainder is the new seed. The following property expresses the multiplicative linear congruential method mathematically.

Property 2.1. ***The MLCG for Random Numbers:*** *Let S_0 be an initial random number seed with a fixed multiplier denoted by a and fixed modulus m. Both the initial random number seed and the multiplier must be less than the modulus. The stream of random number seeds is given by*

$$S_{k+1} = (a S_k) \textbf{ mod } m \text{ for } k = 0, 1, \cdots ,$$

and the associated stream of random numbers is

$$R_{k+1} = \frac{S_{k+1}}{m} \text{ for } k = 0, 1, \cdots .$$

Example 2.3. This first example is clearly absurd, but it should serve to illustrate the mechanics of an MLCG. Consider a generator with multiplier $a = 2$ and modulus $m = 7$. This multiplier will generate the following stream of "random" numbers using an initial random number seed of $S_0 = 3$:

$$\begin{aligned} S_1 &= (2 \times 3) \textbf{ mod } 7 = 6; \quad \text{and } R_1 = 6/7 = 0.8571 \\ S_2 &= (2 \times 6) \textbf{ mod } 7 = 5; \quad \text{and } R_2 = 5/7 = 0.7143 \\ S_3 &= (2 \times 5) \textbf{ mod } 7 = 3; \quad \text{and } R_3 = 3/7 = 0.4286 \\ S_4 &= (2 \times 3) \textbf{ mod } 7 = 6; \quad \text{and } R_4 = 6/7 = 0.8571 \,. \end{aligned}$$

□

There are a few problems with this stream of random numbers. First, the numbers are not very dense. With a modulus of only 7, at most we can generate the number 1/7, 2/7, etc, so that there is a significant amount of "space" between two close random numbers. To fix this problems, we simply increase the modulus. A second problem is that the numbers start repeating themselves after just three values. You should be able to verify that the next random number that will be generated will equal 0.7143; in other words, no new numbers will be generated after the first three. This property is called the period of the generator.

Definition 2.1. The *period* of a random number generator is the number of seeds that can be produced before the exact sequence of seeds repeat themselves. □

You should be able to verify that if the multiplier had been 3, the period would be 6 which is the maximum possible period for an MLCG with modulus of 7. Finally, for a generator to be reasonable, a stream of these numbers need to be able to "fool" a statistician into thinking that the number are from a uniform distribution and independent. The stream of random numbers generated in the above example would obviously fool no one!

The modulus for an MLCG should always be a prime number, in which case it is possible to find a multiplier for which the period of the generator equals the modulus minus one, called a *full period* generator. A popular generator for 32-bit machines used for many years was based on an algorithm by Schrage [9] that gave a portable code for the MLCG using a modulus of $m = 2,147,483,647$ and a multiplier of $a = 16,807$. The modulus is $2^{31} - 1$ which equals the largest signed integer that can be stored in a 32-bit machine and it is prime. Another multiplier that may provide better statistical properties for the resulting stream of random numbers is $a = 630,360,016$. Notice that the period of these generators is slightly more than 2.1×10^9 which seems large; however, with large simulations, especially within a multi-processor environment, it would be nice to have longer periods.

2.2.2 A Multiple Recursive Generator

It is now well recognized that linear congruential generators for large streams of random numbers have some weaknesses in their statistical properties, especially as it relates to the independence of successive numbers in three or more dimensions. The multiple recursive generators (abbreviated MRG) is an extension of the MLCG in that instead of the next seed depending only on the previous seed, the next seed depends on two or more of the previous seeds. The simplest of the fast MRGs suggested by Deng and Lin [2] is the following.

Property 2.2. ***A simplified MRG for Random Numbers:*** *Let S_{-1} and S_0 be two initial random number seeds with a fixed multiplier denoted by a and fixed modulus m. Both the initial random number seed and the multiplier must be less than the modulus. The stream of random number seeds is given by the recursion*

$$S_{k+1} = (aS_{k-1} - S_k) \; \mathbf{mod} \; m \quad \text{for } k = 0, 1, \cdots ,$$

and the associated stream of random numbers is

$$R_{k+1} = \frac{S_{k+1}}{m} \quad \text{for } k = 0, 1, \cdots .$$

The advantages of MRGs are that they have better statistical properties than the MLCGs and they can have much longer periods. Notice that the recursion uses the difference of two positive numbers so that it is possible to have a negative number before using the **mod** operator. When this happens, the modulus is added to the number so that a positive number results. (In other words, (-3) **mod** $7 = 4$ and (-13) **mod** $7 = 1$.)

Example 2.4. Form the MRG from Proposition 2.2 by setting the modulus equal to 7 and the multiplier equal to 4. The first four random numbers using $S_{-1} = 2$ and $S_0 = 6$ are obtained by

$$\begin{aligned}
S_1 &= (4 \times 2 - 6) \textbf{ mod } 7 = 2; \quad \text{and } R_1 = 2/7 = 0.2857 \\
S_2 &= (4 \times 6 - 2) \textbf{ mod } 7 = 1; \quad \text{and } R_2 = 1/7 = 0.1429 \\
S_3 &= (4 \times 2 - 1) \textbf{ mod } 7 = 0; \quad \text{and } R_3 = 0/7 = 0 \\
S_4 &= (4 \times 1 - 0) \textbf{ mod } 7 = 4; \quad \text{and } R_4 = 4/7 = 0.5714\,.
\end{aligned}$$

□

Notice that it is possible to have zero for a random number with MRGs. If the stream were continued, you would observe that $R_6 = 0.8571$ as well as $R_7 = 0.8571$. It is impossible to have two identical consecutive seeds when using an MLCG; whereas, it is possible for the MRG. The largest possible period for an MLCG with $m = 7$ is 6. The period of the generator of Example 2.4 is 48.

When the MRG of Property 2.2 is used on a 32-bit machine, the modulus should be $m = 2,147,483,647$. Five multipliers that work well with this modulus, as found in [2], are 26,403, 33,236, 36,673, 40,851, or 43,693. These values will yield a period of slightly more than 2.6×10^{18}. In today's environment where 48 and 64-bit integer sizes are possible, this MRG should be easy to program.

2.2.3 Composite Generators

The final type of generator to be discussed is a composite generator that essential combines two other generators. For example, suppose a random number generator is desired for an 8-bit machine. Since one bit is to be saved to designate the sign of the number, all integers must be less than 2^7. If a multiplicate generator is used, the modulus should be $m = 127$ since $2^7 - 1$ is a prime[1] number. With the proper choice of the multiplier (there are actually 36 possible choices), a random number generator can be defined to have a period of 126.

Example 2.5. Now consider two multiplicative generators with a slightly shorter period for our 8-bit machine. The first generator has modulus $m_1 = 113$ and multiplier $a_1 = 59$. Once an initial seed, S_0 is defined, the stream of seeds produced will be denoted by $S_1, S_2, \cdots$. The second has modulus $m_2 = 107$ and multiplier $a_2 = 17$. The initial seed for the second generator will be denoted by T_0, with the steam of seeds given by $T_1, T_2, \cdots$. These two generators can be combined by taking the difference of the two seeds to form a new composite seed given by

[1] As a point of interest, 127 is called a *Mersenne Prime* because it is of the form $2^n - 1$. Due to the prevalence of 32-bit machines, another important Mersenne prime is $2^{31} - 1$.

$$U_k = (S_k - T_k) \textbf{ mod } \max\{m_1, m_2\} .$$

(For the mod operation on the negative number, see p. 52.) If the first initial seed is given as $S_0 = 75$ and the second seed is given as $T_0 = 19$, the composite generator will have a period of 5,936 and the first few random numbers generated are as follows.

Table 2.5 Results from the composite generator of Example 2.5

k	S_k	T_k	U_k	RN
0	75	19		
1	18	2	16	0.1416
2	45	34	11	0.0973
3	56	43	13	0.1150
4	27	89	51	0.4513
5	11	15	109	0.9646
6	84	41	43	0.3805

□

A random number generator, due to L'Ecuyer [4, 6], with excellent statistical properties and with a period of about 3.1×10^{57} is a combination of two MRGs, each with a different multiplier and a different modulus. Each of the two MRGs use a recursion based on the previous three seeds so that each MRG requires three initial random number seeds and thus the composite generator requires six initial seeds. The specifics of this composite generator are:

Property 2.3. ***The L'Ecuyer Composite MRG:*** *Let S_{-2}, S_{-1}, and S_0 be three initial random number seeds, and let a second set of three initial random number seeds be given by T_{-2}, T_{-1}, and T_0. Whenever a random number is desired, new random number seeds plus the composite seed are obtained from the sequences according to the following*

$$S_{k+1} = (1{,}403{,}580\, S_{k-1} - 810{,}728\, S_{k-2}) \textbf{ mod } 4{,}294{,}967{,}087$$
$$T_{k+1} = (527{,}612\, T_k - 1{,}370{,}589\, T_{k-2}) \textbf{ mod } 4{,}294{,}944{,}443$$
$$U_{k+1} = (S_{k+1} - T_{k+1}) \textbf{ mod } 4{,}294{,}967{,}087 .$$

The random number is then given by

$$R_{k+1} = \frac{U_{k+1}}{4{,}294{,}967{,}087} .$$

Notice that the S-random seed stream does not use the most recent seed in its recursion, and the T-random seed stream does not use the second most recent seed;

however, both recursions use the third most recent seed. It should be noted that the modulus for the S-random seed stream equals $2^{32} - 209$ and the modulus for the T-random seed stream equals $2^{32} - 22,853$. An excellent discussion on L'Ecuyer's composite generator as well as random number generation in general is found in Law [3]. He also gives an implementation of the composite generator for 32-bit machines. Table 2.6 gives some suggestions as found in [3] for initial random number seeds.

Table 2.6 Ten sets of initial random number seeds for L'Ecuyer's composite MRG (numbers copied with permission from p. 419 of the textbook *Simulation Modeling & Analysis* by Averill M. Law, copyright © 2007 by The McGraw-Hill Companies, Inc.)

S_{-2}	S_{-1}	S_0	T_{-2}	T_{-1}	T_0
1772212344	1374954571	2377447708	540628578	1843308759	549575061
2602294560	1764491502	3872775590	4089362440	2683806282	437563332
376810349	1545165407	3443838735	3650079346	1898051052	2606578666
1847817841	3038743716	2014183350	2883836363	3242147124	1955620878
1075987441	3468627582	2694529948	368150488	2026479331	2067041056
134547324	4246812979	1700384422	2358888058	83616724	3045736624
2816844169	885735878	1824365395	2629582008	3405363962	1835381773
675808621	434584068	4021752986	3831444678	4193349505	2833414845
2876117643	1466108979	163986545	1530526354	68578399	1111539974
411040508	544377427	2887694751	702892456	758163486	2462939166

- *Suggestion: Do Problems 2.3 and 2.7.*

2.3 Generation of Random Variates

Realistic simulations obviously involve many different types of probability distributions so it is important to be able to simulate non-uniform random values, called random variates. Fortunately, once random numbers have been obtained, it is possible (at least theoretically) to transform the random numbers into random variates that follow any distribution.

2.3.1 Discrete Random Variates

Let us assume that we want to write a computer simulation model of the microwave salesperson of Example 2.1. The first random variable represents the decision as to whether or not a customer will end up buying a microwave. Mathematically, we can denote the customer's interest by the random variable C and give its probability mass function as

$$P\{C = 0\} = 0.5 \quad \text{and} \quad P\{C = 1\} = 0.5\,,$$

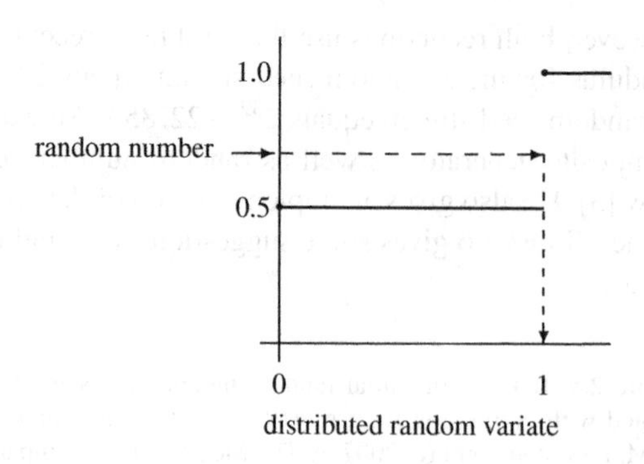

Fig. 2.1 Transformation using the cumulative distribution function to obtain the random variate C

where the random variable being zero ($C = 0$) represents a customer who is not interested in purchasing a microwave, and the random variable being one ($C = 1$) represents the customer who purchases a microwave.

The cumulative probability distribution function (whose values are between zero and one) permits an easy transformation of a random number into a random variate arising from another probability law. The transformation is obtained simply by letting the random number occur along the ordinate (y-axis) and using the cumulative function to form the inverse mapping back to the abscissa (x-axis) as shown in Fig. 2.1 for one possible value of the random number. For the "customer interest" random variable C, the following rule is obtained: If the random number is less than or equal to 0.5, then $C = 0$, i.e., the customer is not interested in a purchase; otherwise, if the random number is greater than 0.5, then $C = 1$, i.e., the customer is interested.

The second random variable that must be simulated in the microwave salesperson example is the profit resulting from the sale of a microwave. Let the "model" random variable be denoted by M with probability mass function

$$\begin{aligned} P\{M = 30\} &= 0.25 \\ P\{M = 60\} &= 0.50 \\ P\{M = 75\} &= 0.25\,. \end{aligned}$$

In this case, $M = 30$ is the profit from the sale of a basic model, $M = 60$ the profit from a standard model, and $M = 75$ the profit from a deluxe model. Again, it is easiest to simulate this random variable if its cumulative probability distribution function is first determined. The jump points of the cumulative function are

$$\begin{aligned} P\{M \leq 30\} &= 0.25 \\ P\{M \leq 60\} &= 0.75 \\ P\{M \leq 75\} &= 1.0\,. \end{aligned}$$

Fig. 2.2 Transformation using the cumulative distribution function to obtain the random variate M

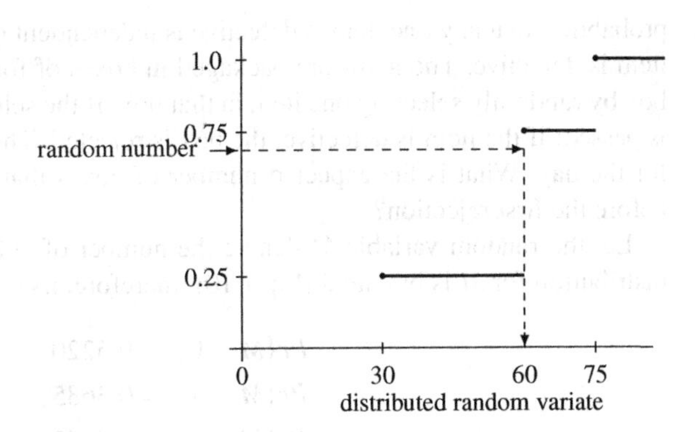

The inverse mapping that gives the transformation from the random number to the random variate representing the profit obtained from the sale of a microwave is illustrated in Fig. 2.2. The results are: If the random number is less than or equal to 0.25, the plain model is sold; if it is greater than 0.25 and less than or equal to 0.75, the standard model is sold; and if it is greater than 0.75, the deluxe model is sold.

The rather intuitive approach for obtaining the random variates C and M as shown in Figs. 2.1 and 2.2 needs to be formalized for use with continuous random variables. Specifically, Figs. 2.1 and 2.2 indicate that the random variates were generated by using the inverse of the cumulative probability distribution function. A mathematical justification for this is now given.

The following property (proven in most introductory mathematical probability and statistics books) leads to the so-called *inverse mapping method* of generating arbitrarily distributed random variates.

Property 2.4. *Let F be an arbitrary cumulative distribution function, and let U be a random variable with a continuous uniform distribution between zero and one. If the inverse of the function F exists, denote it by F^{-1}; otherwise, let $F^{-1}(y) = \min\{t|F(t) \geq y\}$ for $0 \leq y \leq 1$. Then the random variable X defined by*

$$X = F^{-1}(U),$$

has a CDF given by F; that is, $P\{X \leq x\} = F(x)$ for all x.

From the above property, it is seen that the function that relates the continuous uniform zero-one random variable with another random variable is simply the inverse of the cumulative probability distribution function. We utilize this property in the next section. But first, we give some examples simulating discrete random variables.

Example 2.6. A manufacturing process produces items with a defect rate of 15%; in other words, the probability of an individual item being defective is 0.15 and the probability that any one item is defective is independent of whether or not any other item is defective. The items are packaged in boxes of four. An inspector tests each box by randomly selecting one item in that box. If the selected item is good, the box is passed; if the item is defective, the box is rejected. The inspector has just started for the day. What is the expected number of boxes that the inspector will look at before the first rejection?

Let the random variable M denote the number of defective items per box. The distribution for M is binomial (Eq. 1.10); therefore, its pmf is given by

$$\begin{aligned} Pr\{M=0\} &= 0.5220 \\ Pr\{M=1\} &= 0.3685 \\ Pr\{M=2\} &= 0.0975 \\ Pr\{M=3\} &= 0.0115 \\ Pr\{M=4\} &= 0.0005 \end{aligned}$$

The transformation from random numbers to random variates is based on the CDF, so the above probabilities are summed to obtain Table 2.7. (For more efficient techniques than the one demonstrated here for generating some of the common random variates see [3]. Our purpose here is to demonstrate the concept of simulation and not derive the most efficient procedures for large scale programs.) Notice the ranges for the random numbers contained in the table are open on the left and closed on the right. That choice is arbitrary; it is only important to be consistent. In other words, if the random number were equal to 0.8905, the random variate representing the number of defective items in the box would equal 1.

Table 2.7 Mapping to determine the number of defective items per box for Example 2.6

Random number range	number of defective items
(0.0000, 0.5220]	0
(0.5220, 0.8905]	1
(0.8905, 0.9880]	2
(0.9880, 0.9995]	3
(0.9995, 1.0000]	4

The simulation will proceed as follows: First, a box containing a random number of defective items will be generated according to the binomial distribution (Table 2.7). Then, an item will be selected using a Bernoulli random variable, where the probability of selecting a defective item equals the number of defective items divided by four (i.e., the total number of items in a box). To generate random numbers, we can either use a calculator or spreadsheet program. Most calculators have a button labeled "RND" or "Random". Each time such a button is pressed, a (supposedly) random number is generated. If you use Excel, enter `=RAND()` in a cell and a random number will be generated.

Table 2.8 shows the results of the simulation. Notice that Column 4 contains a critical number, which equals the number of defective items divided by four. If the next random number generated is less than or equal to the critical number, the box is rejected; otherwise, the box is accepted. If the critical number is 0, no random number is generated since the box will always be accepted no matter which item the inspector chooses. The simulation is over when the first box containing defective items is selected. Since the question is "how many boxes will be accepted?", it is necessary to make several simulation runs. The final estimate would then be the average length of the runs.

Table 2.8 Three simulation runs for Example 2.6

Box number	Random number	Number of defects	Critical number	Random number	Box accepted?
			Trial # 1		
1	0.2358	0	0	–	yes
2	0.5907	1	0.25	0.3483	yes
3	0.6489	1	0.25	0.9204	yes
4	0.6083	1	0.25	0.2709	yes
5	0.7610	1	0.25	0.5374	yes
6	0.1730	0	0	–	yes
7	0.5044	0	0	–	yes
8	0.6206	1	0.25	0.0474	no
			Trial # 2		
1	0.8403	1	0.25	0.9076	yes
2	0.3143	0	0	–	yes
3	0.0383	0	0	–	yes
4	0.0513	0	0	–	yes
5	0.9537	2	0.50	0.6614	yes
6	0.1637	0	0	–	yes
7	0.4939	0	0	–	yes
8	0.6086	1	0.25	0.1542	no
			Trial # 3		
1	0.2330	0	0	–	yes
2	0.8310	1	0.25	0.8647	yes
3	0.2143	0	0	–	yes
4	0.5414	1	0.25	0.0214	no

Obviously, three trials is not enough to draw any reasonable conclusion, but the idea should be sufficiently illustrated. Based on the results of Table 2.8, we would estimate that the inspector samples an average of 6.67 boxes until finding a box to be rejected. □

Before leaving discrete random variables, a word should be said about generating bivariate random variables. One procedure is to use conditional distributions. For example, if the two dependent random variables, X and Y, are to be generated, we first generate X according to its marginal distribution, then generate Y according to its conditional distribution given the value of X.

Table 2.9 Transformations for determining the contents of phone boxes in Example 2.7

Type of Phone

Random number range	Type phone
(0.00, 0.47]	plain
(0.47, 1.0]	radio

For Plain Phones

RN range	Number defectives
(0.000, 0.787]	0
(0.787, 0.957]	1
(0.957, 1.0]	2

For Radio Phones

RN range	Number defectives
(0.000, 0.849]	0
(0.849, 0.981]	1
(0.981, 1.0]	2

Example 2.7. **Bivariate Discrete Random Variates.** Consider the boxes of phones discussed in Chap. 1. Each box contains two phones, both phones being radio phones or plain phones. The joint pmf describing the probabilities of the phone type and the number of defective phones is given in Example 1.12. Our goal is to estimate, through simulation, the expected number of defective phones per box. (Again, the question we are trying to answer is trivial, but we need to simulate easy systems before attempting complex systems.) Although we could use the joint pmf to generate the bivariate random variables directly, it is often easier to use conditional probabilities when dealing with more complex probability laws. We shall first determine whether the box contains plain or radio phones, and then determine the number of defective phones in the box. Table 2.9 shows the transformations needed to obtain the boxes of phones; namely, the first transformation is from the CDF of the random variable indicating phone type, and the second two transformations come from the *conditional* CDF's for the number of defective phones in a box given the type of phone. For example, the conditional probability that there are no defective phones given that the box contains plain phones is $0.37/0.47 = 0.787$ (see Example 1.12); thus, if a box contains plain phones, a random number in the interval $(0, 0.787]$ will result in no defective phones for that box.

The results obtained after simulating ten boxes is contained in Table 2.10. The random numbers are generated from a calculator. The second random number used for each box (Column 4 of Table 2.10) must be interpreted based on the result of the first random number (Column 3 of Table 2.10). In particular, notice the results for Box 4. The first random number was 0.1454 yielding a box of plain phones. The second random number was 0.9683 yielding two defective phones. However, if the box had been radio phones, the random number of 0.9683 would have yielded only one defective phone. Based on the results of Table 2.10, the expected number of defective phones per box is estimated to be 0.5.

□

- *Suggestion: Do Problems 2.8–2.10.*

Table 2.10 Simulation for boxes of phones in Example 2.7

Box number	Random number	Phone type	Random number	Number defectives	Cumulative defectives
1	0.6594	radio	0.7259	0	0
2	0.6301	radio	0.1797	0	0
3	0.3775	plain	0.5157	0	0
4	0.1454	plain	0.9683	2	2
5	0.7156	radio	0.5140	0	2
6	0.9734	radio	0.9375	1	3
7	0.7269	radio	0.2228	0	3
8	0.1020	plain	0.9991	2	5
9	0.9537	radio	0.6292	0	5
10	0.4340	plain	0.2416	0	5

2.3.2 Continuous Random Variates

Although many physical systems can be modeled using discrete random variables, there are many systems that are more appropriately modeled using continuous random variables. For example, the times between arrivals of customers to a teller window in a bank would be best described by a continuous value instead of a discrete value.

Property 2.4 is used to obtain random variates for any distribution for which the inverse of the cumulative distribution function can be obtained either analytically or numerically. The following example demonstrates the use of the *inverse transformation method* for one of the most commonly used continuous distributions used in stochastic modeling.

Example 2.8. The time between incoming calls to the emergency room between the hours of 10PM and 4AM is according to an exponential distribution with mean of 8 minutes. To simulate the arrival of calls to this emergency room, we first derive a general expression for generating exponential random variates. Let T be an exponentially distributed random variable so that its CDF is given by

$$F(t) = 1 - e^{-t/\theta} \quad \text{for } t \geq 0,$$

where θ is a positive scalar with $E[T] = \theta$. To obtain the inverse, we replace the functional value with the random number U, replace the parameter t with the random variate T, and then solve for the random variate (see Fig. 2.3). Thus,

$$\begin{aligned} U &= 1 - e^{-T/\theta} \\ e^{-T/\theta} &= 1 - U \\ \ln(e^{-T/\theta}) &= \ln(1-U) \\ -\frac{T}{\theta} &= \ln(1-U) \\ T &= -\theta \ln(1-U). \end{aligned}$$

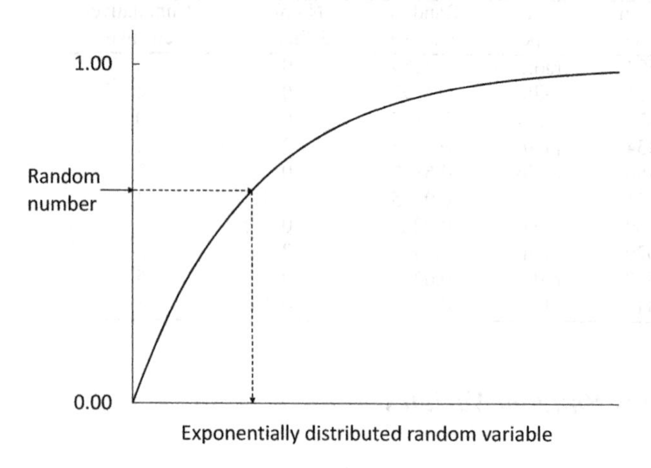

Fig. 2.3 Inverse transformation for an exponentially distributed random variable

One final simplification can be made by observing that if U is uniformly distributed between 0 and 1, then $(1-U)$ must also be uniformly distributed between 0 and 1. Thus, U and $(1-U)$ are equal in distribution which yields

$$T = -E[T]\ln(U)\,. \tag{2.1}$$

To apply Eq. 2.1 to our emergency room call center, suppose we use the "RND" key on our calculator and obtain the following three random numbers: 0.163, 0.742, and 0.448. These three random numbers lead to the following three random variates (i.e., the inter-arrival times): 14.5, 2.4, and 6.4 minutes. If we assume that the simulation starts at 10:00 PM, the arrival times of the first three simulated calls are 10:14.5, 10:16.9, and 10:23.3. □

Since the Erlang distribution (Eq. 1.16) is the sum of exponentials, the above transformation is easily extended for Erlang random variates. Specifically, if X is an Erlang random variate with parameters k and β, then it is the sum of k exponentials, each with mean β/k; thus

$$X = -\frac{\beta}{k}\sum_{i=1}^{k}\ln(U_i) = -\frac{\beta}{k}\ln\left(\prod_{i=1}^{k}U_i\right), \tag{2.2}$$

where U_i are independent random numbers.

The exponential distribution is convenient to use because the expression for its cumulative distribution function is easy to invert. However, many distributions are not so well behaved. A normally distributed random variable (its probability density function has the familiar bell-shaped curve) is an example of a distribution whose cumulative probability distribution function cannot be written in closed form (see Eq. 1.21), much less inverted. Leemis and Park [5] suggest using an approximation for the inverse of the standard normal distribution, and although it may be tedious to use by hand, there should be no problem letting the computer approximate the inverse. The following property, with the constants taken from [8], permits the generation of standard normal random variates that in turn can be used to generate normal random variates with any mean and standard deviation.

Property 2.5. *Let $\Phi(\cdot)$ denote the CDF for a standard normal random variable (i.e., mean 0, standard deviation 1), and let $\Phi^{-1}(u)$ for $0 \le u \le 1$ denote its inverse. Let t be a variable that depends on u as*

$$t = \begin{cases} \sqrt{-2\ln(u)} & \text{for } u < 0.5 \\ \sqrt{-2\ln(1-u)} & \text{for } u \ge 0.5 \,. \end{cases}$$

Then Φ^{-1} can be expressed as the ratio of two polynomials in t as

$$\Phi^{-1}(u) = \begin{cases} -t + \frac{p(t)}{q(t)} & \text{for } u < 0.5 \\ t - \frac{p(t)}{q(t)} & \text{for } u \ge 0.5 \,, \end{cases}$$

where, for all t,

$$p(t) = 0.322232431088 + t + 0.342242088547\,t^2 + 0.0204231210245\,t^3 + 0.0000453642210148\,t^4$$

$$q(t) = 0.099348462606 + 0.588581570495\,t + 0.531103462366\,t^2 + 0.1035377528\,t^3 + 0.0038560700634\,t^4$$

Example 2.9. A manufacturing process produces steel rods. The process is variable and produces rods with a cross sectional area that is distributed according to a normal distribution with a mean of 6.05 cm^2 and standard deviation of 0.1 cm^2. We would like to simulate these rods and, in particular, simulate the cross sectional area of the rod. We note first that if X is normal with mean μ and standard deviation σ, then the random variable

$$Z = \frac{X - \mu}{\sigma} \tag{2.3}$$

is a standard normal random variable so it can be simulated using Property 2.5. Assume that the "RND" function on the calculator produced 0.372 for a random number. From Property 2.5, we get the following: $t = 1.4063$, $p(t) = 2.4624$, $q(t) = 2.2805$, and finally, $\Phi^{-1}(U) = Z = -0.3265$. Thus, the cross sectional area of our first simulated rod is

$$X = \mu + \sigma Z \qquad (2.4)$$
$$= 6.05 + 0.1 \times (-0.3265) = 6.0173\,\text{cm}^2 .$$

□

Example 2.10. In order to simulate the spread of an insect infestation within a forest, a stand of Loblolly pines must be simulated. The height of the trees is distributed according to a log-normal distribution with mean 21.3 m and standard deviation 52.1 m. Since the distribution is log-normal, it can be generated using a normal random variable with mean 2.087 and standard deviation 1.394 (see Example 1.11). Using a calculator to obtain random numbers, the calculations necessary for the first three tree heights are contained in Table 2.11. □

Table 2.11 Simulation for tree heights in Example 2.10

Random Number	Z-value (Property 2.5)	X-value (Eq. 2.4)	Random Height e^X
0.760	0.7063	3.0716	21.576
0.486	-0.03510	2.0381	7.676
0.614	0.2898	2.4909	12.072

There are several other techniques for generating nonuniform random variates when the inverse of the cumulative distribution function cannot be obtained in closed form. However, we will not study these techniques, since either the advanced techniques are buried in the simulation language being used (and thus are transparent to the user) or else an approximation method can be used quite adequately.

- *Suggestion: Do Problems 2.11–2.14.*

2.3.3 Bivariate Continuous Random Variates

As with the discrete bivariate case, a reasonable approach to simulating bivariate continuous random variables is through their conditional probability functions. The disadvantage is that it is often very difficult to obtain the conditional probabilities. One key exception is for bivariate normal random variables, because the conditional distribution is easily obtained from knowledge of the means, variances, and the correlation coefficient. Using the properties given in Problem 1.25, it is not too difficult to obtain an algorithm for generating two bivariate normally distributed random variates. Using the notation from Property 2.5, if we let U be a random number (uniform 0–1), then the standard normal random variate is $X = \Phi^{-1}(U)$.

Property 2.6. *Let U_1 and U_2 denote two independent random numbers and let Φ be the standard normal CDF. Then X_1 and X_2 defined by the following equations will be bivariate normal random variates with means μ_1 and μ_2, variances σ_1^2 and σ_2^2, and correlation coefficient equal to ρ:*

$$X_1 = \mu_1 + \sigma_1 \times \Phi^{-1}(U_1)$$
$$X_2 = \mu_2 + \rho\,\frac{\sigma_2}{\sigma_1}\,(X_1 - \mu_1) + \sigma_2\sqrt{1-\rho^2} \times \Phi^{-1}(U_2)\,.$$

Notice that this property can be simplified if correlated standard normals are desired. Specifically, we again let U_1 and U_2 be two independent random numbers and Φ be the standard normal CDF, then Z_1 and Z_2 will be standard normals with correlation coefficient ρ by the following:

$$Z_1 = \Phi^{-1}(U_1) \tag{2.5}$$
$$Z_2 = \rho\, Z_1 + \sqrt{1-\rho^2} \times \Phi^{-1}(U_2)\,.$$

Obtaining random variates for two random variables with known marginal distributions and a given correlation coefficient is more difficult when the marginals are not normal distributions. For some marginals (like the gamma and log-normal) special algorithms have been developed and are given in Law [3]. For purpose of this review chapter, we present a general procedure that can be used to random variates with arbitrary marginal distributions and with a dependence relationship given by a correlation coefficient. The procedure is to first generate bivariate normals, and then transform these into uniforms, and finally transform the uniforms into the general random variates. We first use Property 2.4 to obtain the uniforms by taking advantage of the distribution Φ and then use Property 2.4 again to obtain the general random variates. The term NORTA (Normal-to-Anything) was coined by Cario and Nelson [1] for this procedure and is given in the following property.

Property 2.7. *Let Z_1 and Z_2 as determined from Eq. (2.5) be two bivariate standard normal random variates with correlation coefficient equal to ρ. Then X_1 and X_2 given by the following equations will be correlated random variates with marginal distributions given by F_1 and F_2.*

$$X_1 = F_1^{-1}(\Phi(Z_1))$$
$$X_2 = F_2^{-1}(\Phi(Z_2)))\,,$$

where Φ is the standard normal CDF.

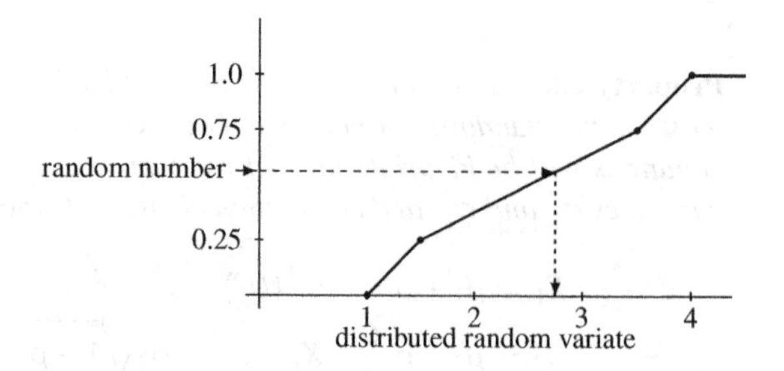

Fig. 2.4 Mapping used to generate a random variate from a distribution based on experimental data

The main difficulty in implementing Property 2.7 is that it does not specify the correlation of the random variables X_1 and X_2. Under many conditions, the correlation of X_1 and X_2 is close to Z_1 and Z_2; however, the exact relationship must be tested empirically. Because this is not difficult to implement in Excel, we give an example illustrating this property in the appendix after discussing the use of several common Excel functions.

2.3.4 Random Variates from Empirical Distributions

Sometimes data are available to approximate a distribution function and so a form of an empirical function is needed. For example, suppose we need to reproduce a continuous random variable we know is always between 1 and 4. Furthermore some data have been collected, and we know that 25% of the data are between 1 and 1.5, 50% of the data are between 1.5 and 3.5, and 25% are between 3.5 and 4. In such a situation we use a piecewise linear function containing three segments to approximate the (unknown) cumulative distribution function. The application of the inverse transformation method is illustrated in Fig. 2.4. Notice from the figure that the cumulative distribution function is graphed, then a random number is created on the y-axis, and finally the random variate of interest is obtained by an inverse mapping from the y-axis to the x-axis.

Example 2.11. Suppose we need to simulate a continuous random variable and all we have are the following ten data points: 5.39, 1.9, 4.62, 2.71, 4.25, 1.11, 2.92, 2.83, 1.88, 2.93. All we know other than the 10 points is that the random variable has some (unknown) upper and lower limits that are positive. Our procedure will be the same as illustrated in the previous example; that is, a cumulative distribution function will be drawn using a piecewise linear approximation and then the inverse transformation method will be used to go from a random number to the random variate. The distribution function should have the property that 10% of the generated points are centered around 5.39, 10% of the generated points are centered around 1.9, etc. To accomplish this, the first step is to order the data, and then calculate the midpoints between adjacent values as shown in Table 2.12.

Since there were 10 data points, there are 10 "cells" with each cell having as its midpoint one of the data points. The major difficulty is the size of the first and last

Table 2.12 Midpoints for Example 2.11

Data point	Midpoint
1.11	
1.88	1.495
1.90	1.890
2.71	2.305
2.83	2.770
2.92	2.875
2.93	2.925
4.25	3.590
4.62	4.435
5.39	5.005

cell; that is, establishing the upper and lower limits. The best way to establish these limits would be through an understanding of the physical limits of whatever it is that is being simulated; however, if the physical limits are not known, one procedure is to assume the lower and upper data points (i.e., 1.11 and 5.39, respectively) are the midpoints of their cells. In other words, the lower and upper limits are established to make this true. Thus, the lower limit is 0.725, and the upper limit is 5.775. (Note that $1.11 = (0.725 + 1.495)/2$ and $5.39 = (5.005 + 5.775)/2$.)

The empirically based distribution function increases 1/10 over each cell because there were ten points. In general, if a distribution function is built with n data points, then it would increase $1/n$ over each data cell. Once the continuous distribution function has been determined, then the inverse mapping technique (similar to Fig. 2.4 except with the function arising from Table 2.12) can be used to obtain the random variate. □

- *Suggestion: Do Problems 2.11–2.14.*

Appendix

As mentioned previously, the Excel function `RAND()` generates random numbers. To generate uniform discrete random variates, use the Excel function

```
RANDBETWEEN( a, b )
```

where a and b are parameters that the user supplies. For example, `RANDBETWEEN(20,30)` will generate discrete uniform random variates between 20 and 30, inclusive. (If you use the 2003 version of Excel, this function is found in the "Data Analysis" tool pack.) For other discrete random variates, the random number should be given in one column and then a nested `If-ELSE` statement used together with the cumulative probabilities of the discrete distribution.

Example 2.12. The discrete random variable N has a probability function given by $\Pr\{N = 2\} = 0.5$, $\Pr\{N = 3\} = 0.3$, and $\Pr\{N = 4\} = 0.2$. In Excel, type the first two rows as

	A	B
1	random number	random variate
2	=RAND()	=IF(A2<0.5,2, IF(A2<0.8,3,4))

Highlight Cells `A2:B10001` and "copy down" (i.e., after highlighting the cells, type `Ctrl-D`). This will give 10,000 replications of the random variate. To determine how well the simulation performed, type 2 in Cell `D1`, type 3 in Cell `E1`, and type 4 in Cell `F1`. Finally, type

```
=COUNTIF($B$2:$B$10001, "=" & D1)/10000
```

in Cell `D2`, and then highlight Cells `D2:F2` and "copy right" (i.e., type `Ctrl-R`). The values in Cells `D2:F2` give the simulated probabilities for the values 2 through 4. Notice that the `RAND()` function is recalculated whenever another cell is changed. If you would like to see the results of a "new" simulation, hit the `F9` key to recalculate all random numbers again. □

For continuous random variates, Excel has several inverse distributions as built-in functions; thus, Property 2.5 can be used to generate the various distributions mentioned in Chap. 1. Table 2.13 lists the associated Excel function that is used for generating random variates from the listed distributions. Some of the listed Excel functions have parameters that must be supplied with numerical values. These parameters are listed using the notation from the corresponding equation as shown in the table.

Table 2.13 Excel functions for some continuous random variates

Distribution	Equation #	Excel Function
Uniform	(1.13)	a + $(b-a)$*RAND()
Exponential	(1.14)	$-(1/\lambda)$*LN(RAND())
Gamma	(1.18)	GAMMAINV(RAND(), α, β)
Weibull	(1.19)	β*(-LN(RAND()))^(1/α)
Standard Normal	(2.3)	NORMSINV(RAND())
Normal	(1.21)	NORMINV(RAND(), μ, σ)
Log Normal	(1.23)	LOGINV(RAND(), μ_N, σ_N)

When using any of these functions within a cell, do not forget to type the equal sign before the function. It might also be noted that in the authors' experience, the random variate for the gamma distribution with shape parameter (α) less than one does not appear to be very accurate with respect to goodness of fit tests.

Example 2.13. **Correlated Random Variables.** Jobs must be processed sequentially through two machines. The processing times for the first machine are distributed according to a gamma distribution with shape parameter 4 and scale parameter 10 minutes. The processing times for jobs on the second machine are distributed according to a continuous uniform distribution from 5 to 25 minutes. After analyzing the times on the two machines for the same job, it is determined that they are not independent and have a correlation coefficient of 0.8. Our goal is to generate pairs of random variates to represent the sequential processing times on the two machines.

We begin the spreadsheet with the constants that are needed to generate the random variates; namely, ρ and $\sqrt{1-\rho^2}$. Type the following in the Cells A1:B2.

	A	B
1	rho	0.8
2	adjusted rho	=SQRT(1-B1*B1)

We shall leave a column to separate the calculations from the random variates and type the following:

	D	E
1	Z1	Z2
2	=NORMSINV(RAND())	=$B$1*D2+$B$2*NORMSINV(RAND())

	F	G
1	X1	X2
2	=GAMMAINV(NORMSDIST(D2),4,10)	=5+20*NORMSDIST(E2)

Now highlight the Cells D2:D30000 and hit the <Ctrl>+D keys to copy the formulas in Row 2 down through Row 30,000, thus producing a large random sample of ordered pairs. In Cell B3 type =CORREL(F2:F30000,G2:G30000). The idea is to change the value in Cell B1 until the value in Cell B3 equals 0.8 (i.e., the time-honored approach of trial and error). Of course, because this is a simulation (i.e., statistical experiment), we cannot claim to have the exact number that will produce our desired value of 0.8. In our experiment, we found that $\rho = 0.842$ yielded a correlation coefficient between the two processing times of 0.8. We also suggest adding some extra data and checking that the mean and standard deviation of the random variates are as expected. □

Problems

2.1. A door-to-door salesman sells pots and pans. He only gets in 50 percent of the houses that he visits. Of the houses that he enters, $1/6$ of the householders are still not interested in purchasing anything, $1/2$ of them end up placing a \$60 order, and $1/3$ of them end up placing a \$100 order. Estimate the average sales receipts per house visit by simulating 25 house visits using a die. Calculate the theoretical value and compare it with the estimate obtained from your simulation.

2.2. Simulate playing the game of Example 2.2, where the contestant uses the "switch" policy.
(a) Which policy would you recommend?

(b) Can you demonstrate mathematically that your suggested policy is the best policy?

2.3. Determine a full-period multiplicative linear congruential generator suitable for a 6-bit computer (i.e., there are 6 bits available to store signed integers). Notice that since one bit is reserved for the sign, the largest integer possible on this machine is 31. Since 31 is a prime number, let $m = 31$ and find a so that the period of the generator is 30.

2.4. Assume that you have a 10-bit computer (i.e., there are 10 bits available to store signed integers). Determine constants a and m for a multiplicative linear congruential generator with the largest possible period on this 10-bit machine. Note that since signed integers are stored in 10-bits and one bit is saved for the sign, all constants must be less than 2^9. (Hint: Use Excel. Make sure the m is a prime and then use trial and error to find a.)

2.5. Use the values of a and m as suggested by Schrage (see p. 52) with an initial random number seed of $S_0 = 101$ and produce the first three random numbers that will be generated with his multiplicative generator. Notice that the seed is first changed and then a random number is produced; in other words, the first random number is derived from S_1, not S_0.

2.6. Assume that you have a 10-bit computer (i.e., there are 10 bits available to store signed integers). Determine constants (multiplier and modulus) less than 2^9, for a simplified multiple recursive generator with a period greater than 2^9. For the constants that you choose, give the period of the generator and the first three numbers generated by your generator.

2.7. Use the third row of Table 2.6 for the initial random number seeds for L'Ecuyer's random number generator and produce the first three random numbers.

2.8. Given the sequence of three uniform random numbers, 0.15, 0.74, 0.57, generate the corresponding sequences of random variates
(a) from a discrete uniform distribution varying between 1 and 4,
(b) for a discrete random variable, with the probability that it is equal to 1, 2, 3, 4 being 0.3, 0.2, 0.1, 0.4, respectively.

2.9. A manufacturing process has a defect rate of 20% and items are placed in boxes of five. An inspector samples two items from each box. If one or both of the selected items are bad, the box is rejected. Simulate this process to answer the following question: If a customer orders 10 boxes, what is the expected number of defective items the customer will receive?

2.10. A binomial random variate with parameters n and p can be obtained by adding n Bernoulli random variates with parameter p (i.e., $\Pr\{X = 0\} = 1 - p$ and $\Pr\{X = 1\} = p$).
(a) Using this fact, use Excel to generate 1,000 binomial random variates with parameters $n = 4$ and $p = 0.35$ and estimate the probabilities for the binomial random

variable. Compare these estimated probabilities to the theoretical values.
(b) Extend the spreadsheet so that 10,000 binomial random variates are created and compare your probability estimates to those obtained in Part (a).

2.11. Given the sequence of four uniform random numbers, 0.23, 0.74, 0.57, 0.07, generate the corresponding sequences of random variates
(a) from an exponential distribution with a mean of 4,
(b) from a Weibull distribution with scale parameter 0.25 and shape parameter 2.5,
(c) from a normal distribution with a mean of 4 and variance of 4, (d) from a continuous random variable designed to reproduce the randomness
observed by the following data: 4.5, 12.6, 13.8. 6.8, 10.3, 12.5, 8.3, 9.2, 15.3, 11.9, 9.3, 8.1, 16.3, 14.0, 7.3, 6.9, 10.5, 12.3, 9.9, 13.6.

2.12. Using Excel, generate 1,000 continuous uniform random variates between 3 and 8.
(a) Estimate the mean and variance from these numbers and compare the estimates to the theoretical values. (b) Repeat Part (a) using 10,000 values.

2.13. Using Excel produce random variates according to a Weibull distribution where the mean and standard deviation are input constants. Generate 1000 random variates and compare the sample mean and standard deviation of the generated data with the input values. Hint: you will need GoalSeek to determine the shape and scale parameters (as in the Appendix of Chap. 1).

2.14. Using Excel, generate 100 steel rods from the shipment of rods described in Problem 1.25. How many rods from your randomly selected sample are outside of the specifications?

2.15. The failure and repair characteristics of a machine are being studied, it has been determined that the mean and standard deviation of time to failure are 80 hours and 50 hours, respectively, and the mean and standard deviation for repair times are 150 minutes and 54 minutes respectively. However, the repair time is not independent of the time to failure and it is estimated that the correlation coefficient of the two is 0.7. We desire to simulate the time to failure using a Weibull distribution and the repair time using a log-normal distribution.
(a) What are the shape (α) and scale (β) parameters for the Weibull distribution that describes the time to failure?
(b) What are the mean (μ_N) and standard deviation (σ_N) for the normally distributed random variable that is associated with the log-normal of the failure times? (See Eq. 1.23.)
(c) What is the (approximate) value of ρ from Property 2.7 that will yield a correlation coefficient of 0.7 for the failure and repair times?
(d) Simulate a few thousands cycles and estimate the fraction of time that the machine is available. Compare this with the theoretical value which is the ratio of the mean time to failure with the sum of the mean time to failure plus the time to repair.

References

1. Cario, M.C., and Nelson, B.L. (1997). Modeling and Generating Random Vectors with Arbitrary Marginal Distributions and Correlation Matrix. Technical Report, Department of Industrial Engineering and Management Sciences, Northwestern University, Evanston, IL.
2. Deng, L.Y., and Lin, D.K.J. (2000). Random Number Generation for the New Century. *The American Statistician*, **54**:145–150.
3. Law, A.M. (2007). *Simulation Modeling and Analysis*, 4th ed., McGraw Hill, Boston.
4. L'Ecuyer, P. (1996). Combined Multiple Recursive Random Number Generators. *Operations Research*, **44**:816–822.
5. Leemis, L.M., and Park, S.K. (2006). *Discrete-Event Simulation: A First Course*, Pearson Prentice Hall, Upper Saddle River, NJ.
6. L'Ecuyer, P. (1999). Good Parameters and Implementations for Combined Multiple Recursive Random Number Generators. *Operations Research*, **47**:159–164.
7. Mlodinow, L. (2008). *The Drunkard's Walk: How Randomness Rules our Lives*, Pantheon Book, New York.
8. Odeh, R.E., Evans, J.O. (1974). The Percentage Points of the Normal Distribution. *Applied Statistics*, **23**:96–97.
9. Schrage, L. (1979). A More Portable Fortran Random Number Generator. *ACM Transactions on Mathematical Software*, **5**:132–138.

Chapter 3
Basic Statistical Review

Statistical analyses are necessary whenever decisions are to be made on the basis of probabilistic data. The conceptual framework for statistics is the assumed existence of an underlying probability space and knowledge is desired regarding the unknown probability law. Furthermore, estimates regarding the probability law are to be made on the basis of the outcome(s) of some random variable(s). This chapter is divided into several major topics. Each topic should be a chapter in itself but the intention is to simply review the material assuming a previous exposure to statistics. The introduction to those statistical techniques that are commonly used in simulation but not necessarily covered in an introductory statistics course will be postponed until the chapter on simulation analysis (Chap. 9) instead of discussed here.

3.1 Collection of Data

It should prove beneficial to briefly comment on some standard notational conventions. Random variables will always be denoted by capital letters (or Greek letters with a "hat" like $\hat{\theta}$); scalars or known numbers will use lower case letters. Since we are going to be dealing with both random variables and *observations* of random variables, the potential to confuse the two is always present. To illustrate, let X_n be a random variable that denotes whether or not a process is "in control" on the n*th* day. For example, we might write $X_n = 1$ to denote the process being in control and $X_n = 0$ to denote the process being out of control (i.e., needing operator intervention). Just before day n starts, we do not know the value of X_n. After the day gets going, we go to the control room and observe whether or not the process is in control. After our observation, we have the realization of the random variable and might write that $x_1 = 1$ if, for example, the process was in control on the first day. The lower case letter emphasizes the fact that the value 1 is now a known, observed value. Although this might seem like a purely technical detail, you will find your understanding of statistics easier if you remember the distinction.

R.M. Feldman, C. Valdez-Flores, *Applied Probability and Stochastic Processes*, 2nd ed., DOI 10.1007/978-3-642-05158-6_3, © Springer-Verlag Berlin Heidelberg 2010

3.1.1 Preliminary Definitions

In statistics there is usually a collection of random variables from which we make an observation and then do something with the observation. The most common situation is when the collection of random variables of interest are mutually independent and with the same distribution. Such a collection is called a random sample.

Definition 3.1. The collection of random variables $X_1, X_2, \cdots, X_n$ is called a *random sample* of size n if they are independent and identically distributed. □

Once we have a random sample, we often want to make some estimate based on a function of the random sample. For example, assume that the underlying distribution function is a normal distribution but the mean and variance are unknown. Suppose we take a random sample of size 2 and then consider an averaging function of those two random variables. In other words, consider the function

$$\frac{X_1 + X_2}{2} .$$

Such a function could be used to estimate the mean of the underlying distribution after an observation is made of the random variables. However, consider the function

$$\frac{(X_1 - \mu)^2 + (X_2 - \mu)^2}{2}$$

where μ is the actual and unknown mean of the underlying distribution. Such a function could not be used to estimate anything because even after the random sample is observed the value of μ is still unknown so it would be impossible to obtain a value for the function. A function for which a numerical value can be obtained after the realization of the random sample is called a statistic.

Definition 3.2. A *statistic* is a function of a random sample that does not contain any unknown parameters. □

The two most common statistics are the sample mean and sample variance. They are the most frequently used measures in describing a random sample. Notice that the definition of the sample variance uses the sample mean instead of μ, the actual mean. Thus the sample variance is a statistic since it can be calculated after an observation has been made of the random sample. The sample standard deviation is the square root of the sample variance.

Definition 3.3. The *sample mean* of a random sample of size n is

$$\overline{X}_n = \frac{X_1 + X_2 + \cdots + X_n}{n}$$

and the *sample variance* is

$$S_n^2 = \frac{(X_1 - \overline{X}_n)^2 + \cdots + (X_n - \overline{X}_n)^2}{n-1} = \frac{(X_1^2 + \cdots + X_n^2) - n\overline{X}_n^2}{n-1} .$$

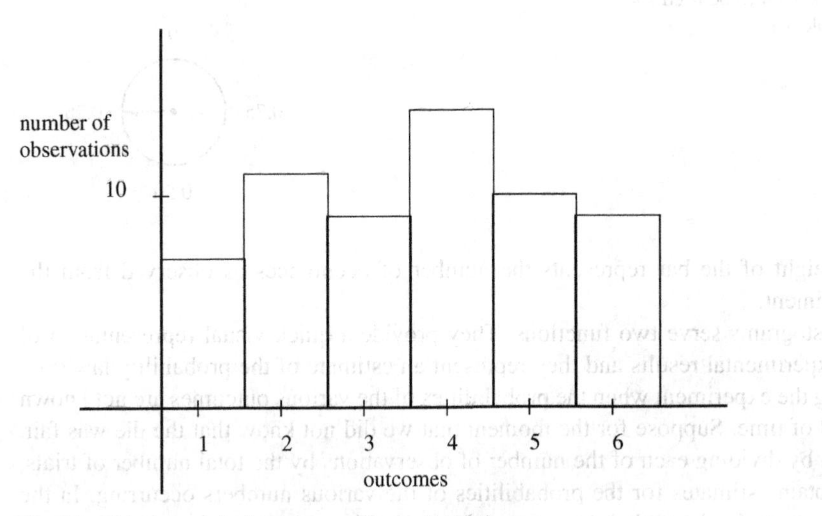

Fig. 3.1 A histogram obtained from rolling a die 60 times

□

Before we discuss the proper statistical procedures for estimating unknown parameters, we need to discuss some of the common ways in which data are represented. (By data we refer to the observations of the random sample.)

3.1.2 Graphical Representations

To reinforce the idea of randomness and the statistical variation of experiments, the reader should consider the results of taking a die and rolling it 60 times. Such an experiment was done with the results (i.e., the number of dots on the upward face) being recorded in Table 3.1. Assuming the die was fair, we would expect each of the

Table 3.1 Experimental results from rolling a die 60 times

Dots on up face	one	two	three	four	five	six
Number of occurrences	7	11	9	14	10	9

six faces to occur 10 times; however that did not happen in the experiment.

Although the list of numbers in Table 3.1 describes the experiment, it is often convenient to present a visual representation of the experimental results (Fig. 3.1). This visual representation is called a *histogram*, and is similar to a bar chart. In the histogram, there is a bar above each number ranging from one to six, where

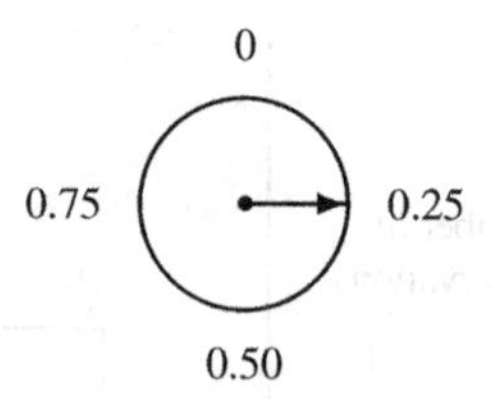

Fig. 3.2 Dial to be used for Example 3.1

the height of the bar represents the number of occurences as observed from the experiment.

Histograms serve two functions. They provide a quick visual representation of the experimental results and they represent an estimate of the probability law governing the experiment when the probabilities of the various outcomes are not known ahead of time. Suppose for the moment that we did not know that the die was fair. Then, by dividing each of the number of observations by the total number of trials, we obtain estimates for the probabilities of the various numbers occurring. In the above example, the probability of a one occurring is estimated to be $7/60$, the probability of a two occurring is estimated to be $11/60$, etc. In this fashion, the histogram of Fig. 3.1 gives the graph for an empirical probability mass function of the die. The notion of randomness is further illustrated by performing the experiment again with the same die and observing that different probability estimates are obtained. Furthermore, if we take more time with the experiment (such as using 600 rolls instead of 60), the probability estimates are better. In other words, for a fair die, the number of observations divided by the total number of rolls becomes closer, on the average, to $1/6$ as the number of rolls gets larger.

Example 3.1. To further illustrate histograms, we perform one more experiment. Consider a free spinning pointer on a dial that is marked zero to one as in Fig. 3.2. The experiment is to (randomly) spin the dial 20 times and record the number to two decimals to which the arrow points each time (assuming the circumference is continuously labeled). Table 3.2 contains the results of this experiment. To draw

Table 3.2 Experimental data from 20 spins of the dial

0.10	0.22	0.24	0.42	0.37
0.06	0.72	0.91	0.14	0.36
0.88	0.48	0.52	0.87	0.71
0.51	0.60	0.94	0.58	0.09

a histogram from this experiment, the interval [0,1) is divided into subintervals or "classes". Specifically, we divide the interval [0,1) into four equal size classes which yields the histogram in Fig. 3.3. □

Another graphical representation of data that is easy and descriptive is a box plot. Before drawing a box plot, the definitions for the median and quartiles of a data set must be given. The median of a distribution is the value such that there is a 50%

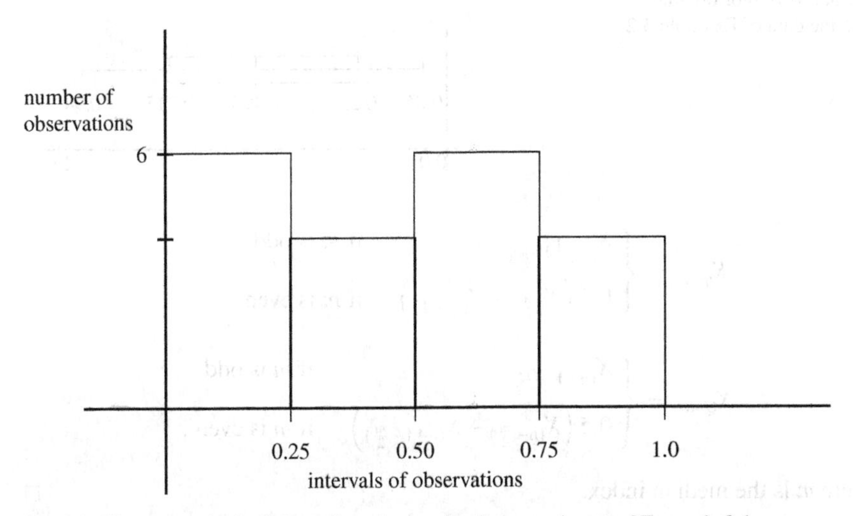

Fig. 3.3 Histogram obtained from the spinning the dial experiment of Example 3.1

probability that a random variable will be less than that value. The quartiles are defined so that there is a 25% probability of the random variable being less than the lower quartile point and a 75% probability of the random variable being less than the upper quartile value. The first step in obtaining the median and quartiles for a random sample is to order the data so that the first data point is the smallest and the last data point is the largest. Indices for the re-ordered data set will be written as subscripts with parentheses. Thus, $X_{(1)}$ denotes the minimum value, $X_{(2)}$ denotes the second lowest value, etc. until $X_{(n)}$ denotes the maximum value. The median value is the midpoint. If there is not a single midpoint, then the average of two midpoints is used. To obtain the quartiles, the data set is divided in half without including the median, and then the quartiles are the medians of each half. Using the ordered random variables, the median and quartile values for a data set are defined as follows:

Definition 3.4. The *median*, X_{med}, of a random sample of size n is

$$X_{med} = \begin{cases} X_{(\frac{n+1}{2})} & \text{if } n \text{ is odd} \\ 0.5\left(X_{(\frac{n}{2})} + X_{(\frac{n}{2}+1)}\right) & \text{if } n \text{ is even ,} \end{cases}$$

and the median index is defined to be $m = \lfloor (n+1)/2 \rfloor$ where the notation $\lfloor a \rfloor$ denotes the truncation of a. The *lower quartile*, X_{quar}, and *upper quartile*, X_{Quar}, of a random sample of size n are defined as

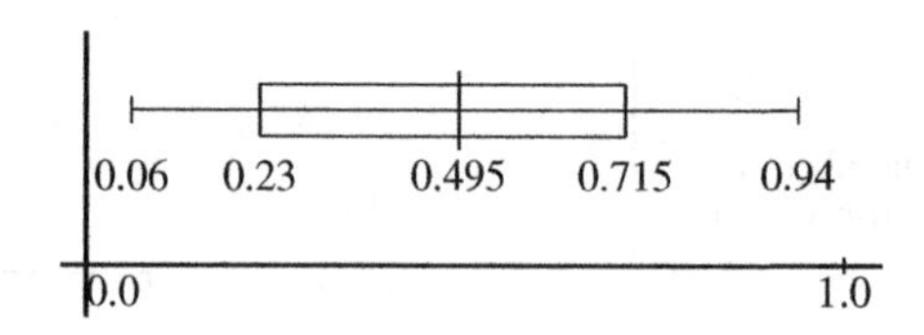

Fig. 3.4 Box plot obtained from the data of Example 3.2

$$X_{quar} = \begin{cases} X_{(\frac{m+1}{2})} & \text{if } m \text{ is odd} \\ 0.5\left(X_{(\frac{m}{2})} + X_{(\frac{m}{2}+1)}\right) & \text{if } m \text{ is even} \end{cases}$$

$$X_{Quar} = \begin{cases} X_{(n+1-\frac{m+1}{2})} & \text{if } m \text{ is odd} \\ 0.5\left(X_{(n-\frac{m}{2})} + X_{(n+1-\frac{m}{2})}\right) & \text{if } m \text{ is even}\,, \end{cases}$$

where m is the median index. □

A box plot uses a line to represent the range of data value and a box to represent the data values between the lower and upper quartiles; thus, the box plot gives a quick visual summary of the distribution of the data points. Note that the box represents the middle 50% of the data values. We should also point out that there are no universally accepted estimators for the upper and lower quartiles. The ones above are suggested because they are are intuitive and easy to compute; however, there are alternate methods also used for determining the quartiles as discussed briefly in the appendix of this chapter.

Example 3.2. The median value from the 20 data points in Table 3.2 is the average of the tenth and eleventh values of the ordered data; thus, $X_{med} = 0.495$. Likewise, $X_{quar} = 0.23$ and $X_{Quar} = 0.715$. A box plot of this in in Fig. 3.4. □

- *Suggestion: Do Problems 3.1–3.3.*

3.2 Parameter Estimation

The purpose of obtaining data is often to estimate the value of one or more parameters associated with the underlying distribution. For example, we may wish to know the average length of time that a job spends within a job shop. In this case, observations of a random sample would be taken and used to estimate the mean value of the underlying distribution.

An *estimator* is a statistic (therefore a random variable) that is used to estimate an unknown parameter. There are two very important properties of estimators (namely, unbiased and efficient) that we illustrate here by way of example. Assume we have a random sample of size 3 from a normal distribution with unknown mean μ and variance σ^2. Consider the following (not necessarily good) estimators for the mean:

$$\hat{\mu}_1 = \frac{X_1 + X_2 + X_3}{3},$$
$$\hat{\mu}_2 = \frac{X_1 + X_2 + X_3}{4},$$
$$\hat{\mu}_3 = \frac{X_1 + 2X_2 + X_3}{4}.$$

(Note that an estimator of a parameter is usually denoted by placing a "hat" over the parameter.)

The questions that need to be asked are "Are these all 'good' estimators?" and "Which of these estimators is the best?" Before answering these questions, let us look at some basic facts regarding the estimators themselves. Since they are random variables, they each have a mean and variance. You should be able to show the following:

$$E[\hat{\mu}_1] = \mu, \qquad V[\hat{\mu}_1] = \tfrac{1}{3}\sigma^2,$$
$$E[\hat{\mu}_2] = 0.75\mu, \quad V[\hat{\mu}_2] = \tfrac{3}{16}\sigma^2,$$
$$E[\hat{\mu}_3] = \mu, \qquad V[\hat{\mu}_3] = \tfrac{3}{8}\sigma^2,$$

where μ and σ^2 are the true (although unknown) mean and variance.

When an observation is made using a particular estimator, the resulting value may or may not be close to the true value of the parameter depending on the characteristics of the estimator. For example, consider the second estimator, $\hat{\mu}_2$, and assume several observations are made using this estimator. Some of the observations may result in an estimate larger than μ and some may result in an estimate smaller than μ; however, in the long-run, the values of the estimates will tend to center around a value equal to 75% of the correct value. Thus, we say that $\hat{\mu}_2$ is biased on the small side (assuming that μ is not zero). Therefore it seems foolish to use $\hat{\mu}_2$. (Note that in this case, the biased estimator is obviously not very good; however, there are other cases in which a biased estimator may, in some sense, be better than an unbiased estimator. It may be interesting to note that the reason that the sample variance of Definition 3.3 has $n-1$ in its denominator instead of n is so that it will be an unbiased estimator for the variance.) If we used $\hat{\mu}_1$ or $\hat{\mu}_3$, our estimates would have in the long-run center around the correct value; thus, we say that $\hat{\mu}_1$ and $\hat{\mu}_3$ are unbiased estimators.

Definition 3.5. An estimator $\hat{\theta}$ for the parameter θ is *unbiased* if

$$E[\hat{\theta}] = \theta.$$

□

We now have two potentially good estimators; namely, $\hat{\mu}_1$ and $\hat{\mu}_3$ since these are both unbiased for estimating μ. Which one is better? We would want to use the estimator that, over the long run, gets closer to the actual value. Therefore, the question is answered by considering the variability of the two estimators. Since the variance of $\hat{\mu}_1$ is less than the variance of $\hat{\mu}_3$, it follows that $\hat{\mu}_1$ will, in general, be closer to the actual value than $\hat{\mu}_3$.

Definition 3.6. If we have two unbiased estimators, $\hat{\theta}_1$ and $\hat{\theta}_2$ for the parameter θ, we say that $\hat{\theta}_1$ is *more efficient* than $\hat{\theta}_2$ if the variance of $\hat{\theta}_1$ is less than the variance of $\hat{\theta}_2$. □

It is important to note that the variance of estimators can only be compared if both estimators are unbiased. For example, the variance of $\hat{\mu}_2$ is less than the variance of $\hat{\mu}_1$ but we still used $\hat{\mu}_1$ because it is an unbiased estimator. As mentioned above, a biased estimator may be preferred over an unbiased one. One possible scenario in which a biased estimator may be better than an unbiased one is when the biased estimator is closer in the long-run to the actual parameter than the unbiased estimator, and one measure of "closeness" is the mean squared error of the estimator.

Definition 3.7. Let $\hat{\theta}$ be an estimator of the parameter θ. The *mean square error* of the estimator is

$$\mathrm{MSE}(\hat{\theta}) = E[(\hat{\theta} - \theta)^2] .$$

□

If we let the bias of an estimator be denoted by $\mathrm{bias}(\hat{\theta}) = E[\hat{\theta}] - \theta$, the mean square error can be calculated as

$$\mathrm{MSE}(\hat{\theta}) = V[\hat{\theta}] + \mathrm{bias}(\hat{\theta})^2 , \quad (3.1)$$

The most common methods for obtaining estimators is by the method of matching moments, by the maximum likelihood method, and by a least squares procedure. A complete discussion regarding these procedures is beyond the scope of this review; however, we will mention the basic idea behind the first two of these procedures in the next subsections.

- *Suggestion: Do Problems 3.4–3.5.*

3.2.1 Method of Moments

A relatively easy method for obtaining parameter estimators is to express the unknown parameters as functions of the moments and then form equalities based on Definition 3.3. If there is one unknown parameter, one equation involving the sample mean is necessary. If there are two unknown parameters, both the sample mean and sample variance would be necessary.

Example 3.3. A random sample of size n is to be collected from a binomial distribution. The binomial distribution has two parameters: m and p, where m is known but p needs to be estimated from the random sample. It should also be remembered that the mean for a binomial distribution is given by $\mu = mp$. Since one parameter is unknown, we form the equation

$$\overline{X}_n = mp .$$

Solving this equation for the unknown parameter yields the estimator as

$$\hat{p} = \frac{\overline{X}_n}{m} .$$

□

Example 3.4. A random sample of size n is to be collected from a gamma distribution. The gamma distribution (Eq. 1.18) has two positive parameters: a shape parameter, α, and a scale parameter, β, where

$$E[T] = \alpha\beta \quad \text{and } V[T] = \alpha\beta^2 .$$

For this example, both parameters are unknown so two equations must be formed; namely,

$$\overline{X}_n = \alpha\beta$$
$$S_n^2 = \alpha\beta^2 .$$

These equations are solved simultaneously to obtain

$$\hat{\alpha} = (\overline{X}_n/S_n)^2$$
$$\hat{\beta} = \overline{X}_n/\hat{\alpha} .$$

□

- *Suggestion: Do Problems 3.8–3.9.*

3.2.2 Maximum Likelihood Estimation

Although the Method of Moments is fairly easy to implement, it is sometimes possible to obtain more efficient estimators using a procedure called the Maximum Likelihood Method. It is beyond the scope of this chapter to give the details of the procedure, and we refer the interested reader to Tables 6.3 and 6.4 of Law [3] for an excellent summary of the maximum likelihood estimators for many commonly used distributions.

The conceptual approach of the Maximum Likelihood Method is to form a likelihood function with the independent variables being the estimators. Intuitively, the function equals that "likelihood" of the observed data set being a result of a distribution having parameters equal to the estimators. The Maximum Likelihood Method then produces estimators based on maximizing the likelihood function. For example, if the underlying distribution was a normal distribution, the maximum likelihood method yields estimators equal to the equations of Definition 3.3 except that the formula for the variance has n in the denominator instead of $n-1$. Maximum likelihood estimators have some very appealing properties; however, in general they are not unbiased, although they are unbiased in the limit. For example, the Maximum likelihood Estimator for the variance of the normal distribution is as follows.

Definition 3.8. The *MLE for the variance* of a random sample of size n from the normal distribution is

$$\hat{\sigma}_n^2 = \frac{(X_1 - \overline{X}_n)^2 + \cdots + (X_n - \overline{X}_n)^2}{n} = \frac{(X_1^2 + \cdots + X_n^2) - n\overline{X}_n^2}{n}.$$

□

The difference between the sample variance (Definition 3.3) and the MLE for the variance is the denominator. The sample variance is unbiased whereas the MLE is not; however, the MLE for the variance has a smaller mean square error (Definition 3.7) than the sample variance.

- *Suggestion: Do Problems 3.11–3.12.*

3.3 Confidence Intervals

The normal distribution plays a key role in statistics, primarily because of the Central Limit Theorem. Therefore, before we start our discussion of confidence intervals, some basic facts regarding the normal distribution are reviewed. For notational ease, when X is normally distributed with mean μ and variance σ^2, we write that X is $N(\mu, \sigma^2)$. The first property deals with the relationship between the normal and chi-square distributions.

Property 3.1. *If Z is $N(0,1)$ (i.e., a standard normal random variable), then Z^2 has a chi-square distribution with one degree of freedom.*

The second property deals with the additive nature of the normal distribution.

Property 3.2. *If X_1 and X_2 are $N(\mu_1, \sigma_1^2)$ and $N(\mu_2, \sigma_2^2)$, respectively, and are independent, then*

$$Y = aX_1 + bX_2$$

is $N(a\mu_1 + b\mu_2,\ a^2\sigma_1^2 + b^2\sigma_2^2)$.

The above properties lead to the following important distributions associated with the sample mean and variance.

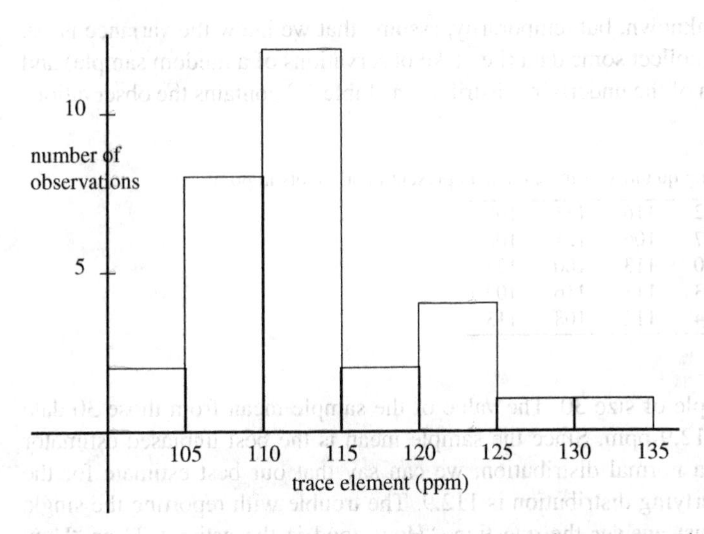

Fig. 3.5 Histogram from 30 ingots giving quantity of trace element present in ppm

Property 3.3. *Let $\overline{X}_n$ and S_n^2 be the sample mean and variance, respectively, from a random sample of size n where the underlying distribution is $N(\mu,\sigma^2)$. Then $\overline{X}_n$ is $N(\mu,\sigma^2/n)$ which is equivalent to saying that*

$$Z = \frac{\overline{X}_n - \mu}{\sigma/\sqrt{n}}$$

is $N(0,1)$. Furthermore

$$V_1 = \frac{(n-1)S_n^2}{\sigma^2}$$

has a chi-square distribution with $n-1$ degrees of freedom.

The chi-square distribution is a special case of a gamma distribution (Eq. 1.18). Specifically, a chi-square distribution with n degrees of freedom is a gamma distribution with shape parameter $n/2$ and scale parameter 2; thus, its mean is n and variance is $2n$.

3.3.1 Means

Suppose that we produce ingots of a metal and are concerned about the presence of a certain trace element within the ingots. The quantity of trace element in parts per million (ppm) within each ingot is random and distributed by the normal distribution whose mean is unknown, but temporarily, assume that we know the variance is 49. We would like to collect some data (i.e., take observations of a random sample) and estimate the mean of the underlying distribution. Table 3.3 contains the observations

Table 3.3 Data giving quantity of trace element present in 30 ingots in ppm

125	110	112	116	131	108
114	121	107	106	121	106
107	113	110	113	100	121
112	109	113	113	116	109
114	123	104	112	108	113

of a random sample of size 30. The value of the sample mean from those 30 data points is $\bar{x}_{30} = 112.9$ ppm. Since the sample mean is the best unbiased estimator for the mean of a normal distribution, we can say that our best estimate for the mean of the underlying distribution is 112.9. The trouble with reporting the single number is we must answer the question, "How good is the estimate?" or "How much confidence do we have in the estimate?" The reason for the question is the understanding that if we took another random sample of 30 points and calculated a new sample mean, its realization would likely be different from 112.9. Because of this variability, we need some way to describe the size of the expected variation in the sample mean.

Therefore, instead of a single value for an estimate, it is more appropriate to give a *confidence interval*. For example, a 95% confidence interval for the mean quantity of the trace element in an ingot is an interval for which we can say with 95% confidence that the interval contains the true mean; that is, we expect to be correct 19 out of 20 times. (The choice of using 95% confidence instead of 97.2% or 88.3% is, for the most part, one of historical precedence.)

The building of confidence intervals for the mean of a normal distribution with known variance makes use of the fact that Z, as defined in Property 3.3, is $N(0,1)$. From the $N(0,1)$ probability law, we need the constant, called z_α, that yields a "right-hand error" of α; that is, z_α is defined to be that value such that

$$P\{Z > z_\alpha\} = \alpha \,.$$

For example, using a table of critical values for the standard normal distribution (see the Normal Distribution Table in the back of the book), we have $z_{0.05} = 1.645$.

Under the assumption of normally distributed data and known variance, σ^2, the $1-\alpha$ confidence interval for the mean is given by

$$\left(\bar{x}_n - z_{\alpha/2}\,\frac{\sigma}{\sqrt{n}},\ \bar{x}_n + z_{\alpha/2}\,\frac{\sigma}{\sqrt{n}}\right) \tag{3.2}$$

where n is the number of data points, $\bar{x}_n$ is the sample mean using the n data points, and $z_{\alpha/2}$ is a critical value based on the standard normal probability distribution. The α in the subscript refers to the amount of error we are willing to risk; thus, a

95% confidence implies a 5% error and we would set $\alpha = 0.05$. To understand why the $1-\alpha$ confidence interval uses a critical value associated with an error of $\alpha/2$, consider a 95% confidence interval. Such a confidence interval has an associated risk of 5% error. Since the error is divided evenly on both sides of the interval, there is a 2.5% chance of error to the left of the interval and a 2.5% chance of error to the right. Statistical table books usually list critical values for statistics associated with only a right-hand error; therefore, for a 5% overall risk, we look up the value associated with a 2.5% "right-hand" risk.

Example 3.5. To illustrate the building of confidence intervals, the data in Table 3.3 is used with the histogram of the data being in Fig. 3.5. (It might appear that it is redundant to plot the histogram since the table of the data gives the information; however, it is recommended that you make a practice of plotting your data in some fashion because quantitative information cannot replace the visual impression for producing an intuitive feel of the data.) The sample mean for the data obtained from the 30 data values is $\bar{x}_{30} = 112.9$, and the critical z-value for a 95% confidence interval is $z_{0.025} = 1.96$. Equation 3.2 thus yields

$$112.9 \pm 1.96 \times \frac{7.0}{\sqrt{30}} = (110.4, 115.4)\,, \tag{3.3}$$

that is, we are 95% confident that the interval (110.4, 115.4) includes the true mean of trace element found in a randomly selected ingot. □

We need to consider again the two assumptions (i.e., independence and normality) under which Eq. (3.2) theoretically holds. The assumption of normality is not critical because of the Central Limit Theorem (Property 1.5). The assumption of independence is very important and special techniques must be developed if the independence assumption is not meet. In particular, simulation data often do not satisfy the independence assumption, therefore, we postpone discussion of what to do in the absence of independence until Chap. 9.

Equation (3.2) assumes that the value of the true variance is known; however, it is much more likely that the variance will not be known and must be estimated. The importance of the variance is in the determination of the variability of the sample mean. Therefore, when an estimate for the variance instead of the variance's actual value is used, an additional degree of variability is introduced. This additional variability increases the size of the confidence interval. Technically, the calculation of the confidence interval (Eq. 3.2) is based on the use of the normal distribution from Property 3.3. Since σ is unknown, we cannot use the standard normal distribution, instead a new distribution called the Student-t distribution is used. Specifically, we have the following.

Property 3.4. *Let $\overline{X}_n$ and S_n^2 be the sample mean and variance, respectively, from a random sample of size n where the underlying distribution is $N(\mu, \sigma^2)$.*

The random variable T_{n-1} defined by

$$T_{n-1} = \frac{\overline{X}_n - \mu}{S_n/\sqrt{n}}$$

is said to have a Student-t with $n-1$ degrees of freedom.

The Student-t distribution is similar to the normal distribution except that it has a slightly larger variance; that is, its pdf is a symmetric bell-shaped curve that is slightly wider than the normal. This increase in variance results in larger confidence intervals. However, as the degrees of freedom increase, the Student-t limits to the normal; thus, for large samples the confidence interval will be the same using the Student-t or the normal. To build confidence intervals we need the critical values associated with the Student-t distribution and these are defined in the same way as the z-values were defined. Namely, $t_{n,\alpha}$ is the constant defined such that yields a right-hand-error of α; that is,

$$Pr\{T_n > t_{n,\alpha}\} = \alpha \,.$$

Under the assumption of normally distributed data and unknown variance, the $1-\alpha$ confidence interval for the mean is given by

$$\left(\overline{x}_n - t_{n-1,\alpha/2}\,\frac{s_n}{\sqrt{n}}\,,\ \overline{x}_n + t_{n-1,\alpha/2}\,\frac{s_n}{\sqrt{n}}\right) \tag{3.4}$$

where n is the number of data points, $\overline{x}_n$ and s_n are the sample mean and standard deviation using the n data points and $t_{n-1,\alpha/2}$ is a critical value based on the Student-t distribution. (In the "old" days when tables were commonly used to obtain critical values, z_α values (standard normal) were used in place of the $t_{n,\alpha}$ values whenever $n > 30$; however, with the use of spreadsheets, there is no need to make this substitution since it is just as easy to obtain a t-statistic with 92 degrees of freedom as it is to obtain the z-statistic.) Statistical tests are usually better as the degrees-of-freedom increases. A statistical test usually loses a degree-of-freedom whenever a parameter must be estimated by the data set; thus, the t-test has only $n-1$ degrees-of-freedom instead of n because we use the data to estimate the variance.

Example 3.6. We return to the data in Table 3.3 and assume the data come from a normal distribution of unknown mean and variance. To calculate the confidence interval, we first note that sample mean and standard deviation are $\overline{x}_n = 112.9$ and $s_n = 6.7$, respectively. The critical t-value for a 95% confidence interval with 29 degrees-of-freedom is 2.045. Equation (3.4) thus yields

$$112.9 \pm 2.045 \times \frac{6.7}{\sqrt{30}} = (110.4, 115.4)\,. \tag{3.5}$$

This confidence interval is slightly larger than the the confidence interval of (3.3), which is what we expected since the estimation of the σ by s_n introduced an additional source of variability (and thus uncertainty). □

3.3.2 Proportions

The second type of confidence interval that we discuss is for proportions. For example, we might be interested in the proportion of items produced by a manufacturing process that falls within specifications, or we might be interested in the proportion of industrial accidents caused by unsafe conditions. The random sample $X_1, X_2, \cdots, X_n$ are now Bernoulli random variables with parameter p, where p is the probability of success; that is X_i is 1 if the specific event of interest occured and is 0 if the specific event of interest did not occur on the ith trial. For ease of discussion, we say there is a "success" when the event occurs; thus the sample mean now represents the proportion of successes out of n trials. If the sample size is large, the confidence interval for proportions is the same for means (Eq. 3.2) where the standard deviation is estimated by $\sqrt{\bar{x}_n(1-\bar{x}_n)}$. In other words, our estimate for p, denoted $\hat{p}_n$, is given by $\bar{x}_n$ with a $1-\alpha$ confidence interval given by

$$\left(\hat{p}_n - z_{\alpha/2}\sqrt{\frac{\hat{p}_n(1-\hat{p}_n)}{n}},\ \hat{p}_n + z_{\alpha/2}\sqrt{\frac{\hat{p}_n(1-\hat{p}_n)}{n}}\right) \tag{3.6}$$

where n is the number of data points, $\bar{x}_n$ is number of successes divided by the number of data points, and $z_{\alpha/2}$ is a critical value based on the standard normal probability distribution.

Example 3.7. Suppose in 4,000 firings of a certain kind of rocket, there were 15 instances in which a rocket exploded upon ignition. We desire to construct a 99% confidence interval for the probability that such a rocket will explode upon ignition. We first note that $\bar{x}_{4000} = 15/4000 = 0.00375$ and $z_{0.005} = 2.576$; thus the 99% confidence interval for the true probability is

$$0.00375 \pm 2.576 \times \sqrt{\frac{0.00375 \times 0.99625}{4000}} = (0.00126, 0.00624)\,.$$

□

3.3.3 Variances

It is sometimes desirable to build a confidence interval for the variance of the underlying distribution. Confidence intervals for variances are built using the Chi-squared distribution (Property 3.3). The critical values from a Chi-squared distribution with

n degrees of freedom and a right-hand error of α is denoted by $\chi^2_{n,\alpha}$; that is,

$$Pr\{V_n > \chi^2_{n,\alpha}\} = \alpha ,$$

where V_n is a Chi-square random variable with n degrees-of-freedom.

Under the assumption of normally distributed data, the $1-\alpha$ confidence interval for the variance is given by

$$\left(\frac{(n-1)s_n^2}{\chi^2_{n-1,\alpha/2}}, \frac{(n-1)s_n^2}{\chi^2_{n-1,1-\alpha/2}} \right) \tag{3.7}$$

where n is the number of data points, s_n^2 is the sample variance, and $\chi^2_{n-1,\alpha/2}$ and $\chi^2_{n-1,1-\alpha/2}$ are critical values based on the Chi-square distribution. (Note that again a degree-of-freedom is lost because one parameter, namely the mean, is estimated in order to obtain the sample variance.) It might also be noted that most table books do not contain chi-squared values for degrees of freedom greater than 30. In such cases, the following can be used as an approximation:

$$\chi^2_{n,\alpha} = n \left(1 - \frac{2}{9n} + z_\alpha \sqrt{\frac{2}{9n}} \right)^3 \quad \text{for } n > 30 ,$$

although with the easy availability of spreadsheet programs, such approximations are not as necessary as they once were.

If we cannot justify the normality assumption of the underlying distribution, we can again use an approximation for large sample sizes. As a rule-of-thumb, whenever the sample size is larger than 30, we can build a confidence interval for the standard deviation by noting that the sample standard deviation is approximately $N(\sigma, \sigma^2/(2n))$ where σ^2 is the true (but unknown) variance. This approximation yields the following confidence interval for the standard deviation for large samples

$$\left(\frac{s_n}{1 + z_{\alpha/2}/\sqrt{2n}}, \frac{s_n}{1 - z_{\alpha/2}/\sqrt{2n}} \right) . \tag{3.8}$$

Example 3.8. Using the data of Table 3.3 for an example, we have $s_{30} = 6.7$, $\chi^2_{29,0.975} = 16.047$ and $\chi^2_{29,0.025} = 45.722$. Thus, the 95% confidence interval for the variance of the population producing the observations seen in Table 3.3 is

$$\left(\frac{29 \times 6.7^2}{45.722}, \frac{29 \times 6.7^2}{16.047} \right) = (28.47, 81.12) .$$

If we stretch the meaning of "large" a little, and use the large sample size approximation to obtain a 95% confidence interval for the standard deviation, the calculations are: $s_{30} = 6.7$ and $z_{0.025} = 1.96$ which yields

$$\left(\frac{6.7}{1+1.96/\sqrt{60}}, \frac{6.7}{1-1.96/\sqrt{60}}\right) = (5.347, 8.970)\,.$$

(Note that the first interval is for the variance and the second interval is for the standard deviation. The equivalent interval for the variance is (28.59, 80.46) which is tighter because of the normality assumption.) □

3.3.4 Correlation Coefficient

When two random variables X and Y are involved, we sometimes wish to estimate their correlation coefficient. Assume that the mean and variance of these random variables are give by μ_x, μ_y, σ_x^2, and σ_y^2, the correlation coefficient is defined by

$$\rho = \frac{E[XY] - \mu_x \mu_y}{\sigma_x \sigma_y}\,. \tag{3.9}$$

An estimator for ρ is given by

$$r = \frac{\sum x_i y_i - n\overline{x}\overline{y}}{(n-1)\, s_x s_y}\,, \tag{3.10}$$

where the random sample is a collection of n ordered pairs $(x_1, y_1), \cdots, (x_n, y_n)$. (This estimator is often called the Pearson correlation coefficient.) The confidence interval is not straight forward due to the fact that r is not normally distributed even when X and Y come from a bivariate normal distribution. However, Fisher has shown that the r value can be transformed into a z^* value that is normally distributed with variance $1/(n-3)$. The process is to first make the z^* transformation by

$$z^* = 0.5 \ln\left(\frac{1+r}{1-r}\right), \tag{3.11}$$

where r is the Pearson correlation coefficient from Eq. (3.10). (Equation 3.11 is called the Fisher transformation.) The $1-\alpha$ confidence interval for z^* is now given by

$$\left(z^* - \frac{z_{\alpha/2}}{\sqrt{n-3}},\ z^* + \frac{z_{\alpha/2}}{\sqrt{n-3}}\right). \tag{3.12}$$

The $1-\alpha$ confidence interval for the correlation coefficient is obtained by using the inverse z^* transformation on the right-hand and left-hand limits of confidence interval given in (3.12). The inverse transformation is

$$r = \frac{e^{2z^*} - 1}{e^{2z^*} + 1}\,. \tag{3.13}$$

Example 3.9. Suppose we have estimated the correlation coefficient based on 100 observations to be $r = 0.62$. The transformation of (3.11) yields

$$z^* = 0.5 \ln \left(\frac{1.62}{0.38} \right) = 0.725 \, .$$

If we desire a 95% confidence interval, the half-width of the interval is $1.96/\sqrt{97} = 0.199$ yielding a z^* confidence interval of $(0.526, 0.924)$. Using the inverse transform of Eq. (3.13) on 0.526 and 0.924 yields the 95% confidence interval for the correlation coefficient of

$$\left(\frac{e^{2\times 0.526} - 1}{e^{2\times 0.526} + 1}, \frac{e^{2\times 0.924} - 1}{e^{2\times 0.924} + 1} \right) = (0.482, 0.728) \, .$$

□

- *Suggestion: Do Problems 3.6–3.7 and 3.10.*

3.4 Fitting Distributions

It is often necessary to estimate not only the mean and variance of a data set, but also the underlying distribution function from which the data set was generated. For example, suppose we wish to model the arrival stream to a drive-in window facility at a bank. In order to develop a model, the modeler must know the distribution of inter-arrival times; therefore, one approach is to assume that the inter-arrival times are independent and are from some specified distribution, like for example, an exponential distribution. We would then estimate the mean of the exponential from the data set and then perform a statistical hypothesis test to determine if the data are such that the exponential assumption cannot be supported. Such a statistical hypothesis test is always susceptible to two different types of errors, called Type I and Type II errors. A Type I error occurs when the hypothesis is rejected when in fact it is true. A Type II error occurs when the hypothesis is not rejected when in fact it is false. The probability of a Type I error (i.e., the probability of rejecting the hypothesis given that it is true) is usually denoted by α. The probability of a Type I error is usually not difficult to establish; however, the probability of a Type II error (in the context of testing distributions) is almost impossible to establish. The reason for the difficulty with the Type II error is that to determine its probability, it is necessary to know the true condition about which the hypothesis is being made; however, if the hypothesis is not true, then the range of possible alternatives is extremely large.

For example, suppose we roll a die and record the number of dots on the face that is up for the purpose of determining if the die is "fair". In other words, we hypothesize that the distribution for the die is a discrete uniform distribution from one to six. Mathematically, the hypothesis is written as

$$H_0 : f(k) = \begin{cases} \frac{1}{6} & \text{for } k = 1, \cdots, 6, \\ 0 & \text{otherwise}. \end{cases}$$

The stated hypothesis of interest is called the *null hypothesis* and is denoted by H_0. If we end up rejecting H_0, then α (which is usually set to be five percent or ten percent) denotes the probability of an error. If we end up not rejecting H_0, then the probability of error depends on the true underlying distribution function. The Type II error is a complicated function, depending on all possible alternative distributions. For our purposes, it is not important to understand any quantitative descriptions of the Type II error except to know that its maximum value is equal to $1 - \alpha$. Since we normally make the probability of a Type I error small, the maximum probability of a Type II error is large. When the hypothesis is not rejected, we do not have much statistical assurance that a correct decision is made, but we assume it is true for lack of anything better. (As the amount of data increases in a random sample, the Type II error decreases; thus, the size of the data sample is often chosen to satisfy requirements of a Type II acceptable error.)

Statistical hypothesis tests are usually based on some established statistic (like the sample mean) that is calculated from the data (called the *test statistic*) and then compared with some value (called the *critical value*) usually obtained from some statistical tables. The general procedure for most hypothesis tests is as follows:

1. State the null hypothesis explicitly.
2. Decide on which statistic to use (discussed in the paragraphs below).
3. Decide on the desired value for α, called the significance level of the test, and then determine the critical value for the test (usually by looking it up in a table book).
4. Collect the data.
5. Calculate the test statistic.
6. Reject or fail to reject the null hypothesis after comparing the test statistic with the critical value of the statistic.

There are two common statistical tests for hypotheses involving equality of distributions: the chi-square goodness-of-fit test and the Kolmogorov-Smirnov test.

3.4.1 The Chi-Square Test

The chi-square test is performed on aggregated data. In other words, the range for the random variable is divided into k intervals, and the data consist of the number of observations that fall within each interval. Thus, the data are the same as results from a histogram. The test compares the actual number of observations within each interval with the (theoretically) expected number of observations, and if the difference is "too" large, the hypothesis is rejected.

To state the test more precisely, let o_i denote the observed number of data points that fall within the ith interval and let e_i denote the expected number of data points that should fall within the ith interval. The test statistic is then given by

$$\chi^2_{test} = \sum_{i=1}^{k} \frac{(o_i - e_i)^2}{e_i}$$

$$= \sum_{i=1}^{k} \frac{(o_i - np_i)^2}{np_i}, \tag{3.14}$$

where k is the number of intervals, n is the number of data points, and p_i is the probability of interval i occuring (from one trial). The intervals do not have to be of equal length or of equal probability, but they should be large enough so that the expected number of data points in each interval is at least 5; that is, $e_i = np_i \geq 5$ for $i = 1, \cdots, k$.

The chi-square random variable has one parameter, its degrees of freedom. Most introductory text books in statistics will contain a table of probabilities associated with chi-square random variables of various degrees of freedom. For the goodness-of-fit test involving a distribution whose parameters are assumed known, the degrees of freedom are equal to the number of intervals minus one. (If there are unknown parameters, one degree of freedom is lost for each unknown parameter. An example of this involving the normal distribution is given at the end of this subsection.) Typically, the value for α (i.e., the probability of an error when the null hypothesis is rejected) is taken to be five percent or ten percent. The critical value for χ^2_{k-1} is found by looking in the appropriate table under $k-1$ degrees of freedom for the specified α. (Sometimes, tables do not list the statistic under α but instead list it under the cumulative probability, which equals $1-\alpha$. In other words, if the table is for the distribution function F of random variable X, usually the table will give x where $\alpha = P\{X > x\} = 1 - F(x)$.) If the test statistic is greater than or equal to the critical value, the null hypothesis is rejected (because the large test statistic showed that the process had more variation than would be expected under the null hypothesis). If the test statistic is less than the critical value, the conclusion of the test is that there is not enough statistical variation within the data set to reject the null hypothesis.

Example 3.10. Again, the data of Table 3.1 are used to illustrate the chi-square goodness-of-fit test. The null hypothesis is the one stated above—that a fair die was used in obtaining the data. Using a significance level of five percent (i.e., $\alpha = 0.05$) along with the six intervals (i.e., 5 degrees of freedom) yields a critical value of $\chi^2_{critical} = 11.1$. The calculations for the test statistic are

$$\chi^2_{test} = \frac{(7-10)^2}{10} + \frac{(11-10)^2}{10} + \frac{(9-10)^2}{10} + \frac{(14-10)^2}{10} + \frac{(10-10)^2}{10} + \frac{(9-10)^2}{10} = 2.8\,.$$

Since $2.8 < 11.1$, we cannot reject the hypothesis that the die was fair. □

Example 3.11. The use of the chi-square test for a continuous distribution is illustrated using the "spinning dial" experiment along with the data contained in his-

Table 3.4 Experimental data, possibly from a normal distribution

99	93	82	95	89
114	77	94	91	99
98	78	93	108	97
85	100	92	104	84
96	92	121	106	107
95	124	122	102	102

togram of Fig. 3.3. The hypothesis is that the pointer produces a uniformly distributed random variable. Recognizing that the cumulative distribution for a continuous uniform zero-one random variable is a simple linear function from zero to one, the null hypothesis is written as

$$H_0 : F(s) = \begin{cases} 0 & \text{for } s < 0, \\ s & \text{for } 0 \le s < 1, \\ 1 & \text{for } s \ge 1. \end{cases} \quad (3.15)$$

In our experiment, there were 20 data points, so if four equal sized intervals are used, each interval would be expected to contain 5 data points. Since the test requires the expected number to be at least 5, it is permissible to use the four intervals shown in the histogram. Using a significance level of five percent along with four intervals yields a critical value of $\chi^2_{critical} = 7.815$. The calculations for the test statistic are

$$\chi^2_{test} = \frac{(6-5)^2}{5} + \frac{(4-5)^2}{5} + \frac{(6-5)^2}{5} + \frac{(4-5)^2}{5}$$
$$= 0.8 .$$

Because $\chi^2_{test} = 0.8 < 7.815$, we cannot reject the hypothesis. □

We often want to hypothesize a distribution, but the relevant parameters (e.g., mean, variance, etc.) are not known. For example, suppose we wish to test whether or not the data in Table 3.4 are from a normal distribution. Our assumption now needs to be verified statistically, so the null hypothesis is formed as

$$H_0 : f(\cdot) \text{ is a normal probability density function} .$$

Before intervals can be decided, the parameters for the normal distribution need to be estimated. The same data used for the chi-square statistic are used to estimate all unknown parameters, but each parameter estimated causes a loss of one degree of freedom in the test statistic. In other words, the chi-square statistic is calculated just as above once the parameters have been estimated; however, the degrees of freedom equal one less than the number of intervals *minus* the number of parameters estimated from the data.

Example 3.12. The data from Table 3.4 yield a mean of 97.97 and a standard deviation of 11.9 (from the equations in Definition 3.3). We will start with six intervals by adding and subtracting 10 from the mean. Thus, our initial intervals are set at

$(-\infty, 77.97]$, $(77.97, 87.97]$, $(87.97, 97.97]$, $(97.97, 107.97]$, $(107.97, 117.97]$, and $(117.97, \infty)$. Using a standard normal table, we find the probabilities associated with these intervals are

$$
\begin{aligned}
P\{X < 77.97\} &= 0.0456 = p_1 \\
P\{77.97 \le X < 87.97\} &= 0.1549 = p_2 \\
P\{87.97 \le X < 97.97\} &= 0.2995 = p_3 \\
P\{97.97 \le X < 107.97\} &= 0.2995 = p_4 \\
P\{107.97 \le X < 117.97\} &= 0.1549 = p_5 \\
P\{X \ge 117.97\} &= 0.0456 = p_6 \,.
\end{aligned}
$$

To obtain the expected number of occurences in each interval, the above numbers are multiplied by the sample size yielding $e_1 = 1.368$, $e_2 = 4.647$, $e_3 = 8.985$, $e_4 = 8.995$, $e_5 = 4.647$, $e_6 = 1.368$. Since not all intervals have an expected number of occurences greater than or equal to 5, some of the intervals must be combined; therefore, before doing the test, we combine the first two intervals into one interval and the last two intervals into one interval. Thus, our chi-squared test will end up with the following four intervals: $(-\infty, 87.97]$, $(87.97, 97.97]$, $(97.97, 107.97]$, and $(107.97, \infty)$.

This grouping meets the criterion of at least 5 expected observations per interval. (The determination of the intervals is arbitrary as long as it meets the criterion, and as long as it is done without reference to the data so it is not influenced by the data.) Thus, the critical value using only one degree of freedom for the test at a five percent significance level is $\chi^2_{critical} = 3.841$ and the test statistic is

$$
\begin{aligned}
\chi^2_{test} &= \frac{(5-6.015)^2}{6.015} + \frac{(11-8.985)^2}{8.985} + \frac{(9-8.985)^2}{8.985} + \frac{(5-6.015)^2}{6.015} \\
&= 0.794 \,.
\end{aligned}
$$

The hypothesis that the data are normally distributed is not rejected since $\chi^2_{test} < 3.841$. However, again we emphasize that such a test does not "prove" the hypothesis is true; it indicates only that with the available data there is not sufficient evidence to reject the hypothesis. □

The test for normality in the above example used only one degree of freedom. This is bothersome because the greater the degrees of freedom the "more powerful" the test. (A test is said to be more powerful than another if the Type II error for the first test is less than the Type II error for the second.) However, there is a trade-off in the chi-square test because it is an approximate test, with the condition $e_i \ge 5$ being the limit of its approximation. The test is more accurate if the intervals are large enough so that the expected number of observations is larger than the minimum of five. However, a chi-square goodness-of-fit test with few degrees of freedom is somewhat weak, and the user should not put much confidence in the test results if the null hypothesis was not rejected. If these statements seem more qualitative than quantitative, you are right. At times, statistics is closer to art than to mathematics!

Another measure of how well the empirical data fit the hypothesized distribution is to measure the "plausibility" that the data set arose from the given distribution. The measure of "plausibility" is called the *p-value* and is defined, for the Chi-Square Goodness-of-Fit test as

$$\text{p-value} = \Pr\{V_n > \chi^2_{test}\}\,, \tag{3.16}$$

where V_n is a chi-square random variable with n degrees, and where n is equal to the number of degrees for freedom needed to determine the chi-square critical value (i.e., n is the size of the data set minus one minus the number of parameters estimated). Thus, the higher the p-value, the higher the plausibility that the data actually came from the hypothesized distribution. Technically, if the p-value is larger than the critical value, the test is not rejected; otherwise, it is not. It is also common (see [2]) to not reject the hypothesis if the p-value is larger than 10%, reject the hypothesis if the p-value is less than 1%, and when the p-value is between 1% and 10% to indicate that the test results make it unclear whether to reject or not reject the hypothesis.

The Goodness-of-Fit Test yielded a p-value of 0.7308 in Example 3.10, a p-value of 0.8495 in Example 3.11, and a p-value of 0.3729 in Example 3.12. Thus, it is plausible to assume that all data sets come from the hypothesized distributions. (See the appendix of this chapter for using Excel to obtain the p-value.)

3.4.2 The Kolmogorov-Smirnov Test

Two disadvantages of the chi-square test is that it cannot be used with small data samples because of the minimum number of expected observations required per interval and a decision must be made regarding how to structure the intervals. For continuous random variables, the Kolmogorov-Smirnov (K-S) test is an alternative approach that can be used with small data sets as well as with large ones. The statistic is based on a comparison between the hypothesized cumulative probability distribution function and the empirical cumulative probability distribution function.

The empirical distribution function is a CDF that arises directly from the data; in other words, it is a step function that increases at each data point proportional to the number of observations at the data value. In other words, the empirical distribution is a step function the increases by an amount equal to $1/n$ at each data point where n is the sample size. For example, assume we have the following 5 data values: 2.4, 5.2, 3.4, 2.4, 1.5. The first step is to order the data and then plot $\hat{F}_5$ so that it increases by $1/5$ at every data value. Note that the step size for $\hat{F}_5(x)$ at $x = 2.4$ is $2/5$ since there were two data points equal to 2.4. (See Fig. 3.6 for the graph from this data set.)

We now give a more precise statement of the K-S test. Let F_0 denote the hypothesized cumulative probability distribution function, and let $\hat{F}_n$ denote the empirical

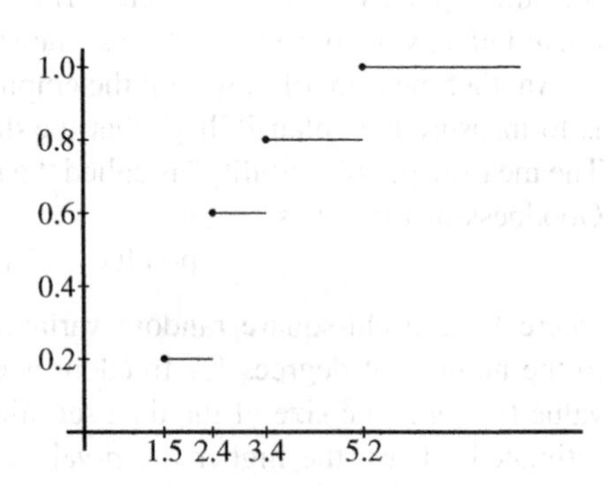

Fig. 3.6 *An empirical distribution function for 5 data points, with two data values occuring at 2.4*

cumulative probability distribution function based on a sample size of n. Then let D_n be a random variable denoting the maximum difference in absolute value between the two distributions; that is,

$$D_n = \sup_x |F_0(x) - \hat{F}_n(x)| \,. \tag{3.17}$$

(The *supremum* or "sup" of a set is the same as the maximum of the set with the addition of limiting values.) The operational procedure for determining the value of D_n is to represent the random sample in its ordered form, so that $X_{(i)}$ is the ith smallest value, which then yields

$$D_n = \max\{|\frac{i}{n} - F_0(X_{(i)})| \,,\, |F_0(X_{(i)}) - \frac{i-1}{n}| \;:\, i = 1, \cdots, n\} \tag{3.18}$$

where n is the sample size. Note that the value of D_n is the maximum of the difference in absolute value between the empirical distribution and hypothesized distribution evaluated at both the left-hand limit and the right-hand limit of the data points. Assuming that all parameters are known or hypothesized (i.e., no parameter estimation is made using the data set), then the test statistic is defined by

$$D_{test} = \left(\sqrt{n} + 0.12 + \frac{0.11}{\sqrt{n}}\right) D_n \text{ for } n \geq 2 \,. \tag{3.19}$$

Thus, the hypothesis is rejected if $D_{test} > D_\alpha$. Some of these critical values, taken from [3], are $D_{0.10} = 1.224$, $D_{0.05} = 1.358$, and $D_{0.01} = 1.628$.

Example 3.13. Returning once again to the spinning dial experiment, we hypothesize that the continuous uniform zero-one distribution is appropriate for that data set (Eq. 3.15). Both the empirical CDF and the hypothesized CDF are shown in Fig. 3.7 and it is seen that the largest difference between the two functions is $D_{20} = 0.10$ which yields a test statistic of $D_{test} = 0.4616$ which is less that the 5% critical value of 1.358; therefore, we conclude that there is not any statistical evidence to reject the null hypothesis. □

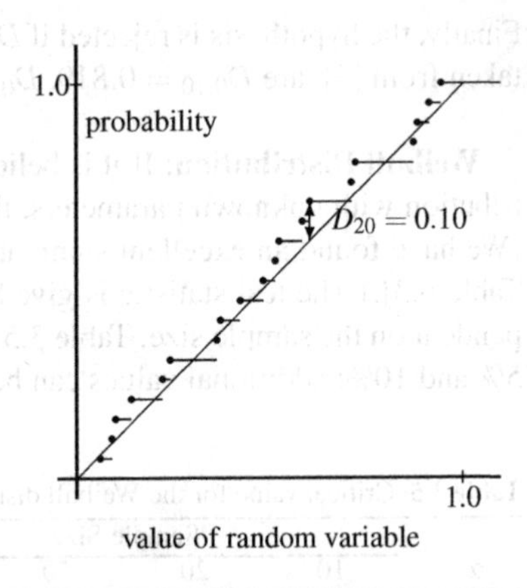

Fig. 3.7 The determination of the K-S statistic for the spinning dial experiment; the straight line is the hypothesized distribution, F_0, and the step function is the empirical distribution, $\hat{F}_{20}$, based on the 20 data points in Table 3.2

A major disadvantage of the K-S test in its original design is that it requires all parameters of the hypothesized distribution to be known, i.e., not determined from the data set. It is difficult to apply the K-S test to a general distribution when some of the parameters are unknown; however, for a few special distributions, easily implemented procedures have been developed that incorporate parameter estimation into the statistical test.

Exponential Distribution.

If it is believed that a data set comes from an exponential distribution with an unknown parameter, both the test statistic and the critical values have simple adjustments. First, the sample mean of Definition 3.3 is used to estimate the mean of the distribution. Second, the test statistic is calculated as

$$D_{test} = \left(\sqrt{n}+0.26+\frac{0.5}{\sqrt{n}}\right)\left(D_n - \frac{0.2}{n}\right) \text{ for } n \geq 2\,. \tag{3.20}$$

Finally, the hypothesis is rejected if $D_{test} > D_\alpha$, where some of these critical values, taken from [3], are $D_{0.10} = 0.990$, $D_{0.05} = 1.094$, and $D_{0.01} = 1.308$.

Normal Distribution: If it is believed that a data set comes from a normal distribution with unknown parameters, we again have simple adjustments to the test statistic and the critical values. The sample mean (Definition 3.3) and MLE for the variance (Definition 3.8) are used to obtain estimates for the unknown parameters, and the test statistic is calculated as

$$D_{test} = \left(\sqrt{n}-0.01+\frac{0.85}{\sqrt{n}}\right)D_n \text{ for } n \geq 2\,. \tag{3.21}$$

Finally, the hypothesis is rejected if $D_{test} > D_{\alpha}$, where some of these critical values, taken from [3], are $D_{0.10} = 0.819$, $D_{0.05} = 0.895$, and $D_{0.01} = 1.035$.

Weibull Distribution: If it is believed that a data set comes from a Weibull distribution with unknown parameters, the Maximum Likelihood Estimators are used. (We have found an excellent summary of distributions and MLE procedures in [3, Table 6.3].) The test statistic is give by $D_{test} = \sqrt{n}D_n$, and the critical value is dependent on the sample size. Table 3.5 gives some of the critical value for an error of 5% and 10%; additional values can be found in [3].

Table 3.5 Critical value for the Weibull distribution

	Sample Size			
α	10	20	50	∞
5%	0.819	0.843	0.856	0.874
10%	0.760	0.779	0.790	0.803

Gamma Distribution: If it is believed that a data set comes from a Gamma distribution with unknown parameters, a much more complicated test is required. This test uses a squared difference between empirical CDF and the hypothesized CDF. The Gamma distribution has two parameters: a shape parameter, a, and a scale parameter, β. As with the previous distributions, we use the Maximum Likelihood Method to estimate both unknown parameters and again refer the reader to [3, Table 6.3] for a description of MLE procedures. The test statistic, denoted by U^2 is given by

$$U^2_{test} = \frac{1}{12n} + \sum_{i=1}^{n}\left(F(X_{(i)}) - \frac{2i-1}{\sqrt{2n}}\right)^2 - n\left(\left(\frac{1}{n}\sum_{i=1}^{n}F(X_{(i)})\right) - 0.5\right)^2 . \quad (3.22)$$

If the test statistic is larger than the critical value, the null hypothesis is rejected. The critical values, taken from [4], depend on the shape parameter, a, according to the Table 3.6. The difficulty with the Gamma distribution is that the test statistic (Eq. 3.22) is more complicated than the other test statistics that arise from the empirical CDF and the table of critical values are for the asymptotic results only. However, it is reported in [4, p. 156] that results are still quite good for small n.

Table 3.6 Critical values for the gamma distribution

	Sample Size						
α	1	2	3	4	5	10	∞
5%	0.119	0.118	0.118	0.117	0.117	0.117	0.117
10%	0.098	0.097	0.097	0.097	0.097	0.096	0.096

- *Suggestion: Do Problems 3.13–3.14.*

3.5 Fitting Models

Oftentimes data collected from random samples need to be characterized in response to more specific questions regarding the relationship of the random sample to another variable that may not be random. We have already seen some of those in Sect. 3.3.4 where the correlation coefficient is used to measure the degree of linear relation between variables collected as pairs. Suppose, however, that the interest is not only about how the two variables behave as a pair, but also how a variable can be used to predict the value for the other variable. For example, the efficiency of an additive to gasoline has been evaluated by adding different amounts of the additive to the gasoline and counting the number of times that clogged injectors have failed to pass performance criteria. Airplane engines have been found to work more efficiently (i.e., fly more miles per gallon) with higher concentrations of the additive. Twenty airplanes of the same type and model are used for two years to test the additive at each of five concentrations of the additive in the gasoline. At the end of the experiment, all the airplanes that needed maintenance to replace clogged injectors caused by the additive were recorded. The following table shows the number of airplanes that needed new injectors at each of the different concentrations of the additive in the gasoline.

Table 3.7 Experimental results regarding risk from gasoline additives

Concentration (ml/l) of additive	0	5	10	50	100
Number Airplanes Tested	20	20	20	20	20
Number Injectors Failed	2	3	5	6	8

The first step in model fitting is to determine an appropriate model for the system under consideration. Determining the best model to fit is a very important decision and should be driven by knowledge of the phenomenon being modeled. Here, we restrict our attention to linear one-variable models; in other words, the relationship between the fraction of fuel injectors that fail and the concentration levels will be assumed to be linear. The variable for the outcomes of the experiment (i.e., the fraction of injectors that fail) is called the dependent variable and is usually denoted by the random variable Y while the variable representing the known values (i.e., the concentration level) is called the independent variable and is denoted by x. Mathematically, this linear relationship can be expressed as follows,

$$Y_i = \alpha + \beta x_i + \varepsilon_i \,, \tag{3.23}$$

where x_i is the ith value of the independent variable, Y_i is the ith value of the dependent variable, ε_i is the random error of Y_i, α is the intercept of the line (i.e., the expected value of Y_i when the value of x_i is zero), and β is the slope of the line (i.e., the rate of change of the dependent variable Y per each unit of change of the independent variable x). Note that if Y and x were linearly related and there were no error (randomness) in the values of Y then the problem would be trivial and one would only need the values of x and Y at two different points to define the straight line. The randomness of the Y values is what makes the task of finding the linear relationship between x and Y a more interesting problem.

Once the type of model to be used to fit observed data has been determined, another important consideration is to select the best method to fit the model to the data. This decision is largely based on the statistical characteristics of the data and the desired statistical behavior of the fitted model. Here, we are going to review two methods to fit a model to the data. Each of the methods has different statistical properties and each is more appropriate to use than others in specific circumstances. The estimation method simply finds the best value for the intercept and slope of Eq. (3.23) that satisfy known statistical properties. These statistical properties depend on what is assumed about the distribution of the unknown error term; namely, the assumption regarding ε_i in Eq. (3.23).

There are several important aspects of regression models that will not be covered in this brief review. Almost all introductory books in mathematical probability and statistics will cover these topics, and we leave it to the reader to pursue further study in this area as needed.

3.5.1 Ordinary Least Squares Regression

For this section, we make three assumptions: (1) the collection of random variables, $\{Y_i;\ i = 1,\cdots,n\}$ are pairwise uncorrelated; (2) each random variable within the collection has the same variance, denoted by σ^2; and, (3) $E[Y_i] = \alpha + \beta x_i$, for some constants α and β. (Pairwise uncorrelated means that $E[Y_i Y_j] = E[Y_i]E[Y_j]$ for $i \neq j$.)

Definition 3.9. The least-squares estimators of α and β of the linear relationship given by Eq. (3.23) between n pairs of observations $(y_i, x_i), i = 1, 2, \cdots, n$ is given by the α and β values that minimize the following sum of squares

$$SSQ(\alpha,\beta) = \sum_{i=1}^{n} (\alpha + \beta x_i - Y_i)^2 .$$

□

The least-squares estimators of α and β can be obtained using a closed-form solution of the least-squares equation given in Definition 3.9. There are no statistical assumptions about the distribution or properties of the unknown error term to obtain ordinary least squares estimates of the parameters. The estimates of the parameters are obtained by taking the partial derivative of the sum of squares equation in Definition 3.9 with respect to α and β, setting the equations equal to zero, and solving for the unknown values of α and β. That is, the solution to the following equations is the least-squares estimates of the parameters

$$\frac{\partial SSQ(\alpha,\beta)}{\partial \alpha} = \sum_{i=1}^{n} 2(\alpha + \beta x_i - Y_i) = 0 \qquad (3.24)$$
$$\frac{\partial SSQ(\alpha,\beta)}{\partial \beta} = \sum_{i=1}^{n} 2x_i(\alpha + \beta x_i - Y_i) = 0 .$$

Property 3.5. *The least-squares estimators for α and β of the linear relationship given by Eq. (3.23) solve the system of equations of (3.24) and are given by*

$$\hat{\beta} = \frac{\sum_{i=1}^{n}(x_i - \bar{x})(y_i - \bar{y})}{\sum_{i=1}^{n}(x_i - \bar{x})^2} = \frac{\sum_{i=1}^{n} x_i y_i - n\bar{x}\bar{y}}{\sum_{i=1}^{n} x_i^2 - n\bar{x}^2}$$
$$\hat{\alpha} = \bar{y} - \beta\bar{x} ,$$

where $\bar{x} = (\sum_{i=1}^{n} x_i)/n$ and $\bar{y} = (\sum_{i=1}^{n} y_i)/n$.

Notice that a lower case letter (y_i) is used within the definition of the least-squares estimators because the estimators are based on *observed* values of the random variable.

Example 3.14. Our goal is to predict the risk associated with using the gasoline additive in airplanes. Taking the data of Table 3.7, we first calculate the estimated probability of failure of the fuel injectors for each concentration level yielding Table 3.8.

Table 3.8 Values used to estimate parameters for the linear model of Example 3.14

Concentration (ml/l): x-values	0	5	10	50	100
Fraction failed: y-values	0.10	0.15	0.25	0.30	0.40

Using the data of Table 3.8, we obtain the model parameters by

$$\hat{\beta} = \frac{58.25 - 5 \times 33 \times 0.24}{12625 - 5 \times 33^2} = 0.002597$$
$$\hat{\alpha} = 0.24 - 0.002597 \times 33 = 0.1543 .$$

Figure 3.8 shows a graph with the estimated probability of injector failures for different concentrations of the gasoline additive. The straight line on the graph is that line that minimizes the sum of squared errors; that is, sum of squared differences between the estimated line and the observed values of Y (i.e., the difference between y_i and $\hat{y}_i$, where $\hat{y}_i = \hat{\alpha} + \hat{\beta} x_i$). In other words, there is no other straight line with a smaller sum of squares errors. □

The assumption of this section is that each random variable Y_i for $i = 1, \cdots, n$ has the same variance denoted by σ^2. Based on Eq. (3.23), each error term ε_i must

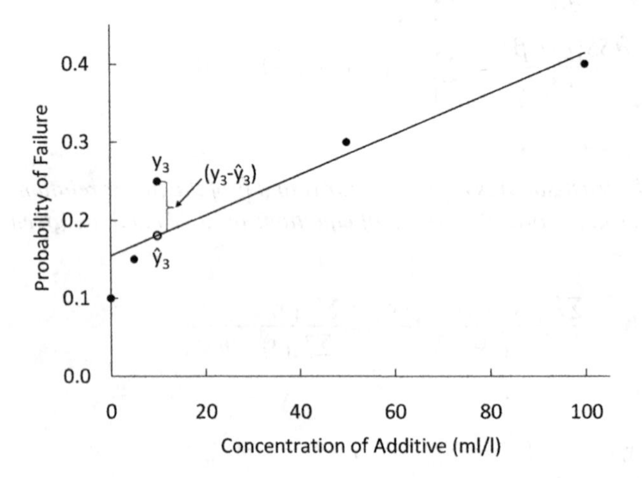

Fig. 3.8 Least squares fit to the probability of injector failure vs. concentration of additive

also have the same variance equal to σ^2. Since the estimates for α and β depend on observations of Y_i, the estimators $\hat{\alpha}$ and $\hat{\beta}$ must be random as well. In particular if the distribution of the errors is assumed to be normal, then the distributions of the estimates are also normal. Even though the ordinary least squares parameter estimators are random, they are unbiased. That is, the mean of the distribution of the parameter estimators equals the true coefficients of Eq. (3.23).

The first step in determining the accuracy of the least-squares estimates given by Property 3.5, is to estimate the value of the variance term σ^2 which is given in the following property.

Property 3.6. *Assume that each ε_i for $i = 1, \cdots, n$ in Eq. (3.23) is normally distributed with mean 0 and variance σ^2. The following estimator for σ^2 is unbiased*

$$s^2 = \frac{SSE}{n-2} = \frac{\sum_{i=1}^{n}(y_i - \hat{y}_i)^2}{n-2} .$$

Notice that the sum of squared errors (SSE) is divided by $n-2$ to account for the two degrees of freedom lost because of the two parameters being estimated. The SSE can be computed by

$$SSE = \sum_{i=1}^{n}(y_i - \hat{y}_i)^2 = \sum_{i=1}^{n} y_i^2 - \hat{\alpha}\sum_{i=1}^{n} y_i - \hat{\beta}\sum_{i=1}^{n} x_i y_i . \tag{3.25}$$

Having an estimate of the variance of the error, we are now in a position of fully characterizing the distribution of the parameter estimators. Let σ_α^2 denote the variance of $\hat{\alpha}$ and let σ_β^2 denote the variance of $\hat{\beta}$. The variance of the two least squares estimators are given by

$$\begin{aligned}\sigma_\alpha^2 &= \frac{\sigma^2 \sum_{i=1}^n x_i^2}{n\sum_{i=1}^n (x_i-\overline{x})^2} \\ \sigma_\beta^2 &= \frac{\sigma^2}{\sum_{i=1}^n (x_i-\overline{x})^2}\,,\end{aligned} \tag{3.26}$$

where σ^2 is the variance of the error term. In order to form a statistic that will allow for building confidence intervals, we need to estimate these variances which is a straight forward substitution of Property 3.6 into (3.26). Thus, the estimate for the variance of $\hat{\alpha}$ and $\hat{\beta}$, denoted by s_α^2 and s_β^2, respectively, are given by

$$\begin{aligned}s_\alpha^2 &= \frac{s^2 \sum_{i=1}^n x_i^2}{n\sum_{i=1}^n (x_i-\overline{x})^2} \\ s_\beta^2 &= \frac{s^2}{\sum_{i=1}^n (x_i-\overline{x})^2}\,,\end{aligned} \tag{3.27}$$

where s^2 is the variance estimator from Property 3.6. The variance statistics of Eqs. (3.27) allow for the construction of Student-t random variables that can be used for hypothesis testing and building confidence intervals for the parameter estimators as showing in the following property.

Property 3.7. *Assume that each ε_i for $i = 1,\cdots,n$ in Eq. (3.23) is normally distributed with mean 0 and variance σ^2. Let $\hat{\alpha}$ and $\hat{\beta}$ be the least squares estimators as determined by Property 3.5 with their variance estimators given by Eqs. (3.27). Then*

$$T_\alpha = \frac{\hat{\alpha}-\alpha}{s_\alpha} \quad \textit{and}$$

$$T_\beta = \frac{\hat{\beta}-\beta}{s_\beta}\,.$$

have Student-t distributions with $n-2$ degrees of freedom.

Notice that the degrees of freedom equal the sample size minus two for the two parameters estimated (i.e., α and β).

The t-distribution allows us to easily build confidence intervals for the parameter estimators. Under the assumptions given in Property 3.7, the $1-a$ confidence interval for the intercept estimator of Eq. (3.23) is given by

$$(\hat{\alpha} - t_{n-2,a/2} \times s_\alpha \, , \, \hat{\alpha} + t_{n-2,a/2} \times s_\alpha) \, , \tag{3.28}$$

and the $1-a$ confidence interval for the slope estimator of Eq. (3.23) is

$$(\hat{\beta} - t_{n-2,a/2} \times s_\beta \, , \, \hat{\beta} + t_{n-2,a/2} \times s_\beta) \, . \tag{3.29}$$

Notice that we use a for the amount of error allowed in the confidence interval instead of α so that there is no confusion between the error allowed and the intercept parameter. (These confidence intervals are very similar to the confidence intervals for the mean. It would be worthwhile to compare Property 3.7 with Property 3.4 and to compare Eqs. (3.28) and (3.29) with Eq. (3.4).)

Example 3.15. In Example 3.14, we found that the least-squares estimate for the intercept and slope were $\hat{\alpha} = 0.1543$ and $\hat{\beta} = 0.002597$. We are now able to determine the accuracy of these estimates and indicate whether or not the estimates are statistically different from zero. Using the above equations, the following values are obtained:

$$\begin{aligned}
s^2 &= \frac{0.008557}{3} = 0.002852 \\
s_\beta^2 &= \frac{0.002852}{7180} = 3.973 \times 10^{-7} \\
s_\alpha^2 &= \frac{0.002852 \times 12625}{5 \times 7180} = 0.001003 \, .
\end{aligned}$$

The t-statistic with a 5% (two-tailed) error and with three degrees of freedom equals 3.1824; therefore, with 95% confidence, we have that

$$0.0535 < \alpha < 0.2551 \quad \text{and} \quad 0.000592 < \beta < 0.004603 \, .$$

Because the value of zero is not within the 95% confidence intervals, we can say that both parameters are significantly different than 0.0. Thus, it is safe to claim that the risk of injector failure depends in the concentration level of the additive, and even without any additive, there is a (statistically) significant probability of injector failure. □

We close this section by giving the prediction interval around the regression line. In other words, if several additional experiments were to be conducted with the independent variable set at the value of $x = x^*$, a 95% prediction interval would be an interval for which approximately 95% of the individual observations fall within the interval. Note that this interval gives the occurences of the individual data points, not just the sample mean. Such a prediction interval is given by

$$\hat{\alpha} + \hat{\beta}\, x^* \pm s \sqrt{\frac{n+1}{n} + \frac{s_\beta^2 (x^* - \overline{x})^2}{s^2}} \tag{3.30}$$

where s^2 is from Property 3.6 and s_β^2 is from Eq. (3.27).

3.5.2 Maximum Likelihood Estimates for Linear Models

A application of the Maximum Likelihood Method for linear models is conceptually the same as briefly discussed in Sect. 3.2.2. Again, we leave the explicit details to other authors and simply give some general results here.

To determine the Maximum Likelihood Estimators (MLE), a likelihood function is obtained which represents the "likelihood" that a specific sample will be observed. The parameter estimates are then the values that maximize the likelihood of the sample being observed. It should be recalled that the Method of Moments and the Maximum Likelihood method often yield identical estimates for means of normally distributed random variables but yield different variance estimators. The same is true for linear models.

Property 3.8. *Assume that each ε_i for $i = 1, \cdots, n$ in Eq. (3.23) is normally distributed with mean 0 and variance σ^2. The MLE for the intercept, α, and slope, β, parameters are identical to the least squares estimators of Property 3.5. The MLE for the variance, σ^2, is given by*

$$\hat{\sigma^2} = \frac{SSE}{n} = \frac{\sum_{i=1}^{n}(y_i - \hat{y}_i)^2}{n}.$$

The MLE for the variance of $\hat{\alpha}$ and $\hat{\beta}$ are the same as given in Eqs. (3.27) except that s is replaced by $\hat{\sigma}$. In other words, the MLE for the variance of $\hat{\alpha}$ and $\hat{\beta}$, denoted by $\hat{\sigma}_{\alpha}^2$ and $\hat{\sigma}_{\beta}^2$, respectively, are given by

$$\hat{\sigma}_{\alpha}^2 = \frac{\hat{\sigma^2} \sum_{i=1}^{n} x_i^2}{n \sum_{i=1}^{n} (x_i - \overline{x})^2} \tag{3.31}$$

$$\hat{\sigma}_{\beta}^2 = \frac{\hat{\sigma^2}}{\sum_{i=1}^{n} (x_i - \overline{x})^2},$$

where $\hat{\sigma^2}$ is the variance estimator from Property 3.8.

Appendix

Spreadsheet programs have made statistical analyses much easier and there no longer needs to be a dependence on statistical tables. In this appendix, we show some uses of Excel in statistics and present an algorithm for calculating sample standard deviations that avoids some potential numerical problems with taking the differences of squares.

Excel Functions: The common Excel functions given by `AVERAGE(`*range*`)` and `STDEV(`*range*`)` yield the sample mean and sample standard deviation (Definition 3.3) of the data given within *range*. The MLE for the standard deviation (Definition 3.8) is given by `STDEVP(`*range*`)`. The Excel estimator for the correlation coefficient, namely `CORREL(`*range1, range2*`)`, is the Pearson correlation coefficient (Eq. 3.10) and is equivalent to the Excel covariance estimator divided by the two MLE's for the standard deviations.

The critical values for many statistical tests are also available from Excel. The only problem is remembering which functions are based on right-hand tails and which functions are based on left-hand tails. For example, the Excel function `NORMSINV(`*prob*`)` is based on the left-hand probability. In other words, the function returns a value such that the probability that a standard normal random variable is to the left of the returned value is equal to the input parameter *prob*. However, the function `CHIINV(`*prob, df*`)` is based on right-hand tails. In other words, this second function returns a value such that the probability that a chi-square random variable with *df* degrees of freedom is to the right of the returned value is equal to the input parameter *prob*. To insure maximum confusion, the Excel function for the t-statistic `TINV(`*prob, df*`)` is based on two tails so that this function returns a value such that a Student-*t* random variable with *df* degrees of freedom is to the right of the returned value is equal to one half of the value of the input parameter *prob* (the other half of the probability is in the left-hand tail). Table 3.9 contains a summary of Excel functions that give the inverse of the distribution functions. (Actually, these functions are the inverse of the distribution function when the probability parameter refers to the left-hand tail, and they are the inverse of the complement of the distribution function when the probability parameter refers to the right-hand tail.)

Table 3.9 Excel functions for some common critical statistical values

Excel Function	Distribution	Tails	Equation References
`NORMSINV(`*prob*`)`	Standard Normal	Left-hand	(3.2), (3.6), (3.8)
`TINV(`*prob, df*`)`	Student-*t*	Two-tailed	(3.4)
`CHIINV(`*prob, df*`)`	Chi-square	Right-hand	(3.7)
`FINV(`*prob, df1, df2*`)`	F	Right-hand	(9.12)

For a 5% error with the Student-*t* statistic as in Eq. (3.4), the value of 0.05 is used for the *prob* parameter; thus, the function `TINV(0.05,` *df*`)` would be used. However, if the z statistic were needed with a 5% error as in Eq. (3.6), the Excel function would be `NORMSINV(0.975)`. The difference in the values of the *prob* parameter illustrates the difference between a left-hand tail function and a two-tailed function.

Example 3.16. The use of Excel in calculating confidence limits and the meaning of a confidence interval will be demonstrated in this example. (Note that when using Excel, we follow the practice of never having data without either a header or a cell to the left of the data with a description of the data.) In Cells A1 and A2, type the following:

	A
1	Data
2	=NORMINV(RAND(),100,20)

Notice that Cell A2 contains a random variate that is normally distributed with mean 100 and standard deviation 20 (see Property 2.4). Highlight Cells A2:A101 and hit the <Ctrl>-D keys to copy the formula in Cell A2 down through Cell A101. While Cells A2:A101 are still highlighted, type `DataRng` in the Name Box (i.e., the box in that appears in the upper left portion of the worksheet just above the A and B columns) and then hit the `Enter` key. Now a random sample of size 100 is contained within the range of cells that have been named `DataRng`. Now in Cells D1:E5, type the following:

	D	E
1	Mean	=AVERAGE(DataRng)
2	StDev	=STDEV(DataRng)
3	t-stat	=TINV(0.05,COUNT(DataRng)-1)
4	Lower Limit	=E1-E2*E3/SQRT(COUNT(DataRng))
5	Upper Limit	=E1+E2*E3/SQRT(COUNT(DataRng))

The 95% confidence interval is given in Cells E4:E5. It is a "good" confidence interval if the value of 100 is between the lower and upper limits. Notice that by hitting the `F9` key, a new realization of the random sample is obtained. On a separate sheet of paper, record whether or not the confidence interval is "good". Hit the `F9` key and record again whether or not the confidence interval is "good". Repeat this several times, and count the number of "good" intervals and the number of "bad" intervals. Since a 95% confidence interval was formed, there should be approximately 19 "good" intervals for every 20 different realizations of the random sample. □

As mentioned when quartiles were introduced, there is not a universally recognized definition for quartiles. Excel uses a different definition than the common one that we used in Definition 3.4. The Excel function `QUARTILE(`*data array, quartile number*`)` will yield the minimum, lower quartile, median, upper quartile, and maximum value as the *quartile number* varies from 0 through 4. The Excel function defines the lower quartile by the index given as $(n+3)/4$ where if the index is not an integer, a weighted average of the ordered data values associated with the two indices that bracket the value $(n+3)/4$. The upper quartile is defined by the index given as $(3n+1)/4$ where again if the index is not an integer, a weighted average of the ordered data values associated the two indices that bracket $(3n+1)/4$. In other words, if $n=50$, the index defining the lower quartile would be $53/4=13.25$; thus, the lower quartile would be given by $x_{(13)}+0.25\times(x_{(14)}-x_{(13)})$.

Two other functions that are helpful are the `FISHER()` and `FISHERINV()` functions. These are needed for confidence intervals dealing with the correlation coefficient and are the functions of Eqs. (3.11) and (3.13). (Problem 3.7 uses these functions to redo Example 3.9.)

Example 3.17. Our final example of this appendix demonstrates using Excel for obtaining the least-squares estimates for a simple linear regression. There are actually two methods that can be used. The first method that we demonstrate is to obtain the slope and intercept individually. The second method is to obtain the slope and intercept plus additional statistical values with one function. For demonstration purposes, we return to Example 3.15 and reproduce those results with Excel. To setup the example, type the following in Cells A1:D6 using word warp in the first line to make it readable.

	A	B
1	Concentration	Prob. Failure
2	0	0.10
3	5	0.15
4	10	0.25
5	50	0.30
6	100	0.40

The slope and intercept values can be obtained directly by the Excel functions with same names as follows.

	A	B
8	alpha	=INTERCEPT(B2:B6,A2:A6)
9	beta	=SLOPE(B2:B6,A2:A6)

Notice that in both functions, the y values are listed first followed by the x values. There is also an Excel function, called `LINEST`, that not only yields the slope and intercept estimates, but also obtains additional statistical information. This function does more that we covered in Sect. 3.5 so we do not explain the entire function but only use what was covered in Example 3.15. To obtain the estimates $\hat{\beta}$ and $\hat{\alpha}$ as well as their standard error terms, s_β and s_α, setup the spreadsheet as follows.

	F	G	H
1		Slope	Intercept
2	Estimator		
3	St. Error		
4	r^2 & s		

Now highlight the 3×2 group of cells G2:H3 and type the following

```
=LINEST(B2:B6,A2:A6,TRUE,TRUE)
```

and while holding down the `<shift>` and `<ctrl>` keys hit the `<enter>` key. The least-squares estimators for the slope and intercept values are contained in Cells G2:H2 and the square root of their associated variances defined by Eq. (3.27) are contained within Cells G3:H3. The value in Cell G4 gives the so-called coefficient of determination, or r^2. The r^2 value gives the fraction of the variation that is accounted for in the regression line. In this case $r^2 = 0.8499$ which implies that approximately

85% of the variation of the fuel injector failures was due to the variation in the concentration levels. Finally, the value that is contained in Cell H4 is the value of s which is the square root of the variance estimator defined by Property 3.6.

Notice that there were two additional parameters that were set equal to `TRUE` in the above example of the `LINEST` function. The third parameter indicates that $\hat{\alpha}$ is desired and the fourth parameter indicates that the additional statistical terms are desired. If the third parameter were set to `FALSE`, the best least-squares line that passes through the *origin* would be obtained, and if the fourth parameter were `FALSE`, only the estimators for $\hat{\beta}$ and $\hat{\alpha}$ would be obtained.

A one pass algorithm: There are two equivalent formulas for calculating the sample variance (Definition 3.3). Because the first formula requires the sample mean to be known ahead of time, it is the second formula that is most commonly used. The difficulty with the second formula is that if the data values are large with a relatively small variance, the second formula can produce round-off errors. A One-Pass Algorithm for Variances (Welford [5]) has been developed that allows the calculation of the sample variance by squaring the difference between each data value and the current estimation of the mean thus reducing the potential for round-off error. Assume we have data value $x_1, \cdots, x_n$. Let $\bar{x}_k$ and ssq_k be the mean estimate and the sum of squares value after considering k data points. The One-Pass Algorithm for the variance is

1. Set $\bar{x}_0 = 0$, $ssq_0 = 0$, and $k = 1$,
2. $\Delta = x_k - \bar{x}_{k-1}$,
3. $\bar{x}_k = \bar{x}_{k-1} + \Delta/k$,
4. $ssq_k = ssq_{k-1} + \Delta \times (x_k - \bar{x}_k)^2$.
5. If $k = n$ stop; otherwise, increment k and return to Step 2.

After the algorithm is finished, the sample variance is $s_n^2 = ssq_n/(n-1)$.

Problems

3.1. The output of a simulation of a queuing system indicates that the waiting times of five customers at a bank teller were 5.7, 16.3, 3.0, 4.5, 1.2 minutes. Using these waiting times, estimate:
(a) The sample mean.
(b) The sample standard deviation.
(c) The probability of the waiting time being longer than 10 minutes.
(d) Assume that the true distribution is a normal distribution with the true mean equal to the sample mean and the true standard deviation equal to the sample standard deviation, give the probability of the waiting time being longer than 10 minutes.
(e) Assume that the true distribution is an exponential distribution with the true mean equal to the sample mean, give the probability of the waiting time being longer than 10 minutes.

3.2. The following 25 values were collected from a process believed to follow a normal distribution. Construct a histogram using the following sample:

2.47	4.87	2.85	7.12	2.25
2.17	2.56	0.83	2.03	0.29
4.63	1.28	4.96	1.32	0.48
4.65	-1.52	1.13	2.81	0.48
5.44	0.90	0.32	3.89	5.86

3.3. Using the sample data in the previous exercise,
(a) Calculate the sample median.
(b) Calculate the lower and upper quartiles.
(c) Construct a box plot.

3.4. Several different estimators have been proposed to estimate the mean of a distribution based on a random sample of size three. For each of the estimators listed below, determine their expected value, $E[\hat{\mu}]$, and variance, $V[\hat{\mu}]$, as a function of the true mean, μ, and true variance, σ^2.
(a) $\hat{\mu} = (3X_1 + X_2 + 2X_3)/6$.
(b) $\hat{\mu} = (3X_1 - X_2 + 4X_3)/6$.
(c) $\hat{\mu} = (X_1 + X_3)/2$.
(d) $\hat{\mu} = (X_1 + X_2 + X_3)/3$.
(e) $\hat{\mu} = (X_1 + X_2 + X_3)/10$.

3.5. Of the estimators for the mean in the previous exercise,
(a) Which estimator(s) are unbiased?
(b) Which estimator has the smallest variance?
(c) Which estimator(s) would you suggest using and why?

3.6. A company that makes hard drives, wishes to determine the reliability of its manufacturing process. The hard drives produced in a plant are inspected for 20 consecutive days. The plant makes exactly 100 hard drives every day with the daily number of defective drives given in the following table. (Assuming the manufacturing process of every hard drive is independent of the manufacturing process of other hard drives.)

1	0	3	5	4
2	4	3	8	1
5	6	1	0	3
7	1	4	2	5

(a) Give an estimate for the probability that a given part will be defective?
(b) Give the 95% confidence interval for your estimate?
(c) Estimate the probability that on any given day, there will be fewer than 3 defective drives produced? (There are actually two different methods to obtain this estimate. Can you identify both methods and suggest which may be best?)

3.7. Confirm the calculations for Example 3.9 using the Excel functions for the Fisher transformation.

3.8. Using the method of moments obtain an expression for estimating the values of a and b for a uniform distribution whose lower and upper limits are given by a and b, respectively.
(a) Do you see any practical difficulties with implementing these estimators?
(b) Generate a random sample of size 20 from a continuous uniform distribution that is between 10 and 15. Demonstrate your estimators with your generated random sample. Are the values of your estimators believable?

3.9. The following is a sample from a gamma distribution. Estimate the parameters of the gamma distribution using the method of moments.

8.9	1.9	1.9	1.6	5.2
2.7	8.3	2.9	6.8	7.4
1.8	4.4	2.5	4.3	0.9
4.5	11.1	5.8	5.4	9.4

3.10. The following sample was taken from a normal distribution with unknown mean and unknown standard deviation.

9.20	7.16	9.57	10.30	11.20	12.23
11.23	9.11	10.50	8.58	14.38	8.22
11.18	7.19	9.52	10.90	5.52	10.08
8.15	8.67	10.77	9.52	10.96	13.13

(a) Compute a 95% confidence interval for the mean.
(b) Compute a 95% confidence interval for the variance.
(c) Compute a 95% confidence interval for the probability that random variable from this distribution will be greater than 14.0.
(d) The data above give the length of fabric ribbons measured in inches. Any ribbon that is within 1.5 inches of the target value of 10 inches is considered acceptable. Calculate a confidence interval for the proportion of acceptable ribbons.

3.11. Using Excel (see Example 3.16), generate a random sample of size 30 from a normal distribution with mean 500 and standard deviation 25. Let the random sample be contained in the second *row* of a worksheet, i.e., A2:AD2. (The first row can be used for a header.) In Cell AF2, calculate the sample variance (i.e., denominator of 29); in Cell AG2, calculate the MLE of the variance (i.e., denominator of 30); in Cell AH2, calculate an estimate for the variance using a denominator of 31; and in Cell AI2, calculate an estimate for the variance using a denominator of 32. Copy the second row down (with formulas) for a 1000 or more rows. Based on your set of 1000 or more random samples, each of size 30, do you think that any of the variance estimators are unbiased? Which estimator appears to have the smallest mean square error?

3.12. Repeat Problem 3.11 except determine four estimators for the standard deviation instead of the variance. Based on your set of 1000 or more random samples, each of size 30, do you think that any of the standard deviation estimators are unbiased? Which estimator appears to have the smallest mean square error?

3.13. The number of customers arriving to a drive-thru bank teller in an hour is randomly sampled from 190 hours. The distribution of the number of customers arriving per hour is as follows:

Number of Customers/hr	Observed Frequency
0	1
1	7
2	15
3	29
4	35
5	33
6	27
7	22
8	14
9	4
10	2
11	1

(a) What is the average number of costumers per hour arriving to the bank teller?
(b) Using a chi-squared test with a Type I error of 5%, test the hypothesis that the number of costumers per hour arriving to the bank teller is distributed according to the Poisson distribution.
(c) Using a chi-squared test with a Type I error of 5%, test the hypothesis that the number of costumers per hour arriving to the bank teller is distributed according to the normal distribution.
(d) Which of the two distributions tested above seems more reasonable for modeling the data collected at the drive-thru bank teller? Why?

3.14. A bank teller recorded the length of time to serve 15 customers. The following were the 15 service times in minutes

4.8	0.9	16.1	2.9	28.4
0.6	14.2	7.4	8.6	9.9
11.1	5.8	1.9	22.6	3.3

For the following use an α error of 5%.
(a) Test the hypothesis that the service times are normally distributed with mean 10 and standard deviation 10 using the K-S test.
(b) Test the hypothesis that the service times are exponentially distributed with mean 10 using the K-S test.
(c) Test the hypothesis that the service times are normally distributed with a mean and standard deviation estimated from the data using the K-S test.
(d) Test the hypothesis that the service times are exponentially distributed with a mean estimated from the data using the K-S test.

3.15. The board of directors of the airplane company in Example 3.14 has decided that the maximum acceptable increase in the probability of failure is 0.05 over the estimated probability of failure when no additive is in the gasoline. They want to know what is the maximum concentration of the additive that can be in the gasoline that increases by 0.05 the expected probability of failure?

3.16. The ordinary least squares estimates of the linear model are given by Property 3.5.
(a) Show that the solution given in the property actually solve Eqs. 3.24.
(b) Show that the solution given in the property also maximize the likelihood function given by

$$L(\alpha,\beta) = \prod_{i=1}^{n} \frac{\exp\{-(y_i - \alpha - \beta x_i)^2/(2\sigma^2)\}}{\sqrt{2\pi}\,\sigma}\,.$$

Note that maximizing L is the same as maximizing the $\ln(L)$.

References

1. Abramowitz, M., and Stegun, I.A., editors (1970). *Handbook of Mathematical Functions*, Dover Publications, Inc., New York.
2. Hayter, A.J. (2007). *Probability and Statistics for Engineers and Scientists*, Thomason, Belmont, CA.
3. Law, A.M. (2007). *Simulation Modeling and Analysis*, 4th ed., McGraw Hill, Boston.
4. Stephens, M.A. (1986). *Goodness-of Fit Techniques*, R. B. D'Agostino and M. A. Stephens (eds.), Marcel Dekker, New York, pp. 97–193.
5. Welford, B.P. (1962). Note on a Method for Calculating Corrected Sums of Squares and Products. *Technometrics*, **4**:419–420.

Chapter 4
Poisson Processes

The introduction to stochastic processes begins with a relatively simply type of process called a Poisson process that is essentially a type of counting process. For example as mentioned in Chap. 1, Bortkiewicz found 1898 that the number of deaths due to horse kicks in the Prussian army could be described by a Poisson random variable. A model depicting the cumulative number of deaths over time in the Prussian army would thus be described as a Poisson process. This chapter gives the formal description of such processes and includes some common extensions.

4.1 Basic Definitions

As stated in the first chapter, a random variable is simply a function that assigns a real number to each possible outcome in the sample space. However, we are usually interested in more than just a single random variable. For example, the daily demand for a particular product during a year would form a sequence of 365 random variables, all of which may influence a single decision. For this reason we usually consider sequences of random variables and their properties.

Definition 4.1. A *stochastic process* is a sequence of random variables. □

It is possible for a stochastic process to consist of a countable number of random variables, like the sequence of daily temperatures, in which case it is called a *discrete parameter process*. It is also possible that a stochastic process consists of an uncountable number of random variables in which case it is called a *continuous parameter process*. An example of a continuous parameter process is the continuous monitoring of the working condition of a machine. That is, at every point in time during the day a random variable is defined which designates the machine's working condition by setting the random variable equal to one if the machine is working and equal to zero if the machine is not working. For example, if the machine was working 55.78 minutes after the day had started, then $X_{55.78} = 1$. Thus, this stochastic process would be represented using a continuous parameter as $\{X_t\ ; t \geq 0\}$.

R.M. Feldman, C. Valdez-Flores, *Applied Probability and Stochastic Processes*, 2nd ed.,
DOI 10.1007/978-3-642-05158-6_4, © Springer-Verlag Berlin Heidelberg 2010

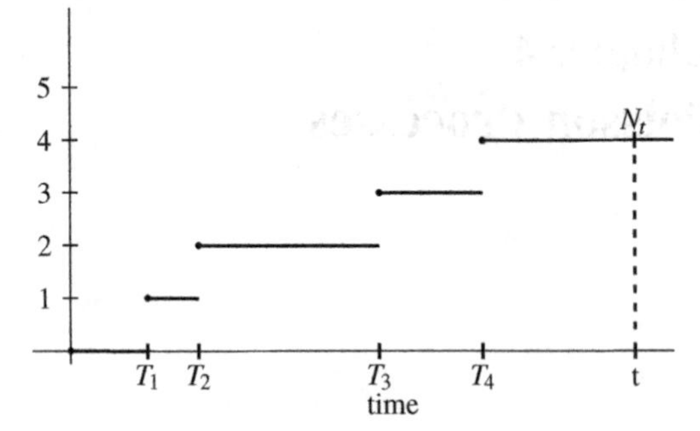

Fig. 4.1 Typical realization for a Poisson process, $\{N_t\,;t \geq 0\}$, with $T_n, n = 1, \cdots$, denoting arrival times

Definition 4.2. The set of real numbers containing the ranges of all the random variables in a stochastic process is called the *state space* of the stochastic process. □

The state space of a process may be either discrete or continuous. The continuous parameter stochastic process mentioned above that represented the continuous monitoring of a machine had a discrete state space of $E = \{0, 1\}$.

Example 4.1. We are interested in the dynamics of the arrival of telephone calls to a call center. To describe these arriving calls, consider a collection of random variables $\{N_t\,;t \geq 0\}$, where each random variable N_t, for a fixed t, denotes the cumulative number of calls coming into the center by time t. Since these calls record a count, the state space is the set of whole numbers $E = \{0, 1, 2, \cdots\}$. In other words, the arrival of telephone calls can be modeled as a continuous parameter stochastic process with a countable state space. □

The arrival process described in the above example is typical of the type of processes introduced in this chapter, namely Poisson processes (see Fig. 4.1). As these are discussed, the terminology of an arrival process will be used generically. For example, a Poisson process may be used to describe the occurrence of errors during the transmission of an electronic message and when an error occurs, it would be described as an "arrival" to the process.

Definition 4.3. Let the continuous parameter stochastic process $\{N_t\,;t \geq 0\}$ with $N_0 = 0$ have a state space equal to the nonnegative integers. It is called a *Poisson process* with rate λ if

(a) $\Pr\{N_t = k\} = \mathrm{e}^{-\lambda t}\,(\lambda t)^k/k!$ for all nonnegative k and $t \geq 0$.
(b) The event $\{N_{s+u} - N_s = i\}$ is independent of the event $\{N_t = j\}$ if $t < s$.
(c) The probability $\Pr\{N_{s+u} - N_s = i\}$ only depends on the value of u.

□

Condition (a) of Definition 4.3 is the defining characteristic of the Poisson process and is the reason for the name of the process. The characteristic specified in Condition (b) indicates that a Poisson process has *independent increments*. In other words, if you consider two non-overlapping intervals, what happens in one interval does not effect what happens in the second interval. The characteristic of Condition

(c) indicates that a Poisson process has *stationary increments*. In other words, the probability that a specified number of arrivals occurs within an interval only depends on the size of the interval not the location of interval.

For example, if the process of Example 4.1 has independent increments, then knowledge of the number of calls that arrived before 11:00AM will not help in predicting the number of calls to arrive between 11:00AM and noon. Likewise, if the process has stationary increments, then the probability that k calls arrive between 10:00AM and 10:15AM will be equal to the probability that k calls arrive between 4:00PM and 4:15PM because the size of both intervals is the same. It turns out that an arrival process with independent and stationary increments in which arrivals can only occur one-at-a-time must be a Poisson process (see [1]) so in that sense Condition (a) is redundant; however, because it is the characteristic that gives the process its name, it seems more descriptive to include it.

Using the definition of a Poisson random variable (Eq. 1.12), it is clear that

$$E[N_t] = \lambda t\,. \tag{4.1}$$

In other words, λ gives the mean arrival rate per unit time for an arrival process that is described by a Poisson process with rate λ. Also, because a Poisson process has independent and stationary increments, we also have, for nonnegative k

$$\Pr\{N_{s+u} - N_s = k\} = \frac{\mathrm{e}^{-\lambda u}(\lambda u)^k}{k!} \quad \text{for } u \geq 0\,. \tag{4.2}$$

Example 4.2. Assume that the arrivals to the call center described in Example 4.1 can be modeled according to a Poisson process with rate 25/hr. The number of calls that are expected within an eight-hour shift is $E[N_{t=8}] = 25 \times 8 = 200$. Assume we had 20 calls that arrived from 8AM through 9AM. What is the probability that there will be no calls that arrive from 9:00AM through 9:06AM? Because of independent increments, the information regarding the 20 calls is irrelevant. Because of stationary increments, we only need to know that the length of the interval is 6 minutes (or 0.1 hours), note that it starts at 9AM; thus, the answer is given as

$$\Pr\{N_{0.1} = 0\} = \frac{\mathrm{e}^{-25\times 0.1}(25 \times 0.1)^0}{0!} = \mathrm{e}^{-2.5} = 0.08208\,.$$

We are also interested in the probability that more than 225 calls arrive during the eight-hour shift. To answer this question, observe that the normal is a good approximation for a Poisson random variable with a large mean. For the approximation, we let X be a normally distributed random variable with the same mean and variance as the Poisson random variable; thus,

$$\begin{aligned}
\Pr\{N_8 > 225\} &\approx \Pr\{X > 225.5\} \\
&= \Pr\{Z > (225.5 - 200)/\sqrt{200}\} \\
&= \Pr\{Z > 1.80\} = 1 - 0.9641 = 0.0359\,.
\end{aligned}$$

(For review, see the discussion regarding the use of a continuous distribution to approximate a discrete random variable on p. 21 and Example 1.10.) □

- *Suggestion: Do Problem 4.1.*

4.2 Properties and Computations

Because of the independent and stationary increment property of a Poisson process, the distribution of inter-arrival times can be easily determined. Consider an interval of length t. The probability that no arrival occurs within that interval is given by

$$\Pr\{N_t = 0\} = e^{-\lambda t} .$$

Let T denote the time that the first arrival occurs. The event $\{T > t\}$ is equivalent to the event $\{N_t = 0\}$; therefore,

$$\Pr\{T \leq 0\} = 1 - \Pr\{T > 0\} = 1 - e^{-\lambda t} \quad \text{for } t \geq 0$$

which is the CDF for the exponential distribution. Extending this by taking advantage of independent and stationary increments, it is not too difficult to show that a Poisson process always has exponential times between arrivals. As mentioned in the first chapter, the exponential distribution has no memory (see Problem 1.23). Using this fact, it is possible to show that a process that has exponential times between arrivals will have independent and stationary increments and thus form a Poisson process.

> **Property 4.1.** *A Poisson process with rate λ has exponentially distributed inter-arrival times with the mean time between arrivals being $1/\lambda$. The converse is also true; namely, an arrival process with exponentially distributed inter-arrival times is a Poisson process.*

Example 4.3. Concern has been expressed regarding the occurrence of accidents at a relatively busy intersection. The time between accidents has been analyzed and it has been determined that this time is random and follows the exponential distribution with a mean time of five days between accidents. It is now noon on Monday and an accident has just occured and we would like to know the probability that the next accident will occur within the next 48 hours. This is given by

$$\Pr\{T_1 \leq 2\} = 1 - e^{-2/5} = 0.3297 .$$

Let us say that it is now noon on Wednesday and no accident happened since Monday and we would like to know what is the probability that the next accident will occur within the next 48 hours. Due to the lack of memory of the exponential, the calculations are exactly the same as above.

Our final calculation for this example is to determine the probability that there will be at least four accidents next week. To determine this, we first observe that the occurrence of accidents form a Poisson process with rate $\lambda = 0.2$ days since the inter-accident times were exponential. (Note that mean rates and mean times are reciprocals of each other.) Since a week is seven days, we have that $\lambda t = 1.4$ so the solution is given by

$$\begin{aligned}\Pr\{N_7 \geq 4\} &= 1 - \Pr\{N_7 \leq 3\} \\ &= 1 - \left(e^{1.4} + 1.4e^{1.4} + \frac{(1.4)^2 e^{1.4}}{2} + \frac{(1.4)^3 e^{1.4}}{3!}\right) \\ &= 1 - (0.2466 + 0.3452 + 0.2417 + 0.1128) = 0.0537\,.\end{aligned}$$

□

Consider a Poisson arrival process with rate λ and denote the successive arrival times by $T_1, T_2, \cdots$. Based on Property 4.1, the time of the nth arrival is composed of n exponential inter-arrival times which implies that the nth arrival time is an Erlang Type-n distribution. Since each inter-arrival time has a mean of $1/\lambda$, the nth arrival time has a mean of n/λ. Using this with Eq. (1.16) gives the following property.

Property 4.2. *Let T_n denote the nth arrival time for a Poisson process with rate λ. The random variable T_n has an Erlang distribution with pdf given by*

$$f(t) = \frac{\lambda(\lambda t)^{k-1} e^{-\lambda t}}{(k-1)!} \quad for\ t \geq 0\,.$$

Assume that we are interested in whether or not T_n has occured; in other words, we would like to evaluate $\Pr\{T_n \leq t\}$ for some fixed value of n and t. In general, to answer a probability questions of a continuous random variable, the pdf must be integrated; however, in this case, it may be easier to sum over a range of discrete probabilities. Note that the event $\{T_n \leq t\}$ is equivalent to the event $\{N_t \geq n\}$; in other words, the nth arrival occurs before time t if and only if there were n or more arrivals that occured in the time interval $[0,t]$. With the equivalence of these events, we can switch from a probability statement dealing with a continuous random variable to a probability statement dealing with a discrete random variable; thus

$$\Pr\{T_n \leq t\} = \Pr\{N_t \geq n\} = 1 - \sum_{k=0}^{n-1} \frac{e^{-\lambda t}(\lambda t)^k}{k!}\,, \tag{4.3}$$

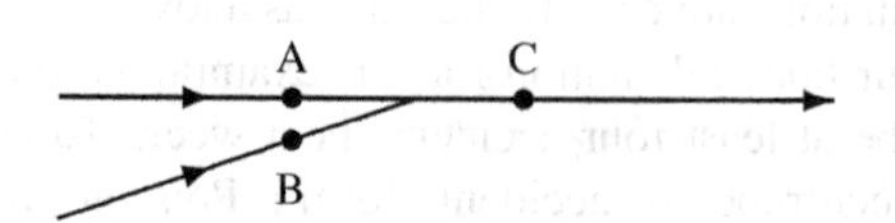

Fig. 4.2 If streams passing Points A and B are Poisson, then the merged stream passing Point C is Poisson as in Example 4.4

where T_n is the nth arrival for a Poisson process with rate λ. For example, the calculations shown in Example 4.3 was to determine $\Pr\{N_7 \geq 4\}$. The same calculations also could be used to find $\Pr\{T_4 \leq 7\}$.

Before leaving the basics of Poisson processes, we mention two other properties that are convenient for modeling purposes; namely Poisson processes maintain their Poisson nature under both superposition and decomposition.

Property 4.3. ***Superposition of Poisson Processes:*** *Let* $\{N_t\ ;t \geq 0\}$ *and* $\{M_t\ ;t \geq 0\}$ *be two independent Poisson processes with rates* λ_1 *and* λ_2. *Form a third process by* $Y_t = N_t + M_t$ *for each* $t \geq 0$. *The process* $\{Y_t\ ;t \geq 0\}$ *is a Poisson process with rate* $\lambda_1 + \lambda_2$.

Example 4.4. Consider the diagram in Fig. 4.2 representing a limited access highway. Assume that the times in which cars pass Point A form a Poisson process with rate $\lambda_1 = 8$/min and cars passing Point B form a Poisson process with rate $\lambda_2 = 2$/min. After these two streams of cars merge, the times at which cars pass Point C form a Poisson process with mean rate 10/min. □

Property 4.4. ***Decomposition of a Poisson Processes:*** *Let* $N = \{N_t\ ;t \geq 0\}$ *be a Poisson processes with rate* λ *and let* $\{X_1, X_2, \cdots\}$ *denote an i.i.d. sequence of Bernoulli random variables (Eq. 1.9) independent of the Poisson process such that* $\Pr\{X_n = 1\} = p$. *Let* $M = \{M_t\ ;t \geq 0\}$ *be a new process formed as follows: for each positive n, consider the* n^{th} *arrival to the N process to also be an arrival to the M process if* $X_n = 1$; *otherwise, the* n^{th} *arrival to the N process is not part of the M process. The resulting M process is a Poisson process with rate* λp.

Example 4.5. Consider traffic coming to a fork in the road, and assume that the arrival times of cars to the fork form a Poisson process with mean rate 2 per minute. In addition, there is a 30% chance that cars will turn left and a 70% chance that cars will turn right. Under the assumption that all cars act independently, the arrival stream on the left-hand fork form a Poisson process with rate 0.6/min and the stream on the right-hand fork form a Poisson process with rate 1.4/min. □

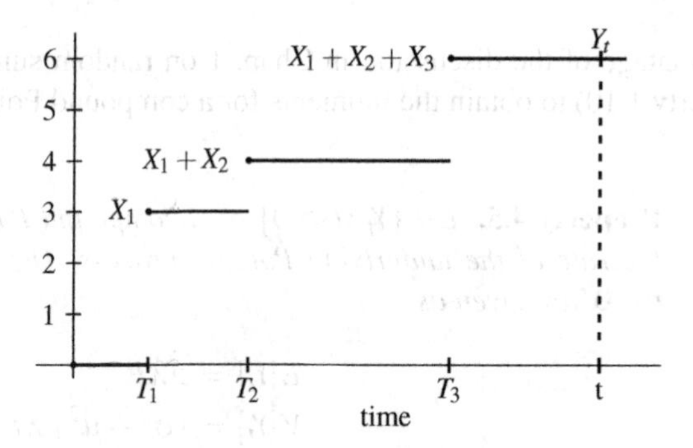

Fig. 4.3 Typical realization for a compound Poisson process, $\{Y_t\,;t \geq 0\}$, with $T_n, n = 1, \cdots$, denoting arrival times and $X_n, n = 1, \cdots$, batch sizes

- *Suggestion: Do Problems 4.2–4.5.*

4.3 Extensions of a Poisson Process

There are two common extensions to the Poisson process that are used for modeling purposes. The first extension is to allow more than one arrival at a specific point in time, and the second is to allow the rate to vary with respect to time.

4.3.1 Compound Poisson Processes

Implicit in Condition (a) of Definition 4.3 for Poisson processes is that arrivals occur one-at-a-time; however, there are many processes for which this assumption needs to be relaxed. Towards this end, consider a Poisson arrival process in which the arrivals are batches of items and the batch sizes form a sequence of independent and identically distributed (*i.i.d.*) positive random variables. The arrival of individual items then form a *compound Poisson process* as illustrated in Fig. 4.3.

Definition 4.4. Let $N = \{N_t\,;t \geq 0\}$ be Poisson process and let $\{X_1, X_2, \cdots\}$ be an *i.i.d.* sequence of random variables independent of N denoting the batch sizes of arrival. Define the process $\{Y_t\,;t \geq 0\}$ by

$$Y_t = \begin{cases} 0 & \text{if } N_t = 0 \\ \sum_{k=0}^{N_t} X_k & \text{if } N_t = 1, 2, \cdots. \end{cases}$$

The process $\{Y_t\,;t \geq 0\}$ is called a *compound Poisson process.* □

Even though the random variables X_n for $n = 1, \cdots$, are described as the batch sizes of the arriving batches, the values of X_n need not be integer and, in fact, may even be negative. For example batch sizes could refer to the amount of water added to a reservoir or removed (a negative quantity) from the reservoir. We can take advantage of the discussion in Chap. 1 on random sums of random variables (Property 1.10) to obtain the moments for a compound Poisson process.

Property 4.5. *Let $\{Y_t\ ;t \geq 0\}$ be a compound Poisson process with λ being the rate of the underlying Poisson process. The mean and variance for Y_t, $t \geq 0$, are given as*

$$E[Y_t] = \mu\lambda t$$
$$V[Y_t] = (\sigma^2 + \mu^2)\,\lambda t\,,$$

where μ and σ^2 are the mean and variance of the batch size, respectively.

Example 4.6. Consider the arrival of passengers to a train station and assume that the individuals come to the train station by car and that cars arrive to the train station according to a Poisson process of 5 per hour. The number of passengers within the nth car to arrive is denoted by the random variable X_n, where $\Pr\{X_n = 1\} = 0.5$, $\Pr\{X_n = 2\} = 0.3$, and $\Pr\{X_n = 3\} = 0.2$ for all positive n; thus, the batch size has mean and variance equal to 1.7 and 0.61, respectively. If the numbers of passengers within the arriving cars form independent random variables, then the arrival process of individual passengers is a compound Poisson process. The mean and standard deviation for the number of individuals that arrive to the station over an eight-hour period is $1.7 \times 40 = 68$ and $\sqrt{(0.61 + 2.89) \times 40} = 11.83$. □

Example 4.7. A monitoring device has been placed on a highway to record the number of vehicles using the highway. Actually, what is recorded are axles, and we would like to model the number of axles passing a particular point in the highway. The time between vehicle arrivals at the monitoring device is exponentially distributed with a mean of 1.2 minutes between vehicles. The probability that a given vehicle has 2, 3, 4, or 5 axles is 0.75, 0.10, 0.07, or 0.08, respectively. We would like to estimate the probability that 150 or more axles will pass the counting point in any given hour interval.

The mean and variance for the number of axles per vehicle are 2.48 and 0.8696; therefore the mean and variance for the number of axles passing over the counter within a given hour interval is 124 and 351. Let Y_1 denote the (random) number of axles passing the counter during an hour interval, and let X denote a normal random variable with mean 124 and standard deviation 18.735, then

$$\begin{aligned}\Pr\{Y_1 \geq 150\} &\approx \Pr\{X > 149.5\}\\ &= \Pr\{Z > (149.5 - 124)/18.735\} = 0.0867\,.\end{aligned}$$

□

This example of a compound Poisson process is, strictly speaking, not a compound Poisson process since the axles do not arrive at exactly the same instant of time; however, this example also illustrates that for modeling purposes, it is okay to make some assumptions that are not exactly true. If traffic conditions are not too heavy, it is clearly reasonable to assume that the time between "axle" arrivals for a single vehicle is insignificant compared to the time between vehicle arrivals.

4.3.2 Non-stationary Poisson Process

Many physical processes that appear at first glance to be a candidate for modeling as a Poisson process fail because of the stationary assumption. For example, consider a traffic process and we are interested in modeling the arrival times of cars to a specific intersection. (Of course, the length of the car prohibits a true Poisson process since inter-arrival times cannot truly approach zero; however, if traffic is not too heavy, the exponential may be an acceptable approximation.) In most places, it is unlikely that the mean arrival rate of cars at 2PM is the same as at 2AM, which means that the process is not stationary and thus the stationary assumption required of Poisson processes would not be true. Therefore, it is convenient to re-consider Definition 4.3 without the stationary assumption of Condition (c).

Definition 4.5. Let the continuous parameter stochastic process $N = \{N_t \ ; t \geq 0\}$ with $N_0 = 0$ have a state space equal to the nonnegative integers. Assume that a non-negative rate function, $\lambda(\cdot)$, has been given and a mean value function, $m(\cdot)$ is defined by

$$m(t) = \int_0^t \lambda(s)\,ds \quad \text{for } t \geq 0\,.$$

The N process is called a *non-stationary Poisson process* if

(a) $\Pr\{N_t = k\} = e^{-m(t)}\,(m(t))^k/k!$ for all nonnegative k and $t \geq 0$,
(b) The event $\{N_{s+u} - N_s = i\}$ is independent of the event $\{N_t = j\}$ if $t < s$.

The process is also called a non-homogenous Poisson process. □

Notice that in the above definition, if the rate function were a constant, Condition (a) in both Definition 4.3 and 4.5 would be the same. The mean value function gives the expected number of arrivals during a time interval, and by using the independent increments property of Condition (b), we have the following property.

Property 4.6. *Let $N = \{N_t \ ; t \geq 0\}$ be a non-stationary Poisson processes with rate function $\lambda(\cdot)$ and let T_n denote the arrival time of the nth arrival. The mean number of arrivals during an interval of time is given by*

$$E[N_{t+s} - N_s] = \int_s^{t+s} \lambda(u)\,du = m(t+s) - m(s) \quad \text{for } t,s \geq 0\,.$$

Example 4.8. A bank has decided to increase its drive-in window capacity, and therefore desires to model the facility. A first step in modeling the drive-in bank facility is to model the arrival process. The drive-in windows open at 7:30AM during the week. It has been determined that the arrival process is a non-stationary Poisson process in which the mean arrival rate slowly increases linearly from 10 per hour to 12 per hour during the first 60 minutes. Thus, we have a rate function defined, for $t \le 1$, by

$$\lambda(t) = 10 + 2t ,$$

where the time units is hours. The mean value function is given as

$$m(t) = \int_0^t (10 + 2s)\,\mathrm{d}s = 10t + t^2 \quad \text{for } t \le 1 .$$

The expected number of arrivals from 8:00AM through 8:30AM is given by

$$E[N_1 - N_{0.5}] = m(1) - m(0.5) = 5.75 .$$

□

- *Suggestion: Do Problems 4.7 and 4.8.*

4.4 Poisson Regression

Ordinary least squares regression (Sect. 3.5.1) of normally distributed random variables is widely used in several areas of study because the observations often have an approximate normal distribution. There are instances, however, that observations can be better described by distributions other than the normal distribution. Such is the case with count data; that is, observations that count the number of events. In this section, we present two examples of Poisson regression: the first is relatively simple involving one independent variable; the second is more realistic (and complex) involving several independent variables.

4.4.1 Poisson Regression with One Independent Variable

To put Poisson regression in the context of risk assessment, assume that failures can be modeled as a Poisson process dependent on some independent variable. That is, for a given value of the independent variable, x, let $\{N_t(x); t \ge 0\}$ be a Poisson process with a mean rate given by $\lambda(x)$ which is defined to have the form

$$\lambda(x) = \alpha(1 + \beta x) , \tag{4.4}$$

where α is the expected number of failures per unit time when $x = 0$ and β is the change in the failure rate per unit increase in x. Notice that the basic model for the regression is again a linear function similar to but not quite identical to the linear model of Eq. (3.23).

Example 4.9. Consider the risk model of Example 3.14 in which fuel injectors are subject to failure, possibly as a function of the concentration of additive within the gasoline. After some additional study, it has been determined that a reasonable assumption would be that the time to failure of the fuel injector is exponentially distributed and that all of the injector failure times within a specific additive concentration are independent and have the same exponential failure time distribution. Our experiment consists of observing each airplane for 100 hours of flight or until failure, whichever occurs first. Thus, if none of the injectors fail, there would be 2,000 hours of flight for each of the concentration levels of the gasoline additive. However, if some fuel injectors fail before the 200 hours elapses, the number of hours of observation will be less than 2,000. Table 4.1 shows the data for this experiment.

Table 4.1 Experimental results regarding risk from gasoline additives for Example 4.9

Concentration (ml/l) of additive	0	5	10	50	100
Number Airplanes Tested	20	20	20	20	20
Number Injectors Failed	2	3	5	6	8
Total Hours for Observations	1,985	1,955	1,925	1,940	1,910

Within the context of injector failures, the parameters of the model of Eq. (4.4) have the following meaning: α is the mean failure rate per unit time of the fuel injectors in the absence of any gasoline additive and β is the increase in mean failure rate per unit time for each additional ml/l of concentration of the additive. □

A maximum likelihood approach is taken to determine the parameters for the Poisson regression (see Problem 4.11). The resulting log-likelihood function and the estimators $\hat{\alpha}$ and $\hat{\beta}$ for the function describing the mean failure rate as a function of the independent variables are given in the following property.

Property 4.7. *Assume there are m groups of observations for which a Poisson regression is desired. Observation i includes the number of failures, n_i, over a time span of t_i, with the independent variable set at a level of x_i. The MLEs for Eq. (4.4) are those values of α and β that maximize the following function:*

$$LL(\alpha,\beta) = \left(\sum_{i=1}^{m} n_i \ln\{\alpha(1+\beta x_i)t_i\}\right) - \left(\sum_{i=1}^{m} \alpha(1+\beta x_i)t_i\right).$$

Notice that the regression models of Sect. 3.5.1 involved a sample of size n and observations denoted by (x_i, y_i) for $i = 1, \cdots, n$. In this section, we are dealing primarily with the *number* of events and their *timing*, so our notation will change slightly. Here we have a sample size of m and observations denoted by (x_i, t_i, n_i) for $i = 1, \cdots, m$.

There are now two approaches that can be taken to find $\hat{\alpha}$ and $\hat{\beta}$. A two-dimensional search routine for maximizing the $LL(\cdot,\cdot)$ function of Property 4.7 could be used. For example, the Solver of Excel works for a problem of this type, although an initial solution that leads to an appropriate solution (i.e., positive value for $\hat{\alpha}$) used for the search may have to be found by trial and error. A second approach is to solve the system of equations obtained by taking the partial derivatives of $LL(\cdot,\cdot)$ and setting these partials equal to zero. Because there are only two variable involved, a substitution can be made to reduce the problem to a single-variable search for which GoalSeek (see the appendix of Chap. 1) can be used. The results of finding the partials and setting them to zero are as shown in the following property.

Property 4.8. *For the Poisson regression problem of Property 4.7, the MLE* $\hat{\beta}$ *is the value of* β *such that*

$$\left(\sum_{i=1}^{m} \frac{n_i x_i}{1+\beta x_i}\right) \sum_{i=1}^{m} (1+\beta x_i)\, t_i = \left(\sum_{i=1}^{m} n_i\right) \sum_{i=1}^{m} x_i t_i \,.$$

The MLE $\hat{\alpha}$ *is then*

$$\hat{\alpha} = \frac{\sum_{i=1}^{m} n_i}{\sum_{i=1}^{m} \left(1+\hat{\beta} x_i\right) t_i} \,.$$

Anytime we have estimators, it is helpful to have some idea of their variability. For maximum likelihood estimators, the standard error can be obtained through the Hessian matrix of the log-likelihood function and then taking the negative of its inverse. As before, the maximum likelihood theory is beyond the scope of this textbook, but we do give the results so that confidence intervals can be written (at least as approximations) if so desired. Consider the following matrix

$$\mathbf{H}' = \begin{bmatrix} \left(\sum_{i=1}^{m} \left(1+\hat{\beta} x_i\right) t_i\right) / \hat{\alpha} & \sum_{i=1}^{m} x_i t_i \\ \sum_{i=1}^{m} x_i t_i & \sum_{i=1}^{m} \left(\hat{\alpha} t_i x_i^2\right) / \left(1+\hat{\beta} x_i\right) \end{bmatrix} . \tag{4.5}$$

It turns out that the inverse has the following form:

$$\mathbf{H}'^{-1} = \begin{bmatrix} \hat{\sigma}_\alpha^2 & \mathrm{cov}(\hat{\alpha},\hat{\beta}) \\ \mathrm{cov}(\hat{\alpha},\hat{\beta}) & \hat{\sigma}_\beta^2 \end{bmatrix} . \tag{4.6}$$

In other words, the inverse of the matrix from Eq. (4.5) yields the standard errors necessary to form confidence intervals using the z-statistic as given in the following property. (Note that the Hessian matrix is actually a function, whereas the matrix in which we are interested is the negative of the Hessian evaluated at the maximum likelihood estimators. We denote this matrix by $\mathbf{H}'$.)

Property 4.9. *Consider a Poisson regression model in which $\hat{\alpha}$ and $\hat{\beta}$ have been determined according to Property 4.8 for a large sample size, and with $\mathbf{H}'$ determined according to Eq. (4.5). Then*

$$Z_\alpha = \frac{\hat{\alpha} - \alpha}{\hat{\sigma}_\alpha} \quad \textit{and}$$

$$Z_\beta = \frac{\hat{\beta} - \beta}{\hat{\sigma}_\beta}.$$

have an approximate standard normal distribution (i.e., mean zero and variance one) where $\hat{\sigma}_\alpha$ is the square root of the upper left element of $\mathbf{H}'^{-1}$ and $\hat{\sigma}_\beta$ is the square root of the lower right element of $\mathbf{H}'^{-1}$.

Example 4.10. Returning to the previous example and Table 4.1, we observe that $\sum_{i=1}^m n_i = 24$ and $\sum_{i=1}^m x_i t_i = 317{,}025$. Using GoalSeek, the value of β that satisfies the first equation of Property 4.8 is $\hat{\beta} = 0.0195$. Since $\sum_{i=1}^m \left(1 + \hat{\beta} x_i\right) t_i = 15894.93$ we have that $\hat{\alpha} = 0.00151$. The negative of the Hessian matrix becomes

$$\mathbf{H}' = \begin{bmatrix} 10{,}527{,}031 & 317{,}025 \\ 317{,}025 & 13{,}797.2 \end{bmatrix}$$

which yields an inverse of

$$\mathbf{H}'^{-1} = \begin{bmatrix} 3.084 \times 10^{-7} & -7.086 \times 10^{-6} \\ -7.086 \times 10^{-6} & 2.353 \times 10^{-4} \end{bmatrix}.$$

Thus, the standard error of the two parameter estimators are

$$\hat{\sigma}_\alpha = \sqrt{3.084 \times 10^{-7}} = 0.000555 \text{ and } \hat{\sigma}_\beta = \sqrt{2.353 \times 10^{-4}} = 0.0153\,.$$

Although this sample size is too small to build confidence intervals, we do the calculations to demonstrate Property 4.9. Using a z statistic value of 1.96, the 95% confidence interval for $\hat{\alpha}$ is 0.00151 ± 0.00109 and for $\hat{\beta}$ is 0.0195 ± 0.0301. □

Maximizing the log-likelihood function as it has been defined may present some difficulties that we have ignored; namely, Eq. (4.4) would make no sense if it turned out that $\hat{\alpha}$ was negative. To insure that this does not happen, it would have been better to define the key regression equation as

$$\lambda(x) = e^{\alpha}(1+\beta x). \tag{4.7}$$

We chose not to take this approach so that the introduction to Poisson regression would stay simple; however, using e^{α} instead of α adds no extra difficulty to the solution procedure. With the parameters defined as Eq. (4.7), the log-likelihood function of Property 4.7 is re-written as

$$LL(\alpha,\beta) = \left(\sum_{i=1}^{m} n_i\left[\alpha + \ln\{(1+\beta x_i)t_i\}\right]\right) - \left(\sum_{i=1}^{m} e^{\alpha}(1+\beta x_i)t_i\right). \tag{4.8}$$

4.4.2 Poisson Regression with Several Independent Variables

Poisson regression modeling can be used in a variety of situations where the occurrence of events is rare. Such is the case in the risk assessment associated with occupational cancer studies where rare events occur over extended periods of observation time. Occupational studies usually include workers employed in a particular industry exposed to a carcinogen and their vital status (whether alive or deceased) and causes of death (if deceased) are known at the end of an observation period. At the end of the observation period workers have been observed for different lengths of time, have different ages, have worked different periods of time, have been exposed to different concentrations of the carcinogen, etc. Poisson regression has been useful in modeling this type of survival data to estimate the relationship between cancer and exposure to the carcinogen after adjusting for factors that also have an effect on the specific cancer mortality. We demonstrate these concepts with a extended example investigating arsenic as a carcinogen.

Example 4.11. A Montana smelter study was to investigate the role of arsenic in respiratory cancer. Motivating factors of the study were that workers were exposed to high concentrations of arsenic and they showed a higher than expected occurrence of respiratory cancer deaths. There are several publications (for example [3, 4]) discussing this particular group of workers. The study included 8,045 workers employed before 1957 for a period of at least 12 months. Company records for each worker included the birth date and the date and duration of each job within the company. The company also had an exposure matrix that matched dates and jobs with arsenic concentrations. Using this matrix, an exposure history for each worker was constructed. The records also included the last date each worker was known alive and the date of death for the decedent along with their cause of death. The data used here included observations up to 1977.

In order to investigate the role of arsenic exposures in respiratory cancers there are several modeling questions that have to be answered. In this example, as it is the case with most survival analyses, age is a well-known factor in the occurrence of respiratory cancer (that is, respiratory cancer rates are higher at older ages than at younger ages). Similarly, in this particular study, the calendar years in which exposures occurred are factors that seem to have affected the respiratory cancer rates. Thus, although the main purpose is to find the relationship between exposure to arsenic and respiratory cancer mortality, there are other factors (age and year of exposure) affecting the observed respiratory cancer mortality rate that should be factored out.

One usual dose measure used in occupational studies is cumulative exposure. Cumulative exposure is defined in terms of concentration-years and is the result of summing the product of durations of exposure and concentrations at different jobs that occurred throughout the work history of each worker. Thus, an exposure to 8 mg/m^3 for one-half a year, or 4 mg/m^3 for one year, or 2 mg/m^3 for two years are all identical cumulative exposures; namely, each is equivalent to 4 mg/m^3-yr.

Instead of age being a continuous variable, four age groups were formed. The four age groups were less than 50, 50 to 59, 60 to 69, and 70 or greater; thus, for example, age group 1 refers to any person-year whose age is less than 50. For purposes of this example, there is some evidence that environmental factors may have been different before 1960 than 1960 and after; therefore, to test whether or not this is a factor for the risk of respiratory cancer, a year indicator will be defined that is either 1 or 2. A year indicator equal to 1 indicates before 1960 and a year indicator of 2 indicates 1960 or after. Likewise, the cumulative exposure to arsenic will be formed into four groups: Group 1 refers to any person-year with a cumulative exposure of less than 1 mg/m^3-yr, Group 2 is between 1 and 5 mg/m^3-yr, Group 3 is between 5 and 15 mg/m^3-yr, and Group 4 is 15 mg/m^3-yr or more. These definitions are summarized in Table 4.2.

Table 4.2 Group definitions associated with the data from the Montana occupational study

Group Definition	Group #1	Group #2	Group #3	Group #4
Age	< 50	50–59	60–69	≥ 70
Calendar Year	< 1960	≥ 1960	—	—
Cumulative Exposure (mg/m^3-yr)	< 1	1–5	5–15	≥ 15
Avg. Exposure within group	0.5	3.0	8.0	21.0

The data from the occupational exposure is divided into the person-years (time at risk) and those person-years are apportioned to different age groups, calendar year groups and exposure groups. Each worker, thus, may contribute person-years at risk to different age groups, different calendar years and different cumulative exposure groups. A worker that dies with respiratory cancer contributes this cancer death to the age interval, calendar year interval and cumulative exposure interval where his last contribution of person years belongs. The data can then be summarized as in Table 4.3.

For example, there were 6 deaths that occurred before 1960 of individuals who were less than 50 years of age and had a cumulative exposure of less than 1 mg/m^3-yr. In addition, the total person-years of exposure summed over all individuals aged 49 or less before 1960 who were exposed to less than 1 mg/m^3-yr was 18,734.3. □

In the above example, as is often the case in dose-response modeling, the interest is on the relationship between cumulative exposure to arsenic and the hazard rate

Table 4.3 Data from the Montana occupational study showing the number of deaths and person-years at risk within each category (adapted from Appendix V of Breslow and Day [2])

Age Group	Year Group	Cumulative Exposure Group 1	2	3	4
1	1	6 : 18,734.3	3 : 3,033.2	1 : 1,703.9	0 : 694.8
1	2	8 : 19,602.3	3 : 3,853.6	0 : 1,288.2	0 : 452.4
2	1	14 : 12,582.7	9 : 1,872.8	3 : 1,122.3	9 : 1,066.6
2	2	24 : 18,436.3	11 : 3,958.0	6 : 1,717.9	4 : 1,054.6
3	1	16 : 6,365.0	3 : 694.0	3 : 691.3	16 : 1,151.3
3	2	42 : 11,130.7	20 : 2,101.7	11 : 1,304.2	6 : 955.4
4	1	10 : 2,279.4	1 : 207.5	2 : 315.1	3 : 556.1
4	2	31 : 4,563.0	5 : 655.1	2 : 397.5	4 : 449.7

of respiratory cancer. However, because age and calendar year are thought to have an impact on the rate of respiratory cancer, the dose-response model includes these variables in order to adjust for their effect on respiratory cancer. The time variable for the Poisson process that describes the event of a death due to the cancer now becomes person-years at risk and the main independent variable is the cumulative exposure, although we will have additional variables indicating the effect of age and calendar year.

In addition to the necessary assumption of exponential times between events, the standard Poisson regression makes the following assumptions: (1) the effects of different factors are independent and multiplicative, (2) the mean rate is constant within each combination of effect intervals, and (3) the observations are independent.

We extend the Poisson regression model of the previous section to include the age and year effects. In particular, we begin with Eq. (4.7) to insure that the extra variables do not yield a negative mean rate. Thus, our basic model is a Poisson process denoted by $\{N_t(i,j,x); t \geq 0\}$ where i denotes the age group, j denotes the calendar year group, x denotes the cumulative exposure in mg/m^3-yr units, and t represents the person-years at risk.

Previously, we had two parameters, α and β associated with the independent variable x. We continue with those two variables being associated with the key independent variable, and also include extra factors that will allow for the possible adjustments due to age and calendar year. Specifically, the parameter a_i will be associated with age group i and the parameter c_j will be associated with calendar group j. These adjustments are relative, so the factors associated with the first age group and the first calendar year group will be set to one.

The mean rate per person-year for our Poisson process is given by

$$\lambda(i,j,x) = \mathrm{e}^{\alpha}\, \mathrm{e}^{a_i}\, \mathrm{e}^{c_j}\, (1 + \beta x)\,, \tag{4.9}$$

where e^{α} is the rate of respiratory cancers per person-year in the absence of exposure to arsenic for the first age group and the first calendar year group, e^{a_i} is the relative effect of age group i (with $a_1 = 0$ and thus $\mathrm{e}^{a_1} = 1$), e^{c_j} is the relative effect of calendar year group j (with $c_1 = 0$ and thus $\mathrm{e}^{c_1} = 1$), and β is the increase in death rate due to respiratory cancers per increase in mg/m^3-yr cumulative exposure. (You may notice that a large negative β could give rise to negative mean rates, but the chance of this seems too small to complicate the model further.) Thus, this problem has six unknowns that need to be determined; namely, $\alpha, \beta, a_2, a_3, a_4$, and c_2. To help simplify the model, the variable x will be set equal to the average cumulative exposure within each of those groups. For example, the first cumulative exposure group would set $x = 0.5\,mg/m^3$-yr as shown by the final row in Table 4.2.

In Sect. 4.4.1, m observations of the form (x_i, t_i, n_i) for $i = 1, \cdots, m$, were used to estimate the regression parameters of Eq. (4.4). For the problem of this section, the observations have a different structure; namely, the observations are associated with specific cells, were a cell refers to a specific combination of age group, calender year group, and cumulative exposure group. To denote the sample size, let m_a be the number of age groups, let m_c be the number of calender year groups, and let m_x be the number of cumulative exposure groups. The sample size is thus the product of the three terms m_a, m_c, and m_x. A specific observation includes n_{ijk} (representing the number of deaths in age group i, calender year j, and cumulative exposure group k), t_{ijk} (representing the number of total person-years of exposure of all individual within age group i, calender year j, and cumulative exposure group k), and $\overline{x}_k$ (representing the average cumulative exposure within cumulative exposure group k). Using the maximum likelihood approach again yields a function in terms of the unknown parameters that is essentially an extension of Eq. (4.8) for the several-variable problem.

Property 4.10. *Assume there are $m = m_a \times m_c \times m_x$ observations for which a Poisson regression is desired. An observation takes the form $(x_{ijk}, t_{ijk}, n_{ijk})$ referring to the cumulative exposure, total person-years, and number of deaths within age group i, calender year j, and cumulative exposure group k. The MLEs for Eq. (4.9) are those values of α, β, a_i, for $i = 2, \cdots, m_a$, and c_j for $j = 2, \cdots, m_c$ that maximize the following function:*

$$LL = \left(\sum_{i=1}^{m_a} \sum_{j=1}^{m_c} \sum_{k=1}^{m_x} n_{ijk} \left[\alpha + a_i + c_j + \ln\{(1 + \beta x_{ijk}) t_{ijk}\} \right] \right) - \left(\sum_{i=1}^{m_a} \sum_{j=1}^{m_c} \sum_{k=1}^{m_x} e^{\alpha} e^{a_i} e^{c_j} (1 + \beta x_{ijk}) t_{ijk} \right),$$

where $x_{ijk} = \overline{x}_k$, $a_1 = 0$, and $c_1 = 0$.

Example 4.12. The application of Property 4.10 to Example 4.11 is not difficult with the use of Excel's Solver. (Solver works better if β is restricted to be greater than zero.) The results of the maximization problem are $\hat{\alpha} = -8.01575$, $\hat{\beta} = 0.09757$, $\hat{a}_2 = 1.43746$, $\hat{a}_3 = 2.29390$, $\hat{a}_4 = 2.52084$, and $\hat{c}_2 = 0.17610$.

These results mean that the estimated background respiratory cancer mortality rate per person-year for the age group of less than 50 years before the year 1960 is 0.00033 (i.e., $e^{-8.01575}$) or approximately a 3.3% chance in a 100-year lifetime. The rate of respiratory cancer in the age group 70 years and older is 12.44 (i.e., $e^{2.52084}$) times greater than the rate of respiratory cancer in the age group of less than 50 years. Similarly, the rate of respiratory cancer after 1959 was 1.19 (i.e., $e^{0.17610}$) times greater than the respiratory cancer before 1960. More importantly, the respiratory cancer rate per person-year increases 9.757% per mg/m3-year of cumulative exposure to arsenic. The estimated number of respiratory cancer deaths using the parameters obtained from the maximization of the log-likelihood function of Property 4.10 in the cohort for each combination of age group, calendar year group, and cumulative arsenic exposure group is given in Table 4.4.

Table 4.4 Predicted number of deaths in each cell using Eq. (4.9) with the parameter estimators of $\hat{\alpha} = -8.01575$, $\hat{\beta} = 0.09757$, $\hat{a}_2 = 1.43746$, $\hat{a}_3 = 2.29390$, $\hat{a}_4 = 2.52084$, and $\hat{c}_2 = 0.17610$

Age Group	Year Group	Cumulative Exposure Group			
		1	2	3	4
1	1	6.49	1.29	1.00	0.70
1	2	8.10	1.96	0.90	0.54
2	1	18.35	3.37	2.78	4.52
2	2	32.09	8.49	5.08	5.34
3	1	21.85	2.94	4.03	11.49
3	2	45.62	10.62	9.07	11.38
4	1	9.82	1.10	2.30	6.97
4	2	23.46	4.15	3.47	6.72

Several observations can be made from Table 4.4. First, the overall sum of expected number of deaths due to respiratory cancer is equal to the total observed number (i.e., 276). Second, within each age group the sum of the expected number is equal to the observed number (i.e., 21, 80, 117 and 58 for age groups less than 50, 50–59, 60–69, and 70+, respectively). Third, within each calendar year group the sum of the expected number is again equal to the observed number of deaths from respiratory cancers (i.e., 99 and 177 for first and second calender year groups, respectively). It turns out that these are, in fact, conditions that have to be satisfied by the maximum likelihood estimates.

Using the MLE of the slope (namely, $\hat{\beta} = 0.09757$) for the linear dose-response relationship between cumulative exposure to arsenic and the hazard rate of respiratory cancer mortality, the estimated risk of respiratory cancer in the U.S. population exposed to a concentration c of arsenic in the air (mg/m^3) can be estimated by multiplying the U.S. age-dependent population's background rate of respiratory cancer by the linear function $(1 + 0.09757x)$, where x is the age-dependent cumulative exposure to arsenic (mg/m^3-yr). To illustrate, assume (1) the rate of respiratory cancer per year in the U.S. population is 0.0001 per year for ages less than 50 years and is 0.0002 after 50 years, (2) there are no other causes of death before age 100 years, (3) the U.S population is exposed to an average of 0.001 mg/m^3 of arsenic. Then, the additional risk of respiratory cancer death due to exposure to arsenic in a 100-year lifetime is approximately given by,

$$\Delta_{\text{risk}} = \Pr\{\text{Respiratory Cancer} \mid c = 0.001\,mg/m^3\} \\ - \Pr\{\text{Respiratory Cancer} \mid c = 0\,mg/m^3\},$$

where

$$\Pr\{\text{Respiratory Cancer} \mid c = 0.001\,mg/m^3\} \\ = \sum_{i=1}^{50} 0.0001 \times (1 + 0.09757 \times 0.001\, i) \\ + \sum_{i=51}^{100} 0.0002 \times (1 + 0.09757 \times 0.001\, i) = 0.015086$$

and

$$\Pr\{\text{Respiratory Cancer} \mid c = 0\,mg/m^3\} = \sum_{i=1}^{50} 0.0001 + \sum_{i=51}^{100} 0.0002 = 0.015\,.$$

Thus, the additional risk is $\Delta_{\text{risk}} = 0.015086 - 0.015 = 0.000086$, or equivalently, an average of 8.6 deaths due to exposure of arsenic in 100,000 people living to age 100 years. □

There are several simplifying assumptions usually made in the above calculations; namely, the dose-response relationship derived from the occupational study holds for the general population, the cumulative exposure for the population is constant during a year of age, and the probability is equal to the cumulative rate of respiratory cancer.

The Poisson regression examples given in this section illustrate how Poisson processes appear in different areas of research. Simplifying assumptions were made here in an effort to help clarify the presentation. Real applications of Poisson regression models to estimate cancer risk for the U.S. population usually entail including the fact that people die before the age of interest due to causes other than respiratory cancer and that background rates of respiratory cancer in the U.S. vary with age, race, gender, etc. Poisson regression has been successfully used by regulatory agencies whenever there are sufficient occupational data to apply this methodology. A more in-depth discussion of Poisson regression applied to cancer risk can be found in [2].

• *Suggestion: Do Problems 4.10 and 4.11.*

Appendix

In this appendix, two simulations are presented using Excel: one for a compound Poisson process and the other for a non-stationary Poisson process. The intent of the examples is to emphasize both the concept of simulation and the dynamic of the two different type of processes. We close the appendix illustrating the use of

Solver in obtaining the maximum likelihood estimates for the Poisson regression of Example 4.9.

Simulation of a Compound Poisson Process: We return to Example 4.7 and simulate that arrival of axles to a monitoring device. As you recall, cars arrive according to a Poisson process with a mean time between arrivals of 1.2 minutes, and the probability that a given vehicle has 2, 3, 4, or 5 axles is 0.75, 0.10, 0.07, or 0.08, respectively. There are two sources of randomness: (1) the inter-arrival times and (2) the number of axles per vehicle. For both of these, random numbers will need to be generated. The first three rows of the Excel spreadsheet would look as follows:

	A	B	C	D	E
1	Inter-arrive Time	Arrival Time	Random Number	Number of Axles	Cumulative Number
2	=-1.2*LN(RAND())	=A2	=RAND()	=IF(C2<0.75,2,IF(C2<0.85,3,IF(C2<0.92,4,5)))	=D2
3	=-1.2*LN(RAND())	=B2+A3	=RAND()	=IF(C3<0.75,2,IF(C3<0.85,3,IF(C3<0.92,4,5)))	=E2+D3

Column A is based on the transformation given by

$$T = -1.2\ln(U)$$

where U is a random number and T is the random variate for an inter-arrival time. Column D is based on the *cumulative* probabilities for the number of axles per vehicle. The simulation is completed by copying Cells A3:E3 down for a thousand or more rows. An estimate for the number of axles per hour is obtained by dividing the number in the final row of Column E by the number in the final row of B and the multiplying by 60 to convert from per minute to per hour. Notice that each time the `F9` is hit, a new realization of the simulation will be obtained. The key dynamics that make this a compound Poisson process are (1) exponential inter-arrival times generated to produce the timing of the arrivals in Column A and (2) independent random numbers were generated to simulate the batch sizes at each arrival time.

Simulation of a Non-stationary Poisson Process: Simulating a non-stationary Poisson process is a little more difficult to simulate than the compound Poisson process. The reason is the simulation of the next arrival time. The simple generation of an exponentially distributed inter-arrival time is not correct because its rate is only instantaneously known. Even if the rate is constant for an interval of time, if the inter-arrival places the next arrival time outside the constant rate interval, its rate would not be correct. One procedure that will produce a non-stationary Poisson process is to generate arrivals at the maximum rate and then "thin" the process by removing some of the arrivals. In particular, let λ^* denote the maximum arrival rate. If an arrival (generated using λ^*) occurs at time t, then it is accepted with a probability equal to $\lambda(t)/\lambda^*$ and rejected with a probability equal to $1 - \lambda(t)/\lambda^*$. When an arrival is rejected, the time is accumulated; however, no event is recorded at the rejected arrival time.

Consider again Example 4.8 that had a rate function defined, for $t \leq 1$, by

$$\lambda(t) = 10 + 2t ,$$

where the time units were in hours.

The Excel setup below is to generate several replicates of the first hour. Note that the maximum rate is $\lambda^* = 12$/hr which is equivalent to an average of 5 minutes between arrivals. The first three rows of the Excel spreadsheet would be as follows:

	A	B	C
1	Inter-arrive Time	Arrival Time	Ratio of Rates
2	=-5*LN(RAND())	=A2	=(10+2*(MOD(B2,60)/60))/12
3	=-5*LN(RAND())	=B2+A3	=(10+2*(MOD(B3,60)/60))/12

	D	E
1	Accept Arrival?	Cumulative Number
2	=IF(RAND()<C2,1,0)	=D2
3	=IF(RAND()<C3,1,0)	=E2+D3

Notice that the unit of time in the simulation is minutes. Column C uses the MOD operator so that the function starts over again at the beginning of each hour. The simulation is completed by copying Cells A3:E3 down for a thousand or more rows. An estimate for the number of arrivals occurring per hour is obtained by dividing the number in the final row of Column E by the number in the final row of B and the multiplying by 60 to convert from per minute to per hour. Notice that each time the F9 is hit, a new realization of the simulation will be obtained.

Maximum Likelihood Estimators for a Poisson Regression: Obtaining the maximum likelihood estimators from Property 4.7 is possible through the use of Excel's Solver which we illustrate through Example 4.9. The first step is to enter the data in Excel as shown below.

	A	B	C	D	E	F
1	x	0	5	10	50	100
2	n	2	3	5	6	8
3	t	1985	1955	1925	1940	1910

Initial guesses for $\hat{\alpha}$ and $\hat{\beta}$ are needed to start the optimization. Since α is the failure rate in the absence of the additive, let the initial guess be the number failed divided by the hours of observation for the case in which there was no additive, in other words, $2/1985 \approx 0.001$. For $\hat{\beta}$, we will start with zero. We enter these initial estimates to the right of the data as follows.

	H	I
1	Alpha	Beta
2	0.001	0

There are two terms on the left-hand side of the log-likelihood function of Property 4.7 which we shall refer to as the "left term" and the "right term". To obtain these expressions, use the following Excel statements.

	A	B
5	right term	=\$H\$2*(1+\$I\$2*B1)*B3
6	left term	=B2*LN(B5)
7	Log-like	=SUM(B6:F6) – SUM(B5:F5)

Highlight the cells B5:F6 and fill to the right (use the <ctrl>-R keys). The spreadsheet is now setup to use Solver. Sometimes, Excel does not install the Solver automatically; however, it is one of the add-ins that comes with Excel. For Excel 2007, Solver will appear on the right-hand part of the Data tab. The older versions of Excel will have Solver as part of the "Tools" menu. When you open Solver, a dialog box will appear. Type `B7` for the "Set Target Cell:" blank and type `H2:I2` for the "By Changing Cells:" blank. Make sure that the "Max" option button is checked and click "Solve" in the upper-right corner of the dialog box.

Problems

4.1. Arrivals to a repair shop occur according to a Poisson process with rate 8 per hour.
(a) What is the probability that there will be 4 or less arrivals during the first 30 minutes?
(b) What is the probability that there will be 6 or more arrivals during the first hour?
(c) Estimate the probability that there will be 70 or more arrivals during an 8-hour day.

4.2. The dispatcher at a central fire station has observed that the time between calls is an exponential random variable with a mean of 32 minutes.
(a) A call has just arrived. What is the probability that the next call will arrive within the next half hour.
(b) It is now noon and the most recent call came in at 11:35AM. What is the probability that the next call will arrive before 12:30PM?
(c) What is the probability that there will be exactly two calls during the next hour?

4.3. A security consultant has observed that the attempts to breach the security of the company's computer system occurs according to a Poisson process with a mean rate of 2.2 attempts per day. (The system is on 24 hours per day.)
(a) What is the probability that there will be 4 breach attempts tomorrow?
(b) What is the probability that there will be no breach attempts during the evening (eight-hour) shift?
(c) It is now midnight and the most recent attempt occured at 10:30PM. What is the probability that the next attempt will occur before 5:30AM?

4.4. Consider again the computer security system from Problem 4.3. Approximately 10% of the attempts to illegally access the computer make it past the first security level.
(a) What is the probability that there will be exactly three attempted illegal entries that successfully make it past the first security level during the next week?
(b) It is now 6AM Monday morning. The most recent security attempt that successfully made it past the first security level occurred at midnight last night. What is the probability that no illegal attempts make it past the first security level during the time period from now (6AM Monday morning) through 6PM Friday?

4.5. There are two common types of failure to a critical electronic element of some machinery: either component A or component B may fail. If either component fails, the machinery goes down. Component A fails according to a Poisson process with mean rate 1.1 failures per shift. (The company operates 24/7 using eight-hour shifts.) Component B fails according to a Poisson process with a mean rate of 1.2 failures per day.
(a) What is the probability that there will be exactly five failures of the machine within a given day?
(b) What is the probability that there will be no more than one failure of the machine during the next shift?
(c) It is now noon and the most recent failure occurred four hours earlier. What the is the probability that the next machine failure will occur before 6PM?

4.6. Using Excel, simulate the machine failures of Problem 4.5 and show within the simulation, which component fails. The assumption is that after a repair, the machine is identical probabilistically to the machine before its failure.
(a) Assume that the time to repair the machine is negligible relative to the failure time.
(b) Assume that the time to repair the machine is uniformly distributed between 5 minutes and 60 minutes.

4.7. Customers arrive to a store according to a Poisson process with mean rate 5 per hour. The probability that the customer leaves without spending anything is 0.20. If the customer spends something, the amount can be approximated by a gamma distribution with scale parameter $100 and shape parameter 2.5.
(a) Give the mean and standard deviation of the revenue generated by customers within any given hour.

(b) Give the mean and standard deviation of the revenue generated by customers within any given 10-hour day.

4.8. Consider again the computer security system from Problem 4.3. Further study of the attempts to breach the security of the computer system has determined that it is a non-stationary Poisson process. To approximate the process, each 24-hour period is been divided into four intervals: 6AM–8PM, 8PM–midnight, midnight–4AM, and 4AM–6AM. Each interval is assumed to have a constant arrival rate for the attempted illegal entries. The mean number of attempts during the time span of each interval is 0.3, 0.6, 1, and 0.3, respectively. Answer the questions of Problem 4.3 where the evening shift is from 3PM until 11PM.

4.9. A mail-order service is open 24 hours per day. Operators receive telephone calls according to a Poisson process. Simulate the process and determine the expected volume (in terms of dollars) of sales during the 15-minute interval from 12:00 noon until 12:15PM under four different assumptions as given below. (Assume each day is probabilistically similar to all other days.)
(a) The Poisson arrival process is stationary with a mean rate of 50 calls per hour. Furthermore, half of the calls do not result in a sale, 30% of the calls result in a sale of $100, and 20% of the calls results in a sale of $200.
(b) Since more people tend to call on their lunch break, the mean arrival rate slowly increases at the start of the noon hour. For the five-minute interval from 12 noon until 12:05 PM, the mean arrival rate is 40 calls per hour; from 12:05 PM until 12:10 PM, the rate is 45 calls per hour, and from 12:10 PM until 12:50 PM, the rate is 50 calls per hour. The probabilities and amounts of a sale are the same as in Part (a).
(c) The arrival rate of calls is approximated by a linear function that equals 40 calls per hour at noon and equals 50 calls per hour at 12:15 PM. The rate is then constant for the next 30 minutes. The probabilities and amounts of a sale are the same as above.
(d) The arrival process is homogenous with a mean rate of 50 calls per hour; however, instead of exponential, the inter-arrival times have Erlang type-three distribution. The probabilities and amounts of a sale are the same as above.

4.10. Obtain the maximum likelihood estimators for α and β of Example 4.10 using Property 4.8

4.11. In a Poisson regression, the likelihood of observing n_i failures at a concentration level of x_i over the time span given by t_i, is

$$\begin{aligned} P\{N_{t_i}(x_i) = n_i\} &= \mathrm{e}^{-\lambda(x_i)t_i} (\lambda(x_i)t_i)^{n_i} / n_i! \\ &= \mathrm{e}^{-[\alpha(1+\beta x_i)]t_i} ([\alpha(1+\beta x_i)]t_i)^{n_i} / n_i! \end{aligned}$$

where i denotes the ith observation. The likelihood function is the product of the individual likelihoods for the entire sample. Use this fact to obtain the results of Properties 4.7 and 4.8.

4.12. The purpose of this problem is to obtain confidence intervals for the estimators of Example 4.12.

(a) The Hessian matrix of a function, **H**, is the square matrix formed by all the second partial derivatives of the function. In other words, if f is a function of x_i for $i = 1, \cdots, n$, then

$$H(i,j) = \begin{cases} \frac{\partial^2 f}{\partial x_i^2} & \text{if } i = j \\ \frac{\partial^2 f}{\partial x_i \partial x_j} & \text{if } i \neq j \end{cases}$$

for $i, j = 1, \cdots, n$. Form the Hessian matrix of the log-likelihood function given in Property 4.10.

(b) Determine the values of $\mathbf{H}'$, where $\mathbf{H}'$ is defined to be the negative of the Hessian matrix evaluated at the maximum likelihood estimates.

(c) Give the variance of each estimator where the variance is the associated diagonal element of $\mathbf{H}'^{-1}$.

(d) Give the 95% confidence interval under the assumption that the estimators are unbiased and are normally distributed with a variance given in Part (c).

References

1. Çinlar, E. (1975). *Introduction to Stochastic Processes*, Prentice-Hall, Inc., Englewood Cliffs, NJ.
2. Breslow, N.S., and Day, N.E. (1987). *Statistical Methods in Cancer Research, Vol. II — The Design and Analysis of Cohort Studies*. International Agency for Research on Cancer, Lyon, France.
3. Lee, A.M., and Fraumeni, J.F., Jr. (1969). Arsenic and Respiratory Cancer in Man: An Occupational Study. *Journal of the National Cancer Institute*, **42**:1045–1052.
4. Lubin, J.H., Moore, L.E., Fraumeni, J.F., Jr., and Cantor, K.P. (2008). Respiratory Cancer and Inhaled Inorganic Arsenic in Copper Smelters Workers: A Linear Relationship with Cumulative Exposure that Increases with Concentration. *Environmental Health Perspectives*, **116**:1661-1665.

Chapter 5
Markov Chains

Many decisions must be made within the context of randomness. Random failures of equipment, fluctuating production rates, and unknown demands are all part of normal decision making processes. In an effort to quantify, understand, and predict the effects of randomness, the mathematical theory of probability and stochastic processes has been developed, and in this chapter, one special type of stochastic process called a Markov chain is introduced. In particular, a Markov chain has the property that the future is independent of the past given the present. These processes are named after the probabilist A. A. Markov who published a series of papers starting in 1907 which laid the theoretical foundations for finite state Markov chains.

An interesting example from the second half of the 19th century is the so-called Galton-Watson process [1]. (Of course, since this was before Markov's time, Galton and Watson did not use Markov chains in their analyses, but the process they studied is a Markov chain and serves as an interesting example.) Galton, a British scientist and first cousin of Charles Darwin, and Watson, a clergyman and mathematician, were interested in answering the question of when and with what probability would a given family name become extinct. In the 19th century, the propagation or extinction of aristocratic family names was important, since land and titles stayed with the name. The process they investigated was as follows: At generation zero, the process starts with a single ancestor. Generation one consists of all the sons of the ancestor (sons were modeled since it was the male that carried the family name). The next generation consists of all the sons of each son from the first generation (i.e., grandsons to the ancestor), generations continuing ad infinitum or until extinction. The assumption is that for each individual in a generation, the probability of having zero, one, two, etc. sons is given by some specified (and unchanging) probability mass function, and that mass function is identical for all individuals at any generation. Such a process might continue to expand or it might become extinct, and Galton and Watson were able to address the questions of whether or not extinction occurred and, if extinction did occur, how many generations would it take. The distinction that makes a Galton-Watson process a Markov chain is the fact that at any generation, the number of individuals in the next generation is completely independent of the number of individuals in previous generations, as long as the

R.M. Feldman, C. Valdez-Flores, *Applied Probability and Stochastic Processes*, 2nd ed., 141
DOI 10.1007/978-3-642-05158-6_5, © Springer-Verlag Berlin Heidelberg 2010

number of individuals in the current generation are known. It is processes with this feature (the future being independent of the past given the present) that will be studied next. They are interesting, not only because of their mathematical elegance, but also because of their practical utilization in probabilistic modeling.

5.1 Basic Definitions

In this chapter we study a special type of discrete parameter stochastic process that is one step more complicated than a sequence of *i.i.d.* random variables; namely, Markov chains. Intuitively, a Markov chain is a discrete parameter process in which the future is independent of the past given the present. For example, suppose that we decided to play a game with a fair, unbiased coin. We each start with five pennies and repeatedly toss the coin. If it turns up heads, then you give me a penny; if tails, I give you a penny. We continue until one of us has none and the other has ten pennies. The sequence of heads and tails from the successive tosses of the coin would form an *i.i.d.* stochastic process; the sequence representing the total number of pennies that you hold would be a Markov chain. To see this, assume that after several tosses, you currently have three pennies. The probability that after the next toss you will have four pennies is 0.5 and knowledge of the past (i.e., how many pennies you had one or two tosses ago) does not help in calculating the probability of 0.5. Thus, the future (how many pennies you will have after the next toss) is independent of the past (how many pennies you had several tosses ago) given the present (you currently have three pennies).

Another example of the Markov property comes from Mendelian genetics. Mendel in 1865 demonstrated that the seed color of peas was a genetically controlled trait. Thus, knowledge about the gene pool of the current generation of peas is sufficient information to predict the seed color for the next generation. In fact, if full information about the current generation's genes are known, then knowing about previous generations does not help in predicting the future; thus, we would say that the future is independent of the past given the present.

Definition 5.1. The stochastic process $X = \{X_n; n = 0, 1, \cdots\}$ with discrete state space E is a *Markov chain* if the following holds for each $j \in E$ and $n = 0, 1, \cdots$

$$\Pr\{X_{n+1} = j | X_0 = i_0, X_1 = i_1, \cdots, X_n = i_n\} = \Pr\{X_{n+1} = j | X_n = i_n\} ,$$

for any set of states $i_0, \cdots, i_n$ in the state space. Furthermore, the Markov chain is said to have *stationary* transition probabilities if

$$\Pr\{X_1 = j | X_0 = i\} = \Pr\{X_{n+1} = j | X_n = i\} .$$

□

The first equation in Definition 5.1 is a mathematical statement of the Markov property. To interpret the equation, think of time n as the present. The left-hand-side

Fig. 5.1 State diagram for the Markov chain of Example 5.1

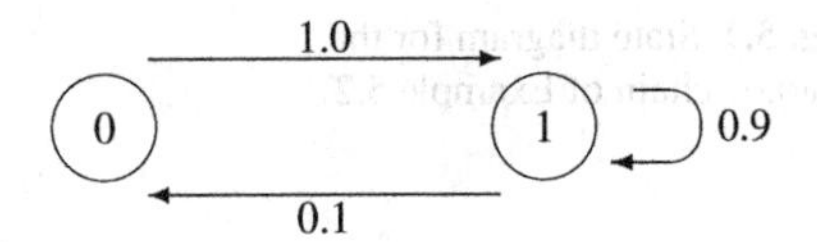

is the probability of going to state j next, given the history of all past states. The right-hand-side is the probability of going to state j next, given only the present state. Because they are equal, we have that the past history of states provides no additional information helpful in predicting the future if the present state is known. The second equation (i.e., the stationary property) simply indicates that the probability of a one-step transition does not change as time increases (in other words, the probabilities are the same in the winter and the summer).

In this chapter it is always assumed that we are working with stationary transition probabilities. Because the probabilities are stationary, the only information needed to describe the process are the initial conditions (a probability mass function for X_0) and the one-step transition probabilities. A square matrix is used for the transition probabilities and is often denoted by the capital letter **P**, where

$$P(i,j) = \Pr\{X_1 = j | X_0 = i\} . \tag{5.1}$$

Since the matrix **P** contains probabilities, it is always nonnegative (a matrix is nonnegative if every element of it is nonnegative) and the sum of the elements within each row equals one. In fact, any nonnegative matrix with row sums equal to one is called a *Markov matrix.*

Example 5.1. Consider a farmer using an old tractor. The tractor is often in the repair shop but it always takes only one day to get it running again. The first day out of the shop it always works but on any given day thereafter, independent of its previous history, there is a 10% chance of it not working and thus being sent back to the shop. Let $X_0, X_1, \cdots$ be random variables denoting the daily condition of the tractor, where a one denotes the working condition and a zero denotes the failed condition. In other words, $X_n = 1$ denotes that the tractor is working on day n and $X_n = 0$ denotes it being in the repair shop on day n. Thus, $X_0, X_1, \cdots$ is a Markov chain with state space $E = \{0, 1\}$ and with Markov matrix

$$\mathbf{P} = \begin{matrix} 0 \\ 1 \end{matrix} \begin{bmatrix} 0 & 1 \\ 0.1 & 0.9 \end{bmatrix} .$$

(Notice that the state space is sometimes written to the left of the matrix to help in keeping track of which row refers to which state.) □

In order to develop a mental image of the Markov chain, it is very helpful to draw a state diagram (Fig. 5.1) of the Markov matrix. In the diagram, each state is represented by a circle and the transitions with positive probabilities are represented by an arrow. Until the student is very familiar with Markov chains, we recommend that state diagrams be drawn for any chain that being discussed.

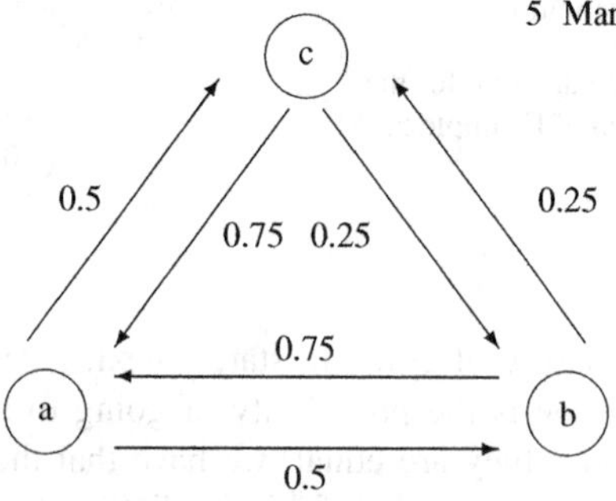

Fig. 5.2 State diagram for the Markov chain of Example 5.2

Example 5.2. A salesman lives in town 'a' and is responsible for towns 'a', 'b', and 'c'. Each week he is required to visit a different town. When he is in his home town, it makes no difference which town he visits next so he flips a coin and if it is heads he goes to 'b' and if tails he goes to 'c'. However, after spending a week away from home he has a slight preference for going home so when he is in either towns 'b' or 'c' he flips two coins. If two heads occur, then he goes to the other town; otherwise he goes to 'a'. The successive towns that he visits form a Markov chain with state space $E = \{a,b,c\}$ where the random variable X_n equals a, b, or c according to his location during week n. The state diagram for this system is given in Fig. 5.2 and the associated Markov matrix is

$$\mathbf{P} = \begin{matrix} a \\ b \\ c \end{matrix} \begin{bmatrix} 0 & 0.50 & 0.50 \\ 0.75 & 0 & 0.25 \\ 0.75 & 0.25 & 0 \end{bmatrix}.$$

□

Example 5.3. Let $X = \{X_n; n = 0,1,\cdots\}$ be a Markov chain with state space $E = \{1,2,3,4\}$ and transition probabilities given by

$$\mathbf{P} = \begin{matrix} 1 \\ 2 \\ 3 \\ 4 \end{matrix} \begin{bmatrix} 1 & 0 & 0 & 0 \\ 0 & 0.3 & 0.7 & 0 \\ 0 & 0.5 & 0.5 & 0 \\ 0.2 & 0 & 0.1 & 0.7 \end{bmatrix}.$$

The chain in this example is structurally different than the other two examples in that you might start in State 4, then go to State 3, and never get to State 1; or you might start in State 4 and go to State 1 and stay there forever. The other two examples, however, involved transitions in which it was always possible to eventually reach every state from every other state. □

Example 5.4. The final example in this section is taken from Parzen [3, p. 191] and illustrates that the parameter n need not refer to time. (It is often true that the "steps" of a Markov chain refer to days, weeks, or months, but that need not be the case.) Consider a page of text and represent vowels by zeros and consonants by ones.

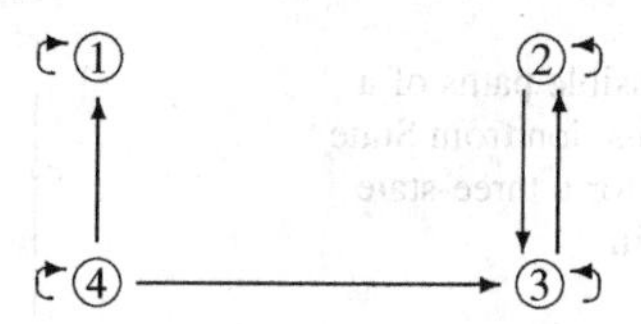

Fig. 5.3 State diagram for the Markov chain of Example 5.3

Thus the page becomes a string of zeros and ones. It has been indicated that the sequence of vowels and consonants in the Samoan language forms a Markov chain, where a vowel always follows a consonant and a vowel follows another vowel with a probability of 0.51 [2]. Thus, the sequence of ones and zeros on a page of Samoan text would evolve according to the Markov matrix

$$\mathbf{P} = \begin{matrix} 0 \\ 1 \end{matrix} \begin{bmatrix} 0.51 & 0.49 \\ 1 & 0 \end{bmatrix} .$$

□

After a Markov chain has been formulated, there are many questions that might be of interest. For example: In what state will the Markov chain be five steps from now? What percent of time is spent in a given state? Starting from one state, will the chain ever reach another fixed state? If a profit is realized for each visit to a particular state, what is the long run average profit per unit of time? The remainder of this chapter is devoted to answering questions of this nature.

- *Suggestion: Do Part (a) of Problems 5.1–5.3 and 5.6–5.9.*

5.2 Multistep Transitions

The Markov matrix provides direct information about one-step transition probabilities and it can also be used in the calculation of probabilities for transitions involving more than one step. Consider the salesman of Example 5.2 starting in town b. The Markov matrix indicates that the probability of being in State a after one step (in one week) is 0.75, but what is the probability that he will be in State a after two steps? Figure 5.4 illustrates the paths that go from b to a in two steps (some of the paths shown have probability zero). Thus, to compute the probability, we need to sum over all possible routes. In other words we would have the following calculations:

$$\begin{aligned} \Pr\{X_2 = a | X_0 = b\} &= \Pr\{X_1 = a | X_0 = b\} \times \Pr\{X_2 = a | X_1 = a\} \\ &\quad + \Pr\{X_1 = b | X_0 = b\} \times \Pr\{X_2 = a | X_1 = b\} \\ &\quad + \Pr\{X_1 = c | X_0 = b\} \times \Pr\{X_2 = a | X_1 = c\} \\ &= P(b,a)P(a,a) + P(b,b)P(b,a) + P(b,c)P(c,a) . \end{aligned}$$

The final equation should be recognized as the definition of matrix multiplication; thus,

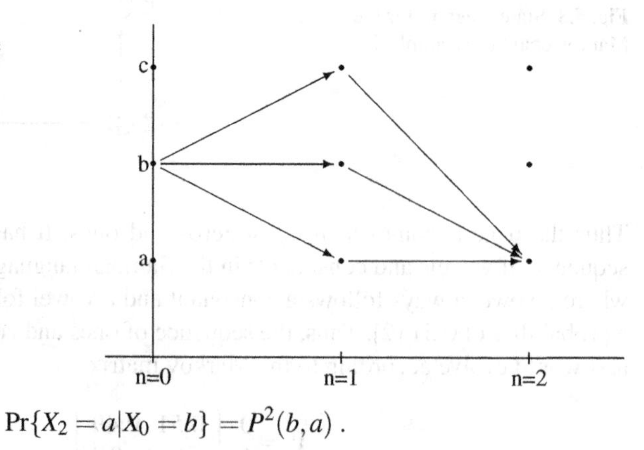

Fig. 5.4 Possible paths of a two-step transition from State b to State a for a three-state Markov chain

$$\Pr\{X_2 = a | X_0 = b\} = P^2(b,a) .$$

This result is easily generalized into the following property:

Property 5.1. *Let $X = \{X_n; n = 0, 1, \cdots\}$ be a Markov chain with state space E and Markov matrix* **P**, *then for $i, j \in E$ and $n = 1, 2, \cdots$*

$$\Pr\{X_n = j | X_0 = i\} = P^n(i,j) ,$$

where the right-hand side represents the $i - j$ element of the matrix $\mathbf{P}^n$.

Property 5.1 indicates that $P^n(i,j)$ is interpreted to be the probability of going from state i to state j in n steps. It is important to remember that the notation $P^n(i,j)$ means that the matrix is *first* raised to the nth power and then the $i - j$ element of the resulting matrix is taken for the answer.

Returning to Example 5.2, the squared matrix is

$$\mathbf{P}^2 = \begin{bmatrix} 0.75 & 0.125 & 0.125 \\ 0.1875 & 0.4375 & 0.375 \\ 0.1875 & 0.375 & 0.4375 \end{bmatrix} .$$

The $b - a$ element of $\mathbf{P}^2$ is 0.1875 and thus there is a 0.1875 probability of being in town a two weeks after being in town b.

Markov chains are often used to analyze the cost or profit of an operation and thus we need to consider a cost or profit function imposed on the process. For example, suppose in Example 5.2 that every week spent in town a resulted in a profit of $1000, every week spent in town b resulted in a profit of $1200, and every week spent in town c resulted in a profit of $1250. We then might ask what would be the expected profit after the first week if the initial town was town a? Or, more generally, what would be the expected profit of the nth week if the initial town was a? It should not be too difficult to convince yourself that the following calculation would be appropriate:

$$E[\text{Profit for week } n] = P^n(a,a) \times 1000 + P^n(a,b) \times 1200 + P^n(a,c) \times 1250 .$$

We thus have the following property.

Property 5.2. *Let $X = \{X_n; n = 0, 1, \cdots\}$ be a Markov chain with state space E, Markov matrix* **P**, *and profit function* **f** *(i.e., each time the chain visits state i, a profit of $f(i)$ is obtained). The expected profit at the nth step is given by*

$$E[f(X_n)|X_0 = i] = \mathbf{P}^n\mathbf{f}(i) .$$

Note again that in the right-hand side of the equation, the matrix $\mathbf{P}^n$ is multiplied by the vector **f** first and then the ith component of the resulting vector is taken as the answer. Thus, in Example 5.2, we have that the expected profit during the second week given that the initial town was a is

$$\begin{aligned} E[f(X_2)|X_0 = a] &= 0.75 \times 1000 + 0.125 \times 1200 + 0.125 \times 1250 \\ &= 1056.25. \end{aligned}$$

Up until now we have always assumed that the initial state was known. However, that is not always the situation. The manager of the traveling salesman might not know for sure the location of the salesman; instead, all that is known is a probability mass function describing his initial location. Again, using Example 5.2, suppose that we do not know for sure the salesman's initial location but know that there is a 50% chance he is in town a, a 30% chance he is in town b, and a 20% chance he is in town c. We now ask, what is the probability that the salesman will be in town a next week? The calculations for that are

$$\Pr\{X_1 = a\} = 0.50 \times 0 + 0.30 \times 0.75 + 0.20 \times 0.75 = 0.375 .$$

These calculations generalize to the following.

Property 5.3. *Let $X = \{X_n; n = 0, 1, \cdots\}$ be a Markov chain with state space E, Markov matrix* **P**, *and initial probability vector* $\boldsymbol{\mu}$ *(i.e., $\mu(i) = \Pr\{X_0 = i\}$). Then*

$$\Pr_{\mu}\{X_n = j\} = \boldsymbol{\mu}\mathbf{P}^n(j) .$$

It should be noticed that μ is a subscript to the probability statement on the left-hand side of the equation. The purpose for the subscript is to insure that there is no confusion over the given conditions. And again when interpreting the right-hand-side of the equation, the vector $\boldsymbol{\mu}$ is multiplied by the matrix $\mathbf{P}^n$ first and then the jth element is taken from the resulting vector.

The last two properties can be combined, when necessary, into one statement.

Property 5.4. *Let $X = \{X_n; n = 0, 1, \cdots\}$ be a Markov chain with state space E, Markov matrix* **P**, *initial probability vector* $\boldsymbol{\mu}$, *and profit function* **f**. *The expected profit at the nth step is given by*

$$E_\mu[f(X_n)] = \boldsymbol{\mu}\mathbf{P}^n\mathbf{f}\,.$$

Returning to Example 5.2 and using the initial probabilities and profit function given above, we have that the expected profit in the second week is calculated to be

$$\boldsymbol{\mu}\mathbf{P}^2\mathbf{f} = (0.50, 0.30, 0.20)\begin{bmatrix} 0.75 & 0.125 & 0.125 \\ 0.1875 & 0.4375 & 0.375 \\ 0.1875 & 0.375 & 0.4375 \end{bmatrix}\begin{pmatrix} 1000 \\ 1200 \\ 1250 \end{pmatrix}$$
$$= 1119.375\,.$$

Example 5.5. A market analysis concerning consumer behavior in auto purchases has been conducted. Body styles traded-in and purchased have been recorded by a particular dealer with the following results:

Number of Customers	Trade
275	sedan for sedan
180	sedan for station wagon
45	sedan for convertible
80	station wagon for sedan
120	station wagon for station wagon
150	convertible for sedan
50	convertible for convertible

These data are believed to be representative of average consumer behavior, and it is assumed that the Markov assumptions are appropriate.

We shall develop a Markov chain to describe the changing body styles over the life of a customer. (Notice, for purposes of instruction, we are simplifying the process by assuming that the age of the customer does not affect the customer's choice of a body style.) We define the Markov chain to be the body-style of the automobile that the customer has immediately after a trade-in, where the state space is $E = \{s, w, c\}$ with s for the sedan, w for the wagon, and c for the convertible. Of 500 customers who have a sedan, 275 will stay with the sedan during the next trade; therefore, the s-s element of the transition probability matrix will be 275/500. Thus, the Markov matrix for the "body style" Markov chain is

$$\mathbf{P} = \begin{matrix} s \\ w \\ c \end{matrix}\begin{bmatrix} 275/500 & 180/500 & 45/500 \\ 80/200 & 120/200 & 0 \\ 150/200 & 0 & 50/200 \end{bmatrix} = \begin{bmatrix} 0.55 & 0.36 & 0.09 \\ 0.40 & 0.60 & 0.0 \\ 0.75 & 0.0 & 0.25 \end{bmatrix}.$$

Let us assume we have a customer whose behavior is described by this model. Furthermore, the customer always buys a new car in January of every year. It is now January 1997, and the customer enters the dealership with a sedan (i.e., $X_{1996} = s$). From the above matrix, $\Pr\{X_{1997} = s | X_{1996} = s\} = 0.55$, or the probability that the customer will leave with another sedan is 55%. We would now like to predict what body-style the customer will have for trade-in during January 2000. Notice that the question involves three transitions of the Markov chain; therefore, the cubed matrix must be determined, which is

$$\mathbf{P}^3 = \begin{bmatrix} 0.5023 & 0.4334 & 0.0643 \\ 0.4816 & 0.4680 & 0.0504 \\ 0.5355 & 0.3780 & 0.0865 \end{bmatrix} .$$

Thus, there is approximately a 50% chance the customer will enter in the year 2000 with a sedan, a 43% chance with a station wagon, and almost a 7% chance of having a convertible. The probability that the customer will leave this year (1997) with a convertible and then leave next year (1998) with a sedan is given by $P(s,c) \times P(c,s) = 0.09 \times 0.75 = 0.0675$. Now determine the probability that the customer who enters the dealership now (1997) with a sedan will leave with a sedan and also leave with a sedan in the year 2000. The mathematical statement is

$$\Pr\{X_{1997} = s, X_{2000} = s | X_{1996} = s\} = P(s,s) \times P^3(s,s) \approx 0.28 .$$

Notice in the above probability statement, that no mention is made of the body style for the intervening years; thus, the customer may or may not switch in the years 1998 and 1999.

Now to illustrate profits. Assume that a sedan yields a profit of \$1200, a station wagon yields \$1500, and a convertible yields \$2500. The expected profit this year from the customer entering with a sedan is \$1425, and the expected profit in the year 1999 from a customer who enters the dealership with a sedan in 1997 is approximately \$1414. Or, to state this mathematically,

$$E[f(X_{1999}) | X_{1996} = s] \approx 1414 ,$$

where $\mathbf{f} = (1200, 1500, 2500)^T$. □

- *Suggestion: Do Problems 5.10a–c, 5.11a–f, and 5.12.*

5.3 Classification of States

There are two types of states possible in a Markov chain, and before most questions can be answered, the individual states for a particular chain must be classified into one of these two types. As we begin the Markov chain study, it is important to cover a significant amount of new notation. The student will discover that the time spent in learning the new terminology will be rewarded with a fuller understanding of

Markov chains. Furthermore, a good understanding of the dynamics involved for Markov chains will greatly aid in the understanding of the entire area of stochastic processes.

Two random variables that will be extremely important denote "first passage times" and "number of visits to a fixed state". To describe these random variables, consider a fixed state, call it State j, in the state space for a Markov chain. The first passage time is a random variable, denoted by T^j, that equals the time (i.e., number of steps) it takes to reach the fixed state *for the first time*. Mathematically, the first passage time random variable is defined by

$$T^j = \min\{n \geq 1 : X_n = j\} , \tag{5.2}$$

where the minimum of the empty set is taken to be $+\infty$. For example, in Example 5.3, if $X_0 = 1$ then $T^2 = \infty$, i.e., if the chain starts in State 1 then it will never reach State 2.

The number of visits to a state is a random variable, denoted by N^j, that equals the total number of visits (including time zero) that the Markov chain makes to the fixed state throughout the life of the chain. Mathematically, the "number of visits" random variable is defined by

$$N^j = \sum_{n=0}^{\infty} I(X_n, j) , \tag{5.3}$$

where I is the identity matrix. The identity matrix is used simply as a "counter". Because the identity has ones on the diagonal and zeroes off the diagonal, it follows that $I(X_n, j) = 1$ if $X_n = j$; otherwise, it is zero. It should be noted that the summation in (5.3) starts at $n = 0$; thus, if $X_0 = j$ then N^j must be at least one.

Example 5.6. Let us consider a realization for the Markov chain of Example 5.3. (By realization, we mean conceptually that an experiment is conducted and we record the random outcomes of the chain.) Assume that the first part of the realization (Fig. 5.5) is $X_0 = 4, X_1 = 4, X_2 = 4, X_3 = 3, X_4 = 2, X_5 = 3, X_6 = 2, \cdots$. The first passage random variables for this realization are $T^1 = \infty, T^2 = 4, T^3 = 3,$ and $T^4 = 1$. To see why $T^1 = \infty$, it is easiest to refer to the state diagram describing the Markov chain (Fig. 5.3). By inspecting the diagram it becomes obvious that once the chain is in either States 2 or 3 that it will never get to State 1 and thus $T^1 = \infty$.

Using the same realization, the number of visits random variables are $N^1 = 0, N^2 = \infty, N^3 = \infty,$ and $N^4 = 3$. Again, the values for $N^1, N^2,$ and N^3 are obtained by inspecting the state diagram and observing that if the chain ever gets to States 2 or 3 it will stay in those two states forever and thus will visit them an infinite number of times.

Let us perform the experiment one more time. Assume that our second realization results in the values $X_0 = 4, X_1 = 4, X_2 = 1, X_3 = 1, X_4 = 1, \cdots$. The new outcomes for the first passage random variables for this second realization are $T^1 = 2, T^2 = \infty, T^3 = \infty,$ and $T^4 = 1$. Furthermore, $N^1 = \infty, N^2 = 0, N^3 = 0,$ and $N^4 = 2$. Thus, you should understand from this example that we may not be able to say what the value

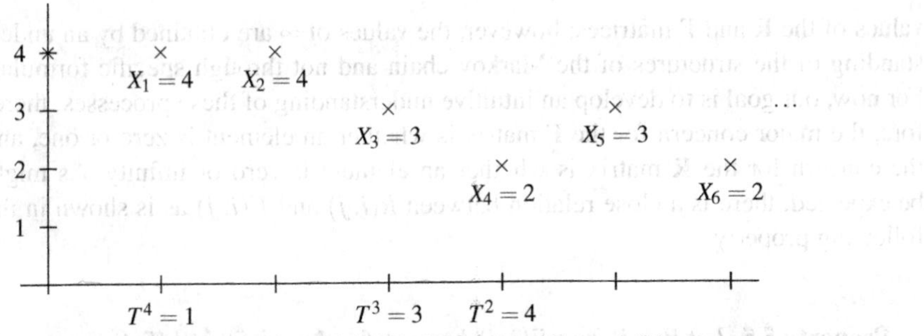

Fig. 5.5 One Possible Realization for the Markov chain of Example5.1

of T^j will be before an experiment, but we should be able to describe its probability mass function. □

The primary quantity of interest regarding the first passage times are the so-called first passage probabilities. The major question of interest is whether or not it is possible to reach a particular state from a given initial state. To answer this question, we determine the first passage probability, denoted by $F(i,j)$, which is the probability of eventually reaching State j at least once given that the initial state was i. The probability $F(i,j)$ ("F" for *first* passage) is defined by

$$F(i,j) = \Pr\{T^j < \infty | X_0 = i\} . \tag{5.4}$$

By inspecting the state diagram in Fig. 5.3, it should be obvious that the first passage probabilities for the chain of Example 5.3 are given by

$$\mathbf{F} = \begin{bmatrix} 1 & 0 & 0 & 0 \\ 0 & 1 & 1 & 0 \\ 0 & 1 & 1 & 0 \\ <1 & <1 & <1 & <1 \end{bmatrix} .$$

The primary quantity of interest for the number of visits random variable is its expected value. The expected number of visits to State j given the initial state was i is denoted by $R(i,j)$ ("R" for *returns*) and is defined by

$$R(i,j) = E[N^j | X_0 = i] . \tag{5.5}$$

Again the state diagram of Fig. 5.3 allows the determination of some of the values of R as follows:

$$\mathbf{R} = \begin{bmatrix} \infty & 0 & 0 & 0 \\ 0 & \infty & \infty & 0 \\ 0 & \infty & \infty & 0 \\ \infty & \infty & \infty & <\infty \end{bmatrix} .$$

The above matrix may appear unusual because of the occurrence of ∞ for elements of the matrix. In Sect. 5.5, numerical methods will be given to calculate the values of the **R** and **F** matrices; however, the values of ∞ are obtained by an understanding of the structures of the Markov chain and not through specific formulas. For now, our goal is to develop an intuitive understanding of these processes; therefore, the major concern for the **F** matrix is whether an element is zero or one, and the concern for the **R** matrix is whether an element is zero or infinity. As might be expected, there is a close relation between $R(i,j)$ and $F(i,j)$ as is shown in the following property.

Property 5.5. *Let $R(i,j)$ and $F(i,j)$ be as defined in (5.5) and (5.4), respectively. Then*

$$R(i,j) = \begin{cases} \frac{1}{1-F(j,j)} & \text{for } i = j, \\ \\ \frac{F(i,j)}{1-F(j,j)} & \text{for } i \neq j; \end{cases}$$

where the convention $0/0 = 0$ is used.

The above discussion utilizing Example 5.3 should help to point out that there is a basic difference between States 1, 2 or 3 and State 4 of the example. Consider that if the chain ever gets to State 1 or to State 2 or to State 3 then that state will continually reoccur. However, even if the chain starts in State 4, it will only stay there a finite number of times and will eventually leave State 4 never to return. These ideas give rise to the terminology of recurrent and transient states. Intuitively, a state is called recurrent if, starting in that state, it will continuously reoccur; and a state is called transient if, starting in that state, it will eventually leave that state never to return. Or equivalently, a state is recurrent if, starting in that state, the chain must (i.e., with probability one) eventually return to the state at least once; and a state is transient if, starting in that state, there is a chance (i.e., with probability greater than zero) that the chain will leave the state and never return. The mathematical definitions are based on these notions developed above.

Definition 5.2. A state j is called *transient* if $F(j,j) < 1$. Equivalently, state j is *transient* if $R(j,j) < \infty$. □

Definition 5.3. A state j is called *recurrent* if $F(j,j) = 1$. Equivalently, state j is *recurrent* if $R(j,j) = \infty$. □

From the above two definitions, a state must either be transient or recurrent. A dictionary[1] definition for the word *transient* is "passing esp. quickly into and out of existence". Thus, the use of the word is justified since transient states will only occur for a finite period of time. For a transient state, there will be a time after which the transient state will never again be visited. A dictionary[1] definition for *recurrent* is "returning or happening time after time". So again the mathematical concept of a

[1] *Webster's Ninth New Collegiate Dictionary*, (Springfield, MA: Merriam-Webster, Inc., 1989)

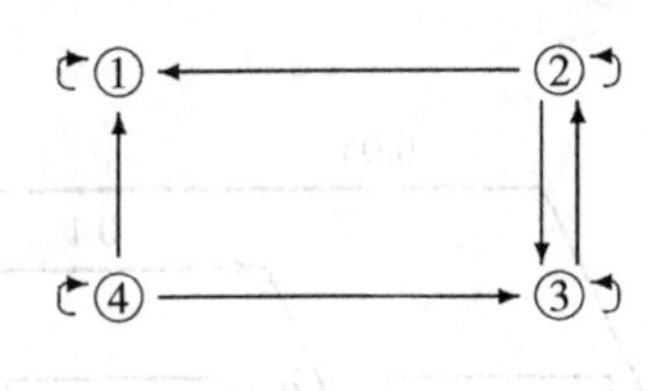

Fig. 5.6 State diagram for the Markov chain of Example 5.7

recurrent state parallels the common English usage: recurrent states are recognized by those states which are continually revisited.

Recurrent states might also be periodic, as with the Markov matrix

$$\mathbf{P} = \begin{bmatrix} 0 & 1 \\ 1 & 0 \end{bmatrix}$$

but discussions regarding periodic chains will be left to other texts.

By drawing the state diagrams for the Markov chains of Examples 5.1 and 5.2, it should become clear that all states in those two cases are recurrent. The only transient state so far illustrated is State 4 of Example 5.3.

Example 5.7. Let $X = \{X_n; n = 0, 1, \cdots\}$ be a Markov chain with state space $E = \{1, 2, 3, 4\}$ and transition probabilities given by

$$\mathbf{P} = \begin{matrix} 1 \\ 2 \\ 3 \\ 4 \end{matrix} \begin{bmatrix} 1 & 0 & 0 & 0 \\ 0.01 & 0.29 & 0.7 & 0 \\ 0 & 0.5 & 0.5 & 0 \\ 0.2 & 0 & 0.1 & 0.7 \end{bmatrix}.$$

Notice the similarity between this Markov chain and the chain in Example 5.3. Although the numerical difference between the two Markov matrices is slight, there is a radical difference between the structures of the two chains. (The difference between a zero term and a nonzero term, no matter how small, can be very significant.) The chain of this example has one recurrent state and three transient states.

By inspecting the state diagram for this Markov chain (Fig. 5.6) the following matrices are obtained (through observation, not through calculation):

$$\mathbf{F} = \begin{bmatrix} 1 & 0 & 0 & 0 \\ 1 & <1 & <1 & 0 \\ 1 & 1 & <1 & 0 \\ 1 & <1 & <1 & <1 \end{bmatrix}.$$

$$\mathbf{R} = \begin{bmatrix} \infty & 0 & 0 & 0 \\ \infty & <\infty & <\infty & 0 \\ \infty & <\infty & <\infty & 0 \\ \infty & <\infty & <\infty & <\infty \end{bmatrix}.$$

It should be observed that $F(3,2) = 1$ even though both State 2 and State 3 are transient. It is only the diagonal elements of the **F** and **R** matrices that determine whether a state is transient or recurrent. □

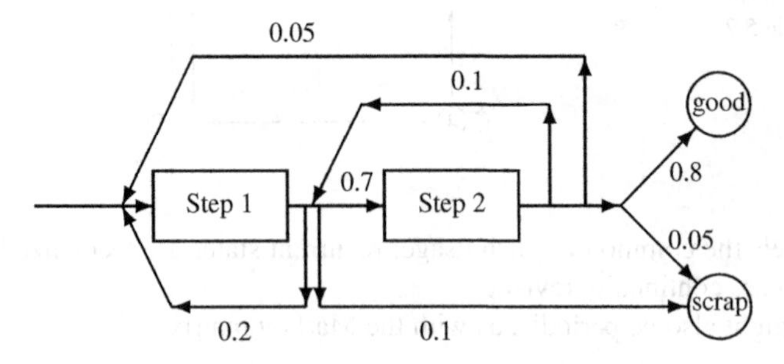

Fig. 5.7 Two step manufacturing process of Example 5.8

Example 5.8. A manufacturing process consists of two processing steps in sequence. After step 1, 20% of the parts must be reworked, 10% of the parts are scrapped, and 70% proceed to the next step. After step 2, 5% of the parts must be returned to the first step, 10% must be reworked and 5% are scrapped; the remainder are sold. The diagram of Fig. 5.7 illustrates the dynamics of the manufacturing process.

The Markov matrix associated with this manufacturing process is given by

$$\mathbf{P} = \begin{array}{c} 1 \\ 2 \\ s \\ g \end{array} \begin{bmatrix} 0.2 & 0.7 & 0.1 & 0.0 \\ 0.05 & 0.1 & 0.05 & 0.8 \\ 0.0 & 0.0 & 1.0 & 0.0 \\ 0.0 & 0.0 & 0.0 & 1.0 \end{bmatrix} . \tag{5.6}$$

Consider the dynamics of the chain. A part to be manufactured will begin the process by entering State 1. After possibly cycling for awhile, the part will end the process by entering either the "g" state or the "s" state. Therefore, States 1 and 2 are transient, and States g and s are recurrent. □

Along with classifying states, we also need to be able to classify sets of states. We will first define a closed set which is a set that once the Markov chain has entered the set, it cannot leave the set.

Definition 5.4. Let $X = \{X_n; n = 0, 1, \cdots\}$ be a Markov chain with Markov matrix $\mathbf{P}$ and let C be a set of states contained in its state space. Then C is *closed* if

$$\sum_{j \in C} P(i, j) = 1 \text{ for all } i \in C .$$

□

To illustrate the concept of closed sets, refer again to Example 5.3. The sets {2,3,4}, {2}, and {3,4} are *not* closed sets. The set {1,2,3,4} is obviously closed, but it can be reduced to a smaller closed set. The set {1,2,3} is also closed, but again it can be further reduced. Both sets {1} and {2,3} are closed and cannot be reduced further. This idea of taking a closed set and trying to reduce it is extremely important and leads to the definition of an irreducible set.

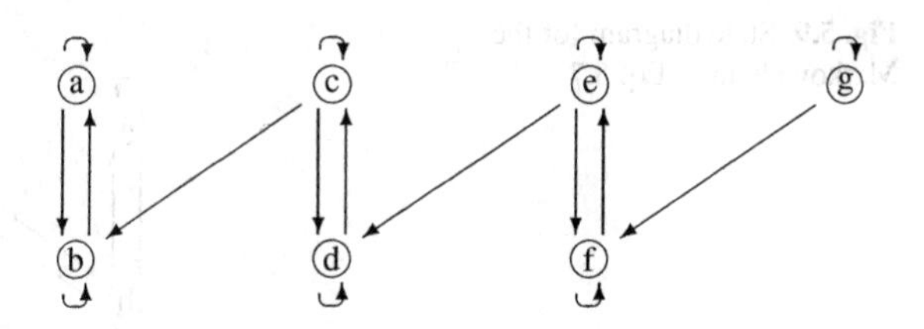

Fig. 5.8 State diagram for the Markov chain of Example 5.9

Definition 5.5. A closed set of states that contains no proper subset which is also closed is called *irreducible*. A state that forms an irreducible set by itself is called an *absorbing* state. □

The Markov chain of Example 5.3 has two irreducible sets: the set $\{1\}$ and the set $\{2,3\}$. Since the first irreducible set contains only one state, that state is an absorbing state.

Example 5.9. Let X be a Markov chain with state space $E = \{a, \cdots, g\}$ and Markov matrix

$$\mathbf{P} = \begin{matrix} a \\ b \\ c \\ d \\ e \\ f \\ g \end{matrix} \begin{bmatrix} 0.3 & 0.7 & 0 & 0 & 0 & 0 & 0 \\ 0.5 & 0.5 & 0 & 0 & 0 & 0 & 0 \\ 0 & 0.2 & 0.4 & 0.4 & 0 & 0 & 0 \\ 0 & 0 & 0.5 & 0.5 & 0 & 0 & 0 \\ 0 & 0 & 0 & 0.8 & 0.1 & 0.1 & 0 \\ 0 & 0 & 0 & 0 & 0.7 & 0.3 & 0 \\ 0 & 0 & 0 & 0 & 0 & 0.4 & 0.6 \end{bmatrix} .$$

By drawing a state diagram (Fig. 5.8), you should observe that states a and b are recurrent and all others are transient. (If that is not obvious, notice that once the chain reaches either a or b, it will stay in those two states, there are no paths leading away from the set $\{a,b\}$; furthermore, all states will eventually reach State b). Obviously, the entire state space is closed. Since no state goes to g, we can exclude g and still have a closed set; namely, the set $\{a,b,c,d,e,f\}$ is closed. However, the set $\{a,b,c,d,e,f\}$ can be reduced to a smaller closed set so it is *not* irreducible. Because there is a path going from e to f, the set $\{a,b,c,d,e\}$ is *not* closed. By excluding the state e, a closed set is again obtained; namely, $\{a,b,c,d\}$ is also closed since there are no paths out of the set $\{a,b,c,d\}$ to another state. Again, however, it can also be reduced so the set $\{a,b,c,d\}$ is *not* irreducible. Finally, consider the set $\{a,b\}$. This two-state set is closed and cannot be further reduced to a smaller closed set; therefore, the set $\{a,b\}$ is an irreducible set. □

The reason that Definitions 5.4 and 5.5 are important is the following property.

Property 5.6. *All states within an irreducible set are of the same classification.*

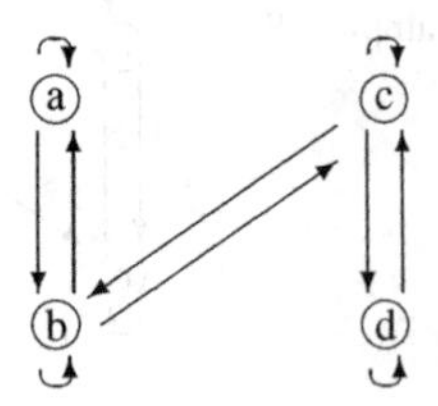

Fig. 5.9 State diagram for the Markov chain of Eq. 5.7

The significance of the above property is that if you can identify one state within an irreducible set as being transient, then all states within the set are transient, and if one state is recurrent, then all states within the set are recurrent. We are helped even more by recognizing that it is impossible to have an irreducible set of transient states if the set contains only a finite number of states; thus we have the following property.

Property 5.7. *Let C be an irreducible set of states such that the number of states within C is finite. Then each state within C is recurrent.*

There is one final concept that will help in identifying irreducible sets; namely, communication between states. Communication between states is like communication between people; there is communication only if messages can go both ways. In other words, two states, i and j, *communicate* if and only if it is possible to eventually reach j from i *and* it is possible to eventually reach i from j. In Example 5.3, States 2 and 3 communicate, but States 4 and 2 do not communicate. Although State 2 can be reached from 4, it does not go both ways because State 4 cannot be reached from 2. The communication must be both ways but it does not have to be in one step. For example, in the Markov chain with state space $\{a,b,c,d\}$ and with Markov matrix

$$\mathbf{P} = \begin{bmatrix} 0.5 & 0.5 & 0 & 0 \\ 0.2 & 0.4 & 0.4 & 0 \\ 0 & 0.1 & 0.8 & 0.1 \\ 0 & 0 & 0.3 & 0.7 \end{bmatrix}, \tag{5.7}$$

all states communicate with every other state. And, in particular, State a communicates with State d even though they can not reach each other in one step (see Fig. 5.9).

The notion of communication is often the concept used to identify irreducible sets as is given in the following property.

Property 5.8. *The closed set of states C is irreducible if and only if every state within C communicates with every other state within C.*

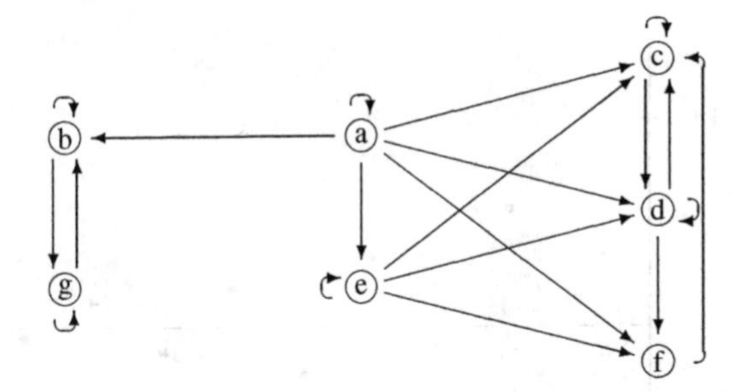

Fig. 5.10 State diagram for the Markov chain of Example 5.3

The procedure for identifying irreducible sets of states for a Markov chain with a finite state space is to first draw the state diagram, then pick a state and identify all states that communicate with it. If the set made up of all those states that communicate with it is closed, then the set is an irreducible, recurrent set of states. If it is not closed, then the originally chosen state and all the states that communicate with it are transient states.

Example 5.10. Let X be a Markov chain with state space $E = \{a, \cdots, g\}$ and Markov matrix

$$\mathbf{P} = \begin{matrix} a \\ b \\ c \\ d \\ e \\ f \\ g \end{matrix} \begin{bmatrix} 0.3 & 0.1 & 0.2 & 0.2 & 0.1 & 0.1 & 0 \\ 0 & 0.5 & 0 & 0 & 0 & 0 & 0.5 \\ 0 & 0 & 0.4 & 0.6 & 0 & 0 & 0 \\ 0 & 0 & 0.3 & 0.2 & 0 & 0.5 & 0 \\ 0 & 0 & 0.2 & 0.3 & 0.4 & 0.1 & 0 \\ 0 & 0 & 1 & 0 & 0 & 0 & 0 \\ 0 & 0.8 & 0 & 0 & 0 & 0 & 0.2 \end{bmatrix} .$$

By drawing a state diagram (Fig. 5.10), it is seen that there are two irreducible recurrent sets of states which are $\{b, g\}$ and $\{c, d, f\}$. States a and e are transient. □

- *Suggestion: Do Problems 5.13a–c and 5.14a–d.*

5.4 Steady-State Behavior

The reason that so much effort has been spent in classifying states and sets of states is that the analysis of the limiting behavior of a chain is dependent on the type of state under consideration. To illustrate long-run behavior, we again return to Example 5.2 and focus attention on State a. Figure 5.11 shows a graph which gives the probabilities of being in State a at time n given that at time zero the chain was in State a. (In other words, the graph gives the values of $P^n(a,a)$ as a function of n.) Although the graph varies dramatically for small n, it reaches an essentially constant value as n becomes large. It is seen from the graph that

$$\lim_{n \to \infty} \Pr\{X_n = a | X_0 = a\} = 0.42857 .$$

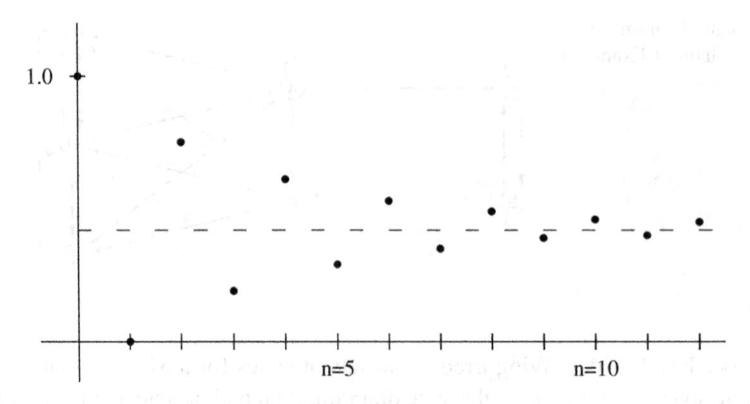

Fig. 5.11 Probabilities from Example 5.2 of being in State a as a function of time

In fact, if you spent the time to graph the probabilities of being in State a starting from State b instead of State a, you would discover the same limiting value, namely,

$$\lim_{n\to\infty} \Pr\{X_n = a|X_0 = b\} = 0.42857\,.$$

When discussing steady-state (or limiting) conditions, this is what is meant, not that the chain stops changing — it is dynamic by definition — but that enough time has elapsed so that the probabilities do not change with respect to time. It is often stated that steady-state results are independent of initial conditions, and this is true for the chain in Example 5.2; however it is not always true. In Example 5.3, it is clear that the steady-state conditions are radically different when the chain starts in State 1 as compared to starting in State 2.

If the entire state space of a Markov chain forms an irreducible recurrent set, the Markov chain is called an irreducible recurrent Markov chain. In this case the steady-state probabilities are independent of the initial state and are not difficult to compute as is seen in the following property.

Property 5.9. *Let $X = \{X_n; n = 0, 1, \cdots\}$ be a Markov chain with finite state space E and Markov matrix* **P**. *Furthermore, assume that the entire state space forms an irreducible, recurrent set, and let*

$$\pi(j) = \lim_{n\to\infty} \Pr\{X_n = j|X_0 = i\}\,.$$

The vector $\boldsymbol{\pi}$ is the solution to the following system

$$\boldsymbol{\pi}\mathbf{P} = \boldsymbol{\pi},$$
$$\sum_{i\in E} \pi(i) = 1\,.$$

To illustrate the determination of $\boldsymbol{\pi}$, observe that the Markov chain of Example 5.2 is irreducible, recurrent. Applying Property 5.9, we obtain

$$\begin{aligned} 0.75\pi_b + 0.75\pi_c &= \pi_a \\ 0.5\pi_a \qquad\qquad + 0.25\pi_c &= \pi_b \\ 0.5\pi_a + 0.25\pi_b \qquad\qquad &= \pi_c \\ \pi_a + \quad \pi_b + \quad \pi_c &= 1\,. \end{aligned}$$

There are four equations and only three variables, so normally there would not be a unique solution; however, for an irreducible Markov matrix there is always exactly one redundant equation from the system formed by $\boldsymbol{\pi}\mathbf{P} = \boldsymbol{\pi}$. Thus, to solve the above system, arbitrarily choose one of the first three equations to discard and solve the remaining 3 by 3 system (*never* discard the final or norming equation) which yields

$$\pi_a = \frac{3}{7},\ \pi_b = \frac{2}{7},\ \pi_c = \frac{2}{7}\,.$$

Property 5.9 cannot be directly applied to the chain of Example 5.3 because the state space is not irreducible. All irreducible, recurrent sets must be identified and grouped together and then the Markov matrix is rearranged so that the irreducible sets are together and transient states are last. In such a manner, the Markov matrix for a chain can always be rewritten in the form

$$\mathbf{P} = \begin{bmatrix} \mathbf{P}_1 & & & & \\ & \mathbf{P}_2 & & & \\ & & \mathbf{P}_3 & & \\ & & & \ddots & \\ \mathbf{B}_1 & \mathbf{B}_2 & \mathbf{B}_3 & \cdots & \mathbf{Q} \end{bmatrix}\,. \tag{5.8}$$

After a chain is in this form, each submatrix $\mathbf{P}_i$ is a Markov matrix and can be considered as an independent Markov chain for which Property 5.9 is applied.

The Markov matrix of Example 5.3 is already in the form of (5.8). Since State 1 is absorbing (i.e., an irreducible set of one state), its associated steady-state probability is easy; namely, $\pi_1 = 1$. States 2 and 3 form an irreducible set so Property 5.9 can be applied to the submatrix from those states resulting in the following system:

$$\begin{aligned} 0.3\pi_2 + 0.5\pi_3 &= \pi_2 \\ 0.7\pi_2 + 0.5\pi_3 &= \pi_3 \\ \pi_2 + \quad \pi_3 &= 1 \end{aligned}$$

which yields (after discarding one of the first two equations)

$$\pi_2 = \frac{5}{12} \text{ and } \pi_3 = \frac{7}{12}\,.$$

The values π_2 and π_3 are interpreted to mean that if a snapshot of the chain is taken a long time after it started and if it started in States 2 or 3, then there is a $5/12$

probability that the picture will show the chain in State 2 and a $7/12$ probability that it will be in State 3. Another interpretation of the steady-state probabilities is that if we recorded the time spent in States 2 and 3, then over the long run the fraction of time spent in State 2 would equal $5/12$ and the fraction of time spent in State 3 would equal $7/12$. These steady-state results are summarized in the following property.

Property 5.10. *Let $X = \{X_n; n = 0, 1, \cdots\}$ be a Markov chain with finite state space with k distinct irreducible sets. Let the ℓth irreducible set be denoted by C_ℓ, and let $\mathbf{P}_\ell$ be the Markov matrix restricted to the ℓth irreducible set (as in Eq. 5.8). Finally, let $\mathbf{F}$ be the matrix of first passage probabilities.*

- *If State j is transient,*

$$\lim_{n\to\infty} \Pr\{X_n = j | X_0 = i\} = 0 .$$

- *If State i and j both belong to the ℓth irreducible set,*

$$\lim_{n\to\infty} \Pr\{X_n = j | X_0 = i\} = \pi(j)$$

where

$$\boldsymbol{\pi}\mathbf{P}_\ell = \boldsymbol{\pi} \text{ and}$$
$$\sum_{i\in C_\ell} \pi(i) = 1 .$$

- *If State j is recurrent and i is not in its irreducible set,*

$$\lim_{n\to\infty} \Pr\{X_n = j | X_0 = i\} = F(i,j)\pi(j) ,$$

where $\boldsymbol{\pi}$ is determined as above.

- *If State j is recurrent and X_0 is in the same irreducible set as j,*

$$\lim_{n\to\infty} \frac{1}{n} \sum_{m=0}^{n-1} I(X_m, j) = \pi(j) ,$$

where $\mathbf{I}$ is the identity matrix.

- *If State j is recurrent,*

$$E[T^j | X_0 = j] = \frac{1}{\pi(j)} .$$

The intuitive idea of the second to last item in the above property is obtained by considering the role that the identity matrix plays in the left-hand-side of the equation. As mentioned previously, the identity matrix acts as a counter so that the summation on the left-hand-side of the equation is the total number of visits that the chain makes to state j. Thus, the equality indicates that the fraction of time spent in State j is equal to the steady-state probability of being in State j. This property is called the *Ergodic Property*. The last property indicates that the reciprocal of the long-run probabilities equals the expected number of steps to return to the state. Intuitively, this is as one would expect it, since the higher the probability, the quicker the return.

Example 5.11. We consider again the Markov chain of Example 5.10. In order to determine its steady-state behavior, the first step is to rearrange the matrix as in Eq. (5.8). The irreducible sets were identified in Example 5.10 and on that basis we order the state space as b, g, c, d, f, a, e. The Markov matrix thus becomes

$$\mathbf{P} = \begin{array}{c} b \\ g \\ c \\ d \\ f \\ a \\ e \end{array} \left[\begin{array}{cc|ccc|cc} 0.5 & 0.5 & & & & & \\ 0.8 & 0.2 & & & & & \\ \hline & & 0.4 & 0.6 & 0 & & \\ & & 0.3 & 0.2 & 0.5 & & \\ & & 1 & 0 & 0 & & \\ \hline 0.1 & 0 & 0.2 & 0.2 & 0.1 & 0.3 & 0.1 \\ 0 & 0 & 0.2 & 0.3 & 0.1 & 0 & 0.4 \end{array} \right].$$

(Blank blocks in a matrix are always interpreted to be zeroes.) The steady-state probabilities of being in States b or g are found by solving

$$\begin{aligned} 0.5\pi_b + 0.8\pi_g &= \pi_b \\ 0.5\pi_b + 0.2\pi_g &= \pi_g \\ \pi_b + \quad \pi_g &= 1\,. \end{aligned}$$

Thus, after deleting one of the first two equations, we solve the two-by-two system and obtain $\pi_b = 8/13$ and $\pi_g = 5/13$. Of course if the chain starts in State c, d, or f the long-run probability of being in State b or g is zero, because it is impossible to reach the irreducible set $\{b, g\}$ from any state in the irreducible set $\{c, d, f\}$. The steady-state probabilities of being in State c, d, or f starting in that set is given by the solution to

$$\begin{aligned} 0.4\pi_c + 0.3\pi_d + \pi_f &= \pi_c \\ 0.6\pi_c + 0.2\pi_d \qquad &= \pi_d \\ 0.5\pi_d \qquad &= \pi_f \\ \pi_c + \quad \pi_d + \pi_f &= 1 \end{aligned}$$

which yields $\pi_c = 8/17$, $\pi_d = 6/17$, and $\pi_f = 3/17$. (It might be noticed that the reasonable person would delete the first equation before solving the above system.) As a final point, assume the chain is in State c, and we wish to know the expected number of steps until the chain returns to State c. Since $\pi_c = 8/17$, it follows that the expected number of steps until the first return is $17/8$. □

Example 5.12. The market analysis discussed in Example 5.5 gave switching probabilities for body styles and associated profits with each body style. The long behavior for a customer can be estimated by calculating the long-run probabilities. The following system of equations

$$\begin{aligned} 0.36\pi_s + 0.60\pi_w \qquad\qquad &= \pi_w \\ 0.09\pi_s \qquad\qquad + 0.25\pi_c &= \pi_c \\ \pi_s + \quad \pi_w + \quad\quad \pi_c &= 1 \end{aligned}$$

is solved to obtain $\pi = (0.495, 0.446, 0.059)$. Therefore, the long-run expected profit per customer trade-in is

$$0.495 \times 1200 + 0.446 \times 1500 + 0.059 \times 2500 = \$1,410.5 \, .$$

□

- *Suggestion: Do Problems 5.10d, 5.11g–i, and 5.15.*

5.5 Computations

The determination of **R** (Eq. 5.5) is straightforward for recurrent states. If States i and j are in the same irreducible set, then $R(i,j) = \infty$. If i is recurrent and j is either transient or in a different irreducible set than i, then $R(i,j) = F(i,j) = 0$. For i transient and j recurrent, then $R(i,j) = \infty$ if $F(i,j) > 0$ and $R(i,j) = 0$ if $F(i,j) = 0$. In the case where i and j are transient, we have to do a little work as is given in the following property.

Property 5.11. *Let $X = \{X_n; n = 0, 1, \cdots\}$ be a Markov chain and let A denote the (finite) set of all transient states. Let* **Q** *be the matrix of transition probabilities restricted to the set A, then, for $i, j \in A$*

$$R(i,j) = (\mathbf{I} - \mathbf{Q})^{-1}(i,j) \, .$$

Continuing with Example 5.11, we first note that **Q** is

$$\mathbf{Q} = \begin{bmatrix} 0.3 & 0.1 \\ 0 & 0.4 \end{bmatrix}$$

and evaluating $(\mathbf{I} - \mathbf{Q})^{-1}$ we have

$$(\mathbf{I} - \mathbf{Q})^{-1} = \begin{bmatrix} 10/7 & 10/42 \\ 0 & 10/6 \end{bmatrix}$$

which yields

$$\mathbf{R} = \left[\begin{array}{cc|ccc|cc} \infty & \infty & 0 & 0 & 0 & 0 & 0 \\ \infty & \infty & 0 & 0 & 0 & 0 & 0 \\ \hline 0 & 0 & \infty & \infty & \infty & 0 & 0 \\ 0 & 0 & \infty & \infty & \infty & 0 & 0 \\ 0 & 0 & \infty & \infty & \infty & 0 & 0 \\ \hline \infty & \infty & \infty & \infty & \infty & 10/7 & 10/42 \\ 0 & 0 & \infty & \infty & \infty & 0 & 10/6 \end{array}\right].$$

The calculations for the matrix **F** are slightly more complicated than for **R**. The matrix **P** must be rewritten so that each irreducible, recurrent set is treated as a single "super" state. Once the Markov chain gets into an irreducible, recurrent set, it will remain in that set forever and all states within the set will be visited infinitely often. Therefore, in determining the probability of reaching a recurrent state from a transient state, it is only necessary to find the probability of reaching the appropriate irreducible set. The transition matrix in which each irreducible set is treated as a single state is denoted by $\hat{\mathbf{P}}$. The matrix $\hat{\mathbf{P}}$ has the form

$$\hat{\mathbf{P}} = \begin{bmatrix} 1 & & & & \\ & 1 & & & \\ & & 1 & & \\ & & & \ddots & \\ \mathbf{b}_1 & \mathbf{b}_2 & \mathbf{b}_3 & \cdots & \mathbf{Q} \end{bmatrix}, \tag{5.9}$$

where $\mathbf{b}_\ell$ is a vector giving the one-step probability of going from transient state i to irreducible set ℓ, that is,

$$b_\ell(i) = \sum_{j \in C_\ell} P(i,j)$$

for i transient and C_ℓ denoting the ℓth irreducible set.

Property 5.12. *Let $\{X_n; n = 0, 1, \cdots\}$ be a Markov chain with a finite state space ordered so that its Markov matrix can be reduced to the form in Eq. (5.9). Then for a transient State i and a recurrent State j, we have*

$$F(i,j) = ((\mathbf{I} - \mathbf{Q})^{-1}\mathbf{b}_\ell)(i)$$

for each State j in the ℓth irreducible set.

We again return to Example 5.11 to illustrate the calculations for **F**. We will also take advantage of Property 5.5 to determine the values in the lower right-hand portion **F**. Note that for i and j both transient, Property 5.5 can be rewritten as

$$F(i,j) = \begin{cases} 1 - \frac{1}{R(j,j)} & \text{for } i = j, \\ \frac{R(i,j)}{R(j,j)} & \text{for } i \neq j. \end{cases} \tag{5.10}$$

Using Property 5.12, the following holds:

$$\hat{\mathbf{P}} = \begin{bmatrix} 1 & 0 & 0 & 0 \\ 0 & 1 & 0 & 0 \\ 0.1 & 0.5 & 0.3 & 0.1 \\ 0 & 0.6 & 0 & 0.4 \end{bmatrix},$$

$$(\mathbf{I}-\mathbf{Q})^{-1}\mathbf{b}_1 = \begin{bmatrix} 1/7 \\ \\ 0 \end{bmatrix},$$

$$(\mathbf{I}-\mathbf{Q})^{-1}\mathbf{b}_2 = \begin{bmatrix} 6/7 \\ \\ 1 \end{bmatrix};$$

thus using Property 5.12 and Eq. (5.10), we have

$$\mathbf{F} = \left[\begin{array}{cc|ccc|cc} 1 & 1 & 0 & 0 & 0 & 0 & 0 \\ 1 & 1 & 0 & 0 & 0 & 0 & 0 \\ \hline 0 & 0 & 1 & 1 & 1 & 0 & 0 \\ 0 & 0 & 1 & 1 & 1 & 0 & 0 \\ 0 & 0 & 1 & 1 & 1 & 0 & 0 \\ \hline 1/7 & 1/7 & 6/7 & 6/7 & 6/7 & 3/10 & 1/7 \\ 0 & 0 & 1 & 1 & 1 & 0 & 4/10 \end{array}\right].$$

We can now use Property 5.10 and the previously computed steady-state probabilities to finish the limiting probability matrix for this example; namely,

$$\lim_{n\to\infty} \mathbf{P}^n = \left[\begin{array}{cc|ccc|cc} 8/13 & 5/13 & 0 & 0 & 0 & 0 & 0 \\ 8/13 & 5/13 & 0 & 0 & 0 & 0 & 0 \\ \hline 0 & 0 & 8/17 & 6/17 & 3/17 & 0 & 0 \\ 0 & 0 & 8/17 & 6/17 & 3/17 & 0 & 0 \\ 0 & 0 & 8/17 & 6/17 & 3/17 & 0 & 0 \\ \hline 8/91 & 5/91 & 48/119 & 36/119 & 18/119 & 0 & 0 \\ 0 & 0 & 8/17 & 6/17 & 3/17 & 0 & 0 \end{array}\right].$$

The goal of many modeling projects is the determination of revenues or costs. The ergodic property (the last item in Property 5.10) of an irreducible recurrent Markov chain gives an easy formula for the long run average return for the process. In Example 5.1, assume that each day the tractor is running, a profit of \$100 is realized; however, each day it is in the repair shop, a cost of \$25 is incurred. The Markov matrix of Example 5.1 yields steady-state results of $\pi_0 = 1/11$ and $\pi_1 = 10/11$; thus, the daily average return in the long run is

$$-25 \times \frac{1}{11} + 100 \times \frac{10}{11} = 88.64 .$$

This intuitive result is given in the following property.

Property 5.13. *Let $X = \{X_n; n = 0, 1, \cdots\}$ be an irreducible Markov chain with finite state space E and with steady-state probabilities given by the vector $\boldsymbol{\pi}$. Let the vector* **f** *be a profit function (i.e., $f(i)$ is the profit received for each visit to State i). Then (with probability one) the long-run average profit per unit of time is*

$$\lim_{n\to\infty} \frac{1}{n} \sum_{k=0}^{n-1} f(X_k) = \sum_{j\in E} \pi(j) f(j) .$$

Example 5.13. We return to the simplified manufacturing system of Example 5.8. Refer again to the diagram in Fig. 5.7 and the Markov matrix of Eq. 5.6. The cost structure for the process is as follows: The cost of the raw material going into Step 1 is \$150; each time a part is processed through Step 1 a cost of \$200 is incurred; and every time a part is processed through Step 2 a cost of \$300 is incurred. (Thus if a part is sold that was reworked once in Step 1 but was not reworked in Step 2, that part would have \$850 of costs associated with it.) Because the raw material is toxic, there is also a disposal cost of \$50 per part sent to scrap.

Each day, we start with enough raw material to make 100 parts so that at the end of each day, 100 parts are finished: some good, some scraped. In order to establish a reasonable strategy for setting the price of the parts to be sold, we first must determine the cost that should be attributed to the parts. In order to answer the relevant questions, we will first need the **F** and **R** matrices, which are

$$\mathbf{F} = \begin{array}{c} 1 \\ 2 \\ s \\ g \end{array} \begin{bmatrix} 0.239 & 0.875 & 0.183 & 0.817 \\ 0.056 & 0.144 & 0.066 & 0.934 \\ 0.0 & 0.0 & 1.0 & 0.0 \\ 0.0 & 0.0 & 0.0 & 1.0 \end{bmatrix}$$

$$\mathbf{R} = \begin{array}{c} 1 \\ 2 \\ s \\ g \end{array} \begin{bmatrix} 1.31 & 1.02 & \infty & \infty \\ 0.07 & 1.17 & \infty & \infty \\ 0.0 & 0.0 & \infty & 0.0 \\ 0.0 & 0.0 & 0.0 & \infty \end{bmatrix} .$$

An underlying assumption for this model is that each part that begins the manufacturing process is an independent entity, so that the actual number of finished "good" parts is a random variable following a binomial distribution. As you recall, the binomial distribution needs two parameters, the number of trials and the probability of success. The number of trials is given in the problem statement as 100; the probability of a success is the probability that a part which starts in State 1 ends in State g. Therefore, the expected number of good parts at the end of the day is

$$100 \times F(1,g) = 81.7\,.$$

The cost per part started is given by

$$150 + 200 \times R(1,1) + 300 \times R(1,2) + 50 \times F(1,s) = 727.15\,.$$

Therefore, the cost which should be associated to each part sold is

$$(100 \times 727.15)/81.7 = \$890/\text{part sold}\,.$$

A rush order for 100 parts from a very important customer has just been received, and there are no parts in inventory. Therefore, we wish to start tomorrow's production with enough raw material to be 95% confident that there will be at least 100 good parts at the end of the day. How many parts should we plan on starting tomorrow morning? Let the random variable N_n denote the number of finished parts that are good given that the day started with enough raw material for n parts. From the above discussion, the random variable N_n has a binomial distribution where n is the number of trials and $F(1,g)$ is the probability of success. Therefore, the question of interest is to find n such that $\Pr\{N_n \geq 100\} = 0.95$. Hopefully, you also recall that the binomial distribution can be approximated by the normal distribution; therefore, define X to be a normally distributed random variable with mean $nF(1,g)$ and variance $nF(1,g)F(1,s)$. We now have the following equation:

$$\begin{aligned}\Pr\{N_n \geq 100\} &\approx \Pr\{X > 99.5\} \\ &= \Pr\{Z > (99.5 - 0.817n)/\sqrt{0.1495n}\} = 0.95\end{aligned}$$

where Z is normally distributed with mean zero and variance one. Thus, using either standard statistical tables or the Excel function `=NORMSINV(0.05)`, we have that

$$\frac{99.5 - 0.817n}{\sqrt{0.1495n}} = -1.645\,.$$

The above equation is solved for n. Since it becomes a quadratic equation, there are two roots: $n_1 = 113.5$ and $n_2 = 130.7$. We must take the second root (why?), and we round up; thus, the day must start with enough raw material for 131 parts. □

Example 5.14. A missile is launched and, as it is tracked, a sequence of course correction signals are sent to it. Suppose that the system has four states that are labeled as follows.

State 0: on-course, no further correction necessary
State 1: minor deviation
State 2: major deviation
State 3: abort, off-course so badly a self-destruct signal is sent

Let X_n represent the state of the system after the nth course correction and assume that the behavior of X can be modeled by a Markov chain with the following probability transition matrix:

$$\mathbf{P} = \begin{matrix} 0 \\ 1 \\ 2 \\ 3 \end{matrix} \begin{bmatrix} 1.0 & 0.0 & 0.0 & 0.0 \\ 0.5 & 0.25 & 0.25 & 0.0 \\ 0.0 & 0.5 & 0.25 & 0.25 \\ 0.0 & 0.0 & 0.0 & 1.0 \end{bmatrix} .$$

As always, the first step is to classify the states. Therefore observe that states 0 and 3 are absorbing, and states 1 and 2 are transient. After a little work, you should be able to obtain the following matrices:

$$\mathbf{F} = \begin{matrix} 0 \\ 1 \\ 2 \\ 3 \end{matrix} \begin{bmatrix} 1.0 & 0.0 & 0.0 & 0.0 \\ 0.857 & 0.417 & 0.333 & 0.143 \\ 0.571 & 0.667 & 0.417 & 0.429 \\ 0.0 & 0.0 & 0.0 & 1.0 \end{bmatrix}$$

$$\mathbf{R} = \begin{matrix} 0 \\ 1 \\ 2 \\ 3 \end{matrix} \begin{bmatrix} \infty & 0.0 & 0.0 & 0.0 \\ \infty & 1.714 & 0.571 & \infty \\ \infty & 1.143 & 1.714 & \infty \\ 0.0 & 0.0 & 0.0 & \infty \end{bmatrix} .$$

Suppose that upon launch, the missile starts in State 2. The probability that it will eventually get on-course is 57.1% (namely, F(2,0)); whereas, the probability that it will eventually have to be destroyed is 42.9%. When a missile is launched, 50,000 pounds of fuel are used. Every time a minor correction is made, 1,000 pounds of fuel are used; and every time a major correction is made, 5,000 pounds of fuel are used. Assume that the missile started in State 2, we wish to determine the expected fuel usage for the mission. This calculation is

$$50000 + 1000 \times R(2,1) + 5000 \times R(2,2) = 59713 .$$

□

The decision maker sometimes uses a discounted cost criterion instead of average costs over an infinite planning horizon. To determine the expected total discounted return for a Markov chain, we have the following.

Property 5.14. *Let $X = \{X_n; n = 0, 1, \cdots\}$ be a Markov chain with Markov matrix* **P**. *Let* **f** *be a return function and let α $(0 < \alpha < 1)$ be a discount factor. Then the expected total discounted cost is given by*

$$E\left[\sum_{n=0}^{\infty} \alpha^n f(X_n) | X_0 = i\right] = ((\mathbf{I} - \alpha\mathbf{P})^{-1}\mathbf{f})(i) .$$

We finish Example 5.11 by considering a profit function associated with it. Let the state space for the example be $E = \{1,2,3,4,5,6,7\}$; that is, State 1 is State b,

State 2 is State g, etc. Let $\mathbf{f}$ be the profit function with $f = (500, 450, 400, 350, 300, 350, 200)$; that is, each visit to State 1 produces \$500, each visit to State 2 produces \$450, etc. If the chain starts in States 1 or 2, then the long-run average profit per period is

$$500 \times \frac{8}{13} + 450 \times \frac{5}{13} = \$480.77\,.$$

If the chain starts in States 3, 4, or 5, then the long-run average profit per period is

$$400 \times \frac{8}{17} + 350 \times \frac{6}{17} + 300 \times \frac{3}{17} = \$364.71\,.$$

If the chain starts in State 6, then the expected value of the long-run average would be a weighted average of the two ergodic results, namely,

$$480.77 \times \frac{1}{7} + 364.71 \times \frac{6}{7} = \$381.29\,,$$

and starting in State 7 gives the same long-run result as starting in States 3, 4, or 5.

Now assume that a discount factor should be used. In other words, our criterion is total discounted profit instead of long-run average profit per period. Let the monetary rate of return be 20% per period which gives $\alpha = 1/1.20 = 5/6$. (In other words, one dollar one period in the future is equivalent to 5/6 dollar, or 83.33 cents, now.) Then

$$(\mathbf{I} - \alpha \mathbf{P})^{-1} = \left[\begin{array}{cc|ccc|cc} 4.0 & 2.0 & 0 & 0 & 0 & 0 & 0 \\ 3.2 & 2.8 & 0 & 0 & 0 & 0 & 0 \\ \hline 0 & 0 & 3.24 & 1.95 & 0.81 & 0 & 0 \\ 0 & 0 & 2.32 & 2.59 & 1.08 & 0 & 0 \\ 0 & 0 & 2.70 & 1.62 & 1.68 & 0 & 0 \\ \hline 0.44 & 0.22 & 1.76 & 1.37 & 0.70 & 1.33 & 0.17 \\ 0 & 0 & 2.02 & 1.66 & 0.82 & 0 & 1.50 \end{array}\right].$$

Denote the total discounted return when the chain starts in state i by $h(i)$. Then, multiplying the above matrix by the profit function $\mathbf{f}$ yields

$$\mathbf{h}^T = (2900, 2860, 2221.5, 2158.5, 2151, 2079, 1935)\,.$$

Thus, if the chain started in State 1, the present worth of all future profits would be \$2,900. If the initial state is unknown and a distribution of probabilities for the starting state is given by $\boldsymbol{\mu}$, the total discounted profit is given by $\boldsymbol{\mu}\mathbf{h}$.

- *Suggestion: Do Problems 5.13d–f, 5.14e–g, and finish 5.1–5.9.*

Appendix

We close the chapter with a brief look at the use of Excel for analyzing Markov chains. We will consider both the matrix operations involved in the computations of Section 5.5 and the use of random numbers for simulating Markov chains.

Matrix Operations in Excel. The application of Property 5.11 to the Markov chain of Example 5.11 requires the determination of $(\mathbf{I}-\mathbf{Q})^{-1}$ as shown on p. 162. There is more than one way to calculate this in Excel, but they all involve the use of the function `MINVERSE(`*array*`)`. In Excel, whenever the output of a function is an array it is important to follow Excel's special instructions for array results: (1) highlight the cells in which you want the array to appear, (2) type in the equal sign (=) followed by the formula, and (3) hold down the `<shift>` and `<ctrl>` keys while hitting the `<enter>` key. To demonstrate this, type the following on a spreadsheet:

	A	B	C	D	E
1	Q matrix			Identity	
2	0.3	0.1		1	0
3	0.0	0.4		0	1

(Notice that as a matter of personal preference, parameters are usually labeled and blank columns or rows are placed between separate quantities.) Now to obtain the inverse of $(\mathbf{I}-\mathbf{Q})$, first, highlight the cells A5:B6; second, type

```
=MINVERSE(D2:E3 - A2:B3)
```

(note that although the range A5:B6 was highlighted, the above formula should occur in Cell A5); finally, while holding down the `<shift>` and `<ctrl>` keys hit the `<enter>` key. After hitting the `<enter>` key, the $(\mathbf{I}-\mathbf{Q})^{-1}$ matrix should appear in the A5:B6 range. To try an alternate method, highlight the cells A8:B9, type

`=MINVERSE({1,0;0,1} - A2:B3)`, and

while holding down the `<shift>` and `<ctrl>` keys hit the `<enter>` key. The obvious advantage of the second method is that the identity matrix need not be explicitly entered on the spreadsheet.

Steady-state probabilities for an irreducible, recurrent Markov chain can also be easily obtained through Excel, but first we re-write the equation $\boldsymbol{\pi}\mathbf{P}=\boldsymbol{\pi}$ as $\boldsymbol{\pi}(\mathbf{I}-\mathbf{P})=\mathbf{0}$. Recall from Property 5.9, that the norming equation must be used and one of the equations from the system $\boldsymbol{\pi}(\mathbf{I}-\mathbf{P})=\mathbf{0}$ must be dropped, which is equivalent to deleting one of the *columns* of $(\mathbf{I}-\mathbf{P})$ and replacing it with a column of all ones. As we do this in Excel, it is the last column that will be dropped.

Example 5.15. Consider a four-state Markov chain with Markov matrix defined by Eq. (5.7). Since it has a finite state space and all states communicate with all other states, it is an irreducible, recurrent Markov chain so that steady-state probabilities can be computed using Property 5.9. Type the following in a spreadsheet:

	A	B	C	D
1		P matrix		
2	0.5	0.5	0	0
3	0.2	0.4	0.4	0
4	0	0.1	0.8	0.1
5	0	0	0.3	0.7
6				
7		Modified I-P matrix		

Next, we need the matrix $(\mathbf{I}-\mathbf{P})$ with the last column replaced by all ones to appear in Cells A8:D11. To accomplish this, highlight Cells A8:C11, type

```
={1,0,0;0,1,0;0,0,1;0,0,0}-A2:C5
```

(again note that although a block of 12 cells were highlighted, the typing occurs in Cell A8), and while holding down the `<shift>` and `<ctrl>` keys hit the `<enter>` key. Finally, place ones in Cells D8:D11.

With the modified $(\mathbf{I}-\mathbf{P})$ matrix, we simply pre-multiply the right-hand side vector by its inverse, where the right-hand side is a vector of all zeros except the last element which equals one (from the norming equation). Center the words `"Steady state probabilities"` across Cells A13:D13 in order to label the quantities to be computed, highlight Cells A14:D14, type

```
=MMULT({0,0,0,1},MINVERSE(A8:D11))
```

(although a block of 4 cells were highlighted, the typing occurs in Cell A14), and while holding down the `<shift>` and `<ctrl>` keys hit the `<enter>` key. Thus, the steady-state probabilities appear in highlighted cells. □

Simulation of a Markov Chain. We close this appendix with an Excel simulation of a Markov chain. The key to simulating a Markov chain is to use the techniques of Sect. 2.3.1 together with conditional probabilities given in the Markov matrix. The Excel implementation of this was illustrated in Example 2.12.

Example 5.16. Consider a Markov chain with state space $\{a,b,c\}$ and with the following transition matrix:

$$\mathbf{P} = \begin{bmatrix} 0.8 & 0.1 & 0.1 \\ 0.4 & 0.6 & 0.0 \\ 0.0 & 0.0 & 1.0 \end{bmatrix}.$$

We know that the expected number of steps taken until the chain is absorbed into State c given that it starts in State a is $R(a,a)+R(a,b) = 12.5$ (Eq. 5.5). However, let us assume that as a new student of Markov chains, our confidence level is low in asserting that the mean absorption time starting from State a is 12.5 (or equivalently, $E[T^c|X_0 = a] = 12.5$). One of the uses of simulation is as a confidence builder; therefore, we wish to simulate this process to help give some evidence that our calculations are correct. In other words, our goal is to simulate this Markov chain starting from State a and recording the number of steps taken until reaching State c.

Prepare a spreadsheet so that the first two rows are as follows:

	A	B	C	D
1	Time Step	Current State	Random Number	Number of Steps
2	0	a	=RAND()	=MATCH("c",B3:B200, 0)

For Cell A3, type =A2+1; for Cell B3, type

```
=IF(B2="a",IF(C2<0.8,"a",IF(C2<0.9,"b","c")),
    IF(B2="b",IF(C2<0.4,"a","b"),"c"))
```

and for Cell C3, type =RAND(). Now copy Cells A3:C3 down through row 200.

Notice that each time the F9 key is pressed, a new value appears in Cell D2. These values represent independent realizations of the random variable T^c given that $X_0 = a$. Obviously, to obtain an estimate for $E[T^c|X_0 = a]$, an average of several realizations are necessary. □

The above example can be modified slightly so that one column produces a random sample containing many realizations of $E[T^c|X_0 = a]$. Change Cell A3 to read =IF(B2="c",0,A2+1) (note that Cell A3 makes reference to B2, not B3), change the last entry in the IF statement of Cell B3 to be "a" instead of "c", change Cell D3 to be =IF(B3="c", A3,"--"), and then copy Cells A3:D3 down through row 30000. The numerical values contained in Column D represent various realizations of the random variable T^c so we now change the formula in D2 to be =AVERAGE(D3:D30000) and thus the sample mean becomes the estimator for $E[T^c|X_0 = a]$.

Problems

The exercises listed below are to help in your understanding of Markov chains and to indicate some of the potential uses for Markov chain analysis. All numbers are fictitious.

5.1. Joe and Pete each have two cents in their pockets. They have decided to match pennies; that is, they will each take one of their own pennies and flip them. If the pennies match (two heads or two tails), Joe gets Pete's penny; if the pennies do not match, Pete gets Joe's penny. They will keep repeating the game until one of them has four cents, and the other one is broke. Although they do not realize it, all four pennies are biased. The probability of tossing a head is 0.6, and the probability of a tail is 0.4. Let X be a Markov chain where X_n denotes the amount that Joe has after the nth play of the game.
(a) Give the Markov matrix for X.
(b) What is the probability that Joe will have four pennies after the second toss?
(c) What is the probability that Pete will be broke after three tosses?
(d) What is the probability that the game will be over before the third toss?
(e) What is the expected amount of money Pete will have after two tosses?

(f) What is the probability that Pete will end up broke?
(g) What is the expected number of tosses until the game is over?

5.2. At the start of each week, the condition of a machine is determined by measuring the amount of electrical current it uses. According to its amperage reading, the machine is categorized as being in one of the following four states: low, medium, high, failed. A machine in the low state has a probability of 0.05, 0.03, and 0.02 of being in the medium, high, or failed state, respectively, at the start of the next week. A machine in the medium state has a probability of 0.09 and 0.06 of being in the high or failed state, respectively, at the start of the next week (it cannot, by itself, go to the low state). And, a machine in the high state has a probability of 0.1 of being in the failed state at the start of the next week (it cannot, by itself, go to the low or medium state). If a machine is in the failed state at the start of a week, repair is immediately begun on the machine so that it will (with probability 1) be in the low state at the start of the following week. Let X be a Markov chain where X_n is the state of the machine at the start of week n.
(a) Give the Markov matrix for X.
(b) A new machine always starts in the low state. What is the probability that the machine is in the failed state three weeks after it is new?
(c) What is the probability that a machine has at least one failure three weeks after it is new?
(d) On the average, how many weeks per year is the machine working?
(e) Each week that the machine is in the low state, a profit of $1,000 is realized; each week that the machine is in the medium state, a profit of $500 is realized; each week that the machine is in the high state, a profit of $400 is realized; and the week in which a failure is fixed, a cost of $700 is incurred. What is the long-run average profit per week realized by the machine?
(f) A suggestion has been made to change the maintenance policy for the machine. If at the start of a week the machine is in the high state, the machine will be taken out of service and repaired so that at the start of the next week it will again be in the low state. When a repair is made due to the machine being in the high state instead of a failed state, a cost of $600 is incurred. Is this new policy worthwhile?

5.3. We are interested in the movement of patients within a hospital. For purposes of our analysis, we shall consider the hospital to have three different types of rooms: general care, special care, and intensive care. Based on past data, 60% of arriving patients are initially admitted into the general care category, 30% in the special care category, and 10% in intensive care. A "general care" patient has a 55% chance of being released healthy the following day, a 30% of remaining in the general care room, and a 15% of being moved to the special care facility. A "special care" patient has a 10% chance of being released the following day, a 20% chance of being moved to general care, a 10% chance of being upgraded to intensive care, and a 5% chance of dying during the day. An "intensive care" patient is never released from the hospital directly from the intensive care unit (ICU), but is always moved to another facility first. The probabilities that the patient is moved to general care, special care, or remains in intensive care are 5%, 30%, or 55%, respectively. Let X

be a Markov chain where X_0 is the type of room that an admitted patient initially uses, and X_n is the room category of that patient at the end of day n.
(a) Give the Markov matrix for X.
(b) What is the probability that a patient admitted into the intensive care room eventually leaves the hospital healthy?
(c) What is the expected number of days that a patient, admitted into intensive care, will spend in the ICU?
(d) What is the expected length of stay for a patient admitted into the hospital as a general care patient?
(e) During a typical day, 100 patients are admitted into the hospital. What is the average number of patients in the ICU?

5.4. Consider again the manufacturing process of Example 5.8. New production plans call for an expected production level of 2000 good parts per month. (In other words, enough raw material must be used so that the expected number of good parts produced each month is 2000.) For a capital investment and an increase in operating costs, all rework and scrap can be eliminated. The sum of the capital investment and operating cost increase is equivalent to an annual cost of $5 million. Is it worthwhile to increase the annual cost by $5 million in order to eliminate the scrap and rework?

5.5. Assume the description of the manufacturing process of Example 5.8 is for the process at Branch A of the manufacturing company. Branch B of the same company has been closed. Branch B had an identical process and they had 1000 items in stock that had been through Step 1 of the process when they were shut down. Because the process was identical, these items can be fed into Branch A's process at the start of Step 2. (However, since the process was identical there is still a 5% chance that after finishing Step 2 the item will have to be reworked at Step 1, a 10% chance the item will have to be reworked at Step 2, and a 5% chance that the item will have to be scrapped.) Branch A purchases these (partially completed) items for a total of $300,000, and they will start processing at Step 2 of Branch A's system. After Branch A finishes processing this batch of items, they must determine the cost of these items so they will know how much to charge customers in order to recover the cost. (They may want to give a discount.) Your task is to determine the cost that would be attributed to each item shipped.

5.6. The manufacture of a certain type of electronic board consists of four steps: tinning, forming, insertion, and solder. After the forming step, 5% of the parts must be retinned; after the insertion step, 20% of the parts are bad and must be scrapped; and after the solder step, 30% of the parts must be returned to insertion and 10% must be scrapped. (We assume that when a part is returned to a processing step, it is treated like any other part entering that step.) Figure 5.12 gives a schematic showing the flow of a job through the manufacturing steps.
(a) Model this process using Markov chains and give its Markov matrix.
(b) If a batch of 100 boards begins this manufacturing process, what is the expected number that will end up scraped?
(c) How many boards should we start with if the goal is to have the expected number

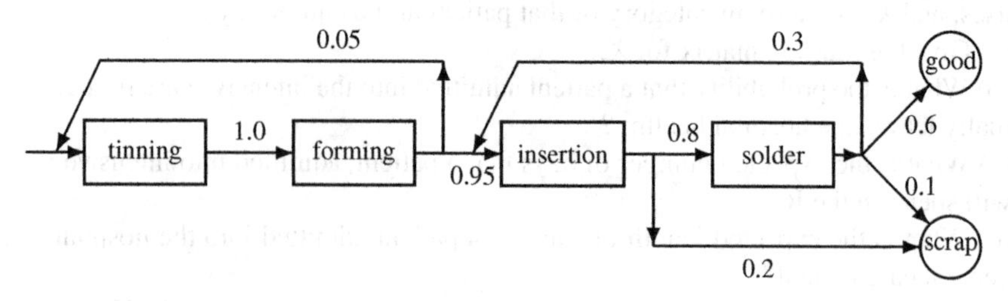

Fig. 5.12 *Manufacturing structure for board processing steps*

of boards that finish in the good category equal to 100?
(d) How many boards should we start with if we want to be 90% sure that we end up with a batch of 100 boards? (Hint: the final status of each board can be considered a Bernoulli random variable, the sum of independent Bernoullis is binomial, and the binomial can be approximated by a normal.)
(e) Each time a board goes through a processing step, direct labor and material costs are \$10 for tinning, \$15 for forming, \$25 for insertion, and \$20 for solder. The raw material costs \$8, and a scrapped board returns \$2. The average overhead rate is \$1,000,000 per year, which includes values for capital recovery. The average processing rate is 5,000 board starts per week. We would like to set a price per board so that expected revenues are 25% higher than expected costs. At what value would you set the price per board?

5.7. The government has done a study of the flow of money among three types of financial institutions: banks, savings associations, and credit unions. We shall assume that the Markov assumptions are valid and model the monthly movement of money among the various institutions. Some recent data are:

June amounts	Units in billions of dollars May bank	savings	credit	other	June totals
bank	10.0	4.5	1.5	3.6	19.6
savings	4.0	6.0	0.5	1.2	11.7
credit	2.0	3.0	6.0	2.4	13.4
other	4.0	1.5	2.0	4.8	12.3
May totals	20	15	10	12	57

For example, of the 20 billion dollars in banks during the month of May, half of it remained in banks for the month of June, and 2 out of the 20 billion left banks to be invested in credit unions during June.
(a) Use the above data to estimate a Markov matrix. What would be the problems with such a model? (That is, be critical of the Markov assumption of the model.)
(b) In the long-run, how much money would you expect to be in credit unions during any given month?

(c) How much money would you expect to be in banks during August of the same year that the above data were collected?

5.8. Within a certain market area, there are two brands of soap that most people use: "super soap" and "cheap soap", with the current market split evenly between the two brands. A company is considering introducing a third brand called "extra clean soap", and they have done some initial studies of market conditions. Their estimates of weekly shopping patterns are as follows: If a customer buys super soap this week, there is a 75% chance that next week the super soap will be used again, a 10% chance that extra clean will be used and a 15% chance that the cheap soap will be used. If a customer buys the extra clean this week, there is a fifty-fifty chance the customer will switch, and if a switch is made it will always be to super soap. If a customer buys cheap soap this week, it is equally likely that next week the customer will buy any of the three brands.
(a) Assuming that the Markov assumptions are good, use the above data to estimate a Markov matrix. What would be the problems with such a model?
(b) What is the long-run market share for the new soap?
(c) What will be the market share of the new soap two weeks after it is introduced?
(d) The market consists of approximately one million customers each week. Each purchase of super soap yields a profit of 15 cents; a purchase of cheap soap yields a profit of 10 cents; and a purchase of extra clean will yield a profit of 25 cents. Assume that the market was at steady-state with the even split between the two products. The initial advertising campaign to introduce the new brand was $100,000. How many weeks will it be until the $100,000 is recovered from the added revenue of the new product?
(e) The company feels that with these three brands, an advertising campaign of $30,000 per week will increase the weekly total market by a quarter of a million customers? Is the campaign worthwhile? (Use a long-term average criterion.)

5.9. A small company sells high quality laser printers and they use a simple periodic inventory ordering policy. If there are two or fewer printers in inventory at the end of the day on Friday, the company will order enough printers so that there will be five printers in stock at the start of Monday. (It only takes the weekend for printers to be delivered from the wholesaler.) If there are more than two printers at the end of the week, no order is placed. Weekly demand data has been analyzed yielding the probability mass function for weekly demand as $\Pr\{D=0\}=0.05$, $\Pr\{D=1\}=0.1$, $\Pr\{D=2\}=0.2$, $\Pr\{D=3\}=0.4$, $\Pr\{D=4\}=0.1$, $\Pr\{D=5\}=0.1$, and $\Pr\{D=6\}=0.05$. Let X be a Markov chain where X_n is the inventory at the end of week n. (Note: backorders are not accepted.)
(a) Give the Markov matrix for X.
(b) If at the end of week 1 there were five items in inventory, what is the probability that there will be five items in inventory at the end of week 2?
(c) If at the end of week 1 there were five items in inventory, what is the probability that there will be five items in inventory at the end of week 3?
(d) What is the expected number of times each year that an order is placed?
(e) Each printer sells for $1,800. Each time an order is placed, it costs $500 plus

\$1,000 times the number of items ordered. At the end of each week, each unsold printer costs \$25 (in terms of keeping them clean, money tied up, and other such inventory type expenses). Whenever a customer wants a printer not in stock, the company buys it retail and sends it by airfreight to the customer; thus the customer spends the \$1,800 sales price but it costs the company \$1,900. In order reduce the lost sales, the company is considering raising the reorder point to three, but still keeping the order up to quantity at five. Would you recommend the change in the reorder point?

5.10. Let X be a Markov chain with state space $\{a,b,c\}$ and transition probabilities given by

$$\mathbf{P} = \begin{bmatrix} 0.3 & 0.4 & 0.3 \\ 1.0 & 0.0 & 0.0 \\ 0.0 & 0.3 & 0.7 \end{bmatrix}.$$

(a) What is $\Pr\{X_2 = a | X_1 = b\}$?
(b) What is $\Pr\{X_2 = a | X_1 = b, X_0 = c\}$?
(c) What is $\Pr\{X_{35} = a | X_{33} = a\}$?
(d) What is $\Pr\{X_{200} = a | X_0 = b\}$? (Use steady-state probability to answer this.)

5.11. Let X be a Markov chain with state space $\{a,b,c\}$ and transition probabilities given by

$$\mathbf{P} = \begin{bmatrix} 0.3 & 0.7 & 0.0 \\ 0.0 & 0.6 & 0.4 \\ 0.4 & 0.1 & 0.5 \end{bmatrix}.$$

Let the initial probabilities be given by the vector $(0.1, 0.3, 0.6)$ with a profit function given by $(10, 20, 30)$. (That means, for example each visit to state a yields a profit of \$10.) Find the following:
(a) $\Pr\{X_2 = b | X_1 = c\}$.
(b) $\Pr\{X_3 = b | X_1 = c\}$.
(c) $\Pr\{X_3 = b | X_1 = c, X_0 = c\}$.
(d) $\Pr\{X_2 = b\}$.
(e) $\Pr\{X_1 = b, X_2 = c | X_0 = c\}$.
(f) $E[f(X_2) | X_1 = c]$.
(g) $\lim_{n\to\infty} \Pr\{X_n = a | X_0 = a\}$.
(h) $\lim_{n\to\infty} \Pr\{X_n = b | X_{10} = c\}$.
(i) $\lim_{n\to\infty} E[f(X_n) | X_0 = c]$.

5.12. Let X be a Markov chain with state space $\{a,b,c,d\}$ and transition probabilities given by

$$\mathbf{P} = \begin{bmatrix} 0.1 & 0.3 & 0.6 & 0.0 \\ 0.0 & 0.2 & 0.5 & 0.3 \\ 0.5 & 0.0 & 0.0 & 0.5 \\ 0.0 & 1.0 & 0.0 & 0.0 \end{bmatrix}.$$

Each time the chain is in state a, a profit of \$20 is made; each visit to state b yields a \$5 profit; each visit to state c yields \$15 profit; and each visit to state d costs \$10.

Find the following:
(a) $E[f(X_5)|X_3 = c, X_4 = d]$.
(b) $E[f(X_1)|X_0 = b]$.
(c) $E[f(X_1)^2|X_0 = b]$.
(d) $V[f(X_1)|X_0 = b]$.
(e) $V[f(X_1)]$ with an initial probability vector of $\boldsymbol{\mu} = (0.2, 0.4, 0.3, 0.1)$.

5.13. Consider the following Markov matrix representing a Markov chain with state space $\{a, b, c, d, e\}$.

$$\mathbf{P} = \begin{bmatrix} 0.3 & 0.0 & 0.0 & 0.7 & 0.0 \\ 0.0 & 1.0 & 0.0 & 0.0 & 0.0 \\ 1.0 & 0.0 & 0.0 & 0.0 & 0.0 \\ 0.0 & 0.0 & 0.5 & 0.5 & 0.0 \\ 0.0 & 0.2 & 0.4 & 0.0 & 0.4 \end{bmatrix}.$$

(a) Draw the state diagram.
(b) List the transient states.
(c) List the irreducible set(s).
(d) Let $F(i, j)$ denote the first passage probabilities of reaching (or returning to) State j given that $X_0 = i$. Calculate the $\mathbf{F}$ matrix.
(e) Let $R(i, j)$ denote the expected number of visits to State j given that $X_0 = i$. Calculate the $\mathbf{R}$ matrix.
(f) Calculate the $\lim_{n\to\infty} \mathbf{P}^n$ matrix.

5.14. Let X be a Markov chain with state space $\{a, b, c, d, e, f\}$ and transition probabilities given by

$$\mathbf{P} = \begin{bmatrix} 0.3 & 0.5 & 0.0 & 0.0 & 0.0 & 0.2 \\ 0.0 & 0.5 & 0.0 & 0.5 & 0.0 & 0.0 \\ 0.0 & 0.0 & 1.0 & 0.0 & 0.0 & 0.0 \\ 0.0 & 0.3 & 0.0 & 0.0 & 0.0 & 0.7 \\ 0.1 & 0.0 & 0.1 & 0.0 & 0.8 & 0.0 \\ 0.0 & 1.0 & 0.0 & 0.0 & 0.0 & 0.0 \end{bmatrix}.$$

(a) Draw the state diagram.
(b) List the recurrent states.
(c) List the irreducible set(s).
(d) List the transient states.
(e) Calculate the $\mathbf{F}$ matrix.
(f) Calculate the $\mathbf{R}$ matrix.
(g) Calculate the $\lim_{n\to\infty} \mathbf{P}^n$ matrix.

5.15. Let X be a Markov chain with state space $\{a, b, c, d\}$ and transition probabilities given by

$$\mathbf{P} = \begin{bmatrix} 0.3 & 0.7 & 0.0 & 0.0 \\ 0.5 & 0.5 & 0.0 & 0.0 \\ 0.0 & 0.0 & 1.0 & 0.0 \\ 0.1 & 0.1 & 0.2 & 0.6 \end{bmatrix}.$$

(a) Find $\lim_{n\to\infty} \Pr\{X_n = a | X_0 = a\}$.
(b) Find $\lim_{n\to\infty} \Pr\{X_n = a | X_0 = b\}$.
(c) Find $\lim_{n\to\infty} \Pr\{X_n = a | X_0 = c\}$.
(d) Find $\lim_{n\to\infty} \Pr\{X_n = a | X_0 = d\}$.

5.16. Consider a Markov chain with state space $\{a,b,c\}$ and with the following transition matrix:

$$P = \begin{bmatrix} 0.35 & 0.27 & 0.38 \\ 0.82 & 0.00 & 0.18 \\ 1.00 & 0.00 & 0.00 \end{bmatrix}.$$

(a) Given that the Markov chain starts in State b, estimate through simulation the expected number of steps until the first return to State b.
(b) The expected return time to a state should equal the reciprocal of the long-run probability of being in that state. Estimate through simulation and analytically the steady-state probability and compare it to your answer to Part (a). Explain any differences.

5.17. Consider a Markov chain with state space $\{a,b,c\}$ and with the following transition matrix:

$$P = \begin{bmatrix} 0.3 & 0.5 & 0.2 \\ 0.1 & 0.2 & 0.7 \\ 0.8 & 0.0 & 0.2 \end{bmatrix}.$$

Each visit to State a results in a profit of \$5, each visit to State b results in a profit of \$10, and each visit to State c results in a profit of \$12. Write a computer program that will simulate the Markov chain so that an estimate of the expected profit per step can be made. Assume that the chain always starts in State a. The simulation should involve accumulating the profit from each step; then the estimate per simulation run is the cumulative profit divided by the number of steps in the run.
(a) Let the estimate be based on 10 replications, where each replication has 25 steps. The estimate is the average value over the 10 replicates. Record both the overall average and the range of the averages.
(b) Let the estimate be based on 10 replications, where each replication has 1000 steps. Compare the estimates and ranges for parts (a) and (b) and explain any differences.

References

1. Galton, F., and Watson, H.W. (1875). On the Probability of the Extinction of Families, *J. Anthropol. Soc. London*, **4**:138–144.

2. Miller, G.A. (1952). Finite Markov Processes in Psychology, *Psychometrika* **17**:149–167.
3. Parzen, E. (1962). *Stochastic Processes*, Holden-Day, Inc., Oakland, CA.

Chapter 6
Markov Processes

The key Markov property introduced with Markov chains is that the future is independent of the past given the present. A Markov process is simply the continuous analog to a chain, namely, the Markov property is maintained and time is measured continuously. For example, suppose we are keeping track of patients within a large hospital, and the state of our process is the specific department housing the patient. If we record the patient's location every day, the resulting process will be a Markov chain assuming that the Markov property holds. If, however, we record the patient's location continuously throughout the day, then the resulting process will be called a Markov process. Thus, instead of transitions only occurring at times 1, 2, 3, etc., transitions may occur at times 2.54 or 7.0193.

Our major reason for covering Markov processes is that they play a key role in queueing theory, and queueing theory has applications in many areas. (In particular, any system in which there may be a build up of waiting lines is a candidate for a queueing model.) Therefore, in this chapter, we give a cursory treatment of Markov processes in preparation for the next chapter dealing with queues.

6.1 Basic Definitions

The definition for Markov processes is similar to that for Markov chains except that the Markov property must be shown to hold for all future times.

Definition 6.1. The process $Y = \{Y_t; t \geq 0\}$ with finite state space E is a *Markov process* if the following holds for all $j \in E$ and $t, s \geq 0$

$$\Pr\{Y_{t+s} = j | Y_u; u \leq t\} = \Pr\{Y_{t+s} = j | Y_t\} .$$

Furthermore, the Markov process is said to be *stationary* if

$$\Pr\{Y_{t+s} = j | Y_t = i\} = \Pr\{Y_s = j | Y_0 = i\} .$$

□

R.M. Feldman, C. Valdez-Flores, *Applied Probability and Stochastic Processes*, 2nd ed., DOI 10.1007/978-3-642-05158-6_6, © Springer-Verlag Berlin Heidelberg 2010

These equations are completely analogous to those in Definition (5.1) except that for a Markov chain it is sufficient to show that the Markov property holds just for the next step; whereas, for Markov processes, we must show that the Markov property holds for all future times. To understand this definition, think of the present time as being time t. The left-hand side of the first equation in Definition (6.1) indicates that a prediction of the future s time units from now is desired given all the past history up to and including the current time t. The right-hand side of the equation indicates that the prediction is the same if the only information available is the state of the system at the present time. (Note that the implications of this definition is that it is a waste of resources to keep historical records of a process if it is Markovian.) The second equation of the definition is the time homogenous condition which indicates that the probability law governing the process does not change during the life of the process.

Example 6.1. A salesman lives in town 'a' and is responsible for towns 'a', 'b', and 'c'. The amount of time he spends in each town is random. After some study it has been determined that the amount of consecutive time spent in any one town follows an exponentially distributed random variable (see Eq. 1.14) with the mean time depending on the town. In his hometown, a mean time of two weeks is spent, in town 'b' a mean time of one week is spent, and in town 'c' a mean time of one and a half weeks is spent. When he leaves town 'a', he flips a coin to determine to which town he goes next; when he leaves either town 'b' or 'c', he flips two coins so that there is a 75% chance of returning to 'a' and a 25% chance of going to the other town. Let Y_t be a random variable denoting the town that the salesman is in at time t. The process $Y = \{Y_t; t \geq 0\}$ is a Markov process due to the lack of memory inherent in the exponential distribution (see Exercise 1.23). □

The construction of the Markov process in Example 6.1 is instructive in that it contains the major characterizations for any finite state Markov process. A Markov process remains in each state for an exponential length of time and then when it jumps, it jumps according to a Markov chain (see Fig. 6.1). To describe this characterization mathematically, we introduce some additional notation that should be familiar due to its similarity to the previous chapter. We first let $T_0 = 0$ and $X_0 = Y_0$. Then, we denote the successive jump times of the process by $T_1, T_2, \cdots$ and denote the state of the process immediately after the jumps as $X_1, X_2, \cdots$. To be more explicit, for $n = 1, 2, \cdots$,

$$T_{n+1} = \min\{t > T_n : Y_t \neq Y_{T_n}\} \text{ , and}$$

$$X_n = Y_{T_n} \text{ .}$$

The time between jumps (namely, $T_{n+1} - T_n$) is called the sojourn time. The Markov property implies that the only information needed for predicting the sojourn time is the current state. Or, to say it mathematically,

$$\Pr\{T_{n+1} - T_n \leq t | X_0, \cdots, X_n, T_0, \cdots, T_n\} = \Pr\{T_{n+1} - T_n \leq t | X_n\} .$$

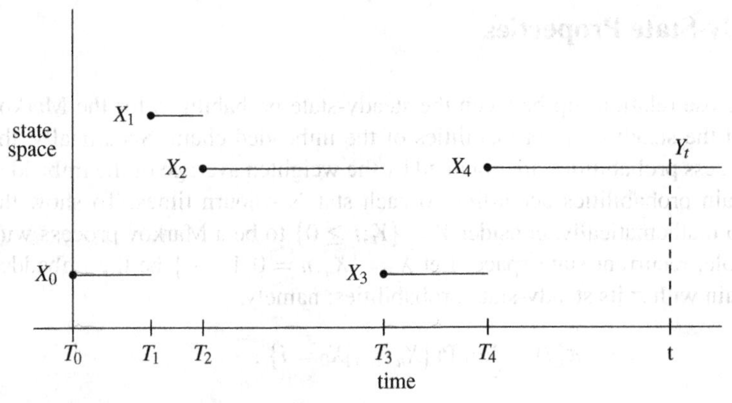

Fig. 6.1 Typical realization for a Markov process

The Markov property also implies that if we focus our attention only on the process $X_0, X_1, \cdots$ then we would see a Markov chain. This Markov chain is called the *imbedded Markov chain.*

Definition 6.2. Let $Y = \{Y_t; t \geq 0\}$ be a Markov process with finite state space E and with jump times denoted by $T_0, T_1, \cdots$ and the imbedded process at the jump times denoted by $X_0, X_1, \cdots$. Then there is a collection of scalars, $\lambda(i)$ for $i \in E$, called the *mean sojourn rates*, and a Markov matrix, $\mathbf{P}$, called the *imbedded Markov matrix*, that satisfy the following:

$$\Pr\{T_{n+1} - T_n \leq t | X_n = i\} = 1 - \mathrm{e}^{-\lambda(i)t},$$

$$\Pr\{X_{n+1} = j | X_n = i\} = P(i,j),$$

where each $\lambda(i)$ is nonnegative and the diagonal elements of $\mathbf{P}$ are zero. □

Example 6.2. Consider again Example 6.1. The imbedded Markov chain described by the above definition is

$$\mathbf{P} = \begin{matrix} a \\ b \\ c \end{matrix} \begin{bmatrix} 0 & 0.50 & 0.50 \\ 0.75 & 0 & 0.25 \\ 0.75 & 0.25 & 0 \end{bmatrix},$$

and the mean sojourn rates are $\lambda(a) = 1/2, \lambda(b) = 1,$ and $\lambda(c) = 2/3$. (It should be noted that mean rates are always the reciprocal of mean times.) □

To avoid pathological cases, we shall adopt the policy in this chapter that the mean time spent in each state be finite and nonzero. The notions of state classification can now be carried over from the previous chapter.

Definition 6.3. Let $Y = \{Y_t; t \geq 0\}$ be a Markov process with finite state space E such that $0 < \lambda(i) < \infty$ for all $i \in E$. A state is called *recurrent* or *transient* according as it is recurrent or transient in the imbedded Markov chain of the process. A set of states is *irreducible* for the process if it is irreducible for the imbedded chain. □

6.2 Steady-State Properties

There is a close relationship between the steady-state probabilities for the Markov process and the steady-state probabilities of the imbedded chain. Specifically, the Markov process probabilities are obtained by the weighted average of the imbedded Markov chain probabilities according to each state's sojourn times. To show the relationship mathematically, consider $Y = \{Y_t; t \geq 0\}$ to be a Markov process with an irreducible, recurrent state space. Let $X = \{X_n; n = 0, 1, \cdots\}$ be the imbedded Markov chain with π its steady-state probabilities; namely,

$$\pi(j) = \lim_{n\to\infty} \Pr\{X_n = j | X_0 = i\} .$$

(In other words, $\boldsymbol{\pi}\mathbf{P} = \boldsymbol{\pi}$, where $\mathbf{P}$ is the Markov matrix of the imbedded chain.) The steady-state probabilities for the Markov process are denoted by the vector $\mathbf{p}$ where

$$p(j) = \lim_{t\to\infty} \Pr\{Y_t = j | Y_0 = i\} . \tag{6.1}$$

Note that the limiting values are independent of the initial state since the state space is assumed irreducible, recurrent. The relationship between the vectors π and $\mathbf{p}$ is given by

$$p(j) = \frac{\pi(j)/\lambda(j)}{\sum_{k\in E} \pi(k)/\lambda(k)} \tag{6.2}$$

where E is the state space and $\lambda(j)$ is the mean sojourn rate for State j.

The use of Eq. (6.2) is easily illustrated with Example 6.2 since the steady-state probabilities π where calculated in Chap. 5. Using these previous results (p. 159), we have that the imbedded chain has steady-state probabilities given by $\pi = (3/7, 2/7, 2/7)$. Combining these with the mean sojourn rates in Example 6.2 and with Eq. (6.2) yields

$$p(a) = \frac{6}{11},\ p(b) = \frac{2}{11}, \text{and}\ p(c) = \frac{3}{11} .$$

The salesman of the example thus spends 54.55% of his time in town 'a', 18.18% of his time in town 'b', and 27.27% of his time in town 'c'.

The information contained in the imbedded Markov matrix and the mean sojourn rates can be combined into one matrix which is called the generator matrix of the Markov process.

Definition 6.4. Let $Y = \{Y_t; t \geq 0\}$ be a Markov process with an imbedded Markov matrix $\mathbf{P}$ and a mean sojourn rate of $\lambda(i)$ for each state i in the state space. The *generator matrix*, $\mathbf{G}$, for the Markov process is given by

$$G(i,j) = \begin{cases} -\lambda(i) & \text{for } i = j, \\ \lambda(i)P(i,j) & \text{for } i \neq j. \end{cases}$$

□

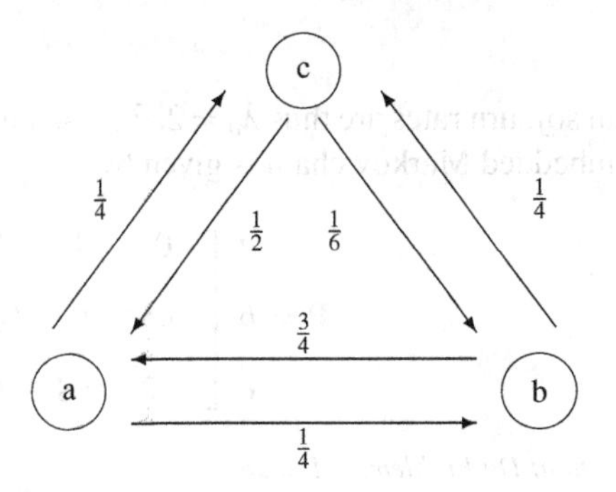

Fig. 6.2 State diagram for the Markov process of Example 6.3

Generator matrices are extremely important in the application of Markov processes. A generator matrix for a Markov process has two properties: (1) each row sum is zero and (2) the off-diagonal elements are nonnegative. These properties can be seen from Definition (6.4) by remembering that the imbedded Markov matrix is nonnegative with row sums of one and has zeroes on the diagonals. The physical interpretation of $\mathbf{G}$ is that $G(i, j)$ is the *rate* at which the process goes from State i to State j.

Example 6.3. We again return to our initial example to illustrate the generator matrix. Using the imbedded Markov chain and sojourn times given in Example 6.2, the generator matrix for that Markov process is

$$\mathbf{G} = \begin{matrix} a \\ b \\ c \end{matrix} \begin{bmatrix} -\frac{1}{2} & \frac{1}{4} & \frac{1}{4} \\ \frac{3}{4} & -1 & \frac{1}{4} \\ \frac{1}{2} & \frac{1}{6} & -\frac{2}{3} \end{bmatrix} .$$

Thus, Definition 6.4 indicates that if the transition matrix for the imbedded Markov chain and the mean sojourn rates are known, the generator matrix can be computed. State diagrams for Markov process are very similar to diagrams for Markov chains, except that the arrows representing transition probabilities represent transition rates (see Fig. 6.2). □

It should also be clear that not only can the transition rate matrix be obtained from the imbedded Markov matrix and mean sojourn rates, but also the reverse is true. From the generator matrix, the absolute value of the diagonal elements give the mean sojourn rates. The transition matrix for the imbedded chain is then obtained by dividing the off-diagonals element by the absolute value of that row's diagonal element. The diagonal elements of the transition matrix for the imbedded Markov chain are zero. For example, suppose a Markov process with state space $\{a, b, c\}$ has a generator matrix given by

$$\mathbf{G} = \begin{matrix} a \\ b \\ c \end{matrix} \begin{bmatrix} -2 & 2 & 0 \\ 2 & -4 & 2 \\ 1 & 4 & -5 \end{bmatrix} .$$

The mean sojourn rates are thus $\lambda_a = 2$, $\lambda_b = 4$, and $\lambda_c = 5$, and the transition matrix for the imbedded Markov chain is given by

$$\mathbf{P} = \begin{array}{c} a \\ b \\ c \end{array} \begin{bmatrix} 0 & 1 & 0 \\ 0.5 & 0 & 0.5 \\ 0.2 & 0.8 & 0 \end{bmatrix} .$$

- *Suggestion: Do Problems 6.1–6.2a.*

It is customary when describing a Markov process to obtain the generator matrix directly and never specifically look at the imbedded Markov matrix. The usual technique is to obtain the off-diagonal elements first and then let the diagonal element equal the negative of the sum of the off-diagonal elements on the row. The generator is used to obtain the steady-state probabilities directly as follows.

Property 6.1. *Let $Y = \{Y_t; t \geq 0\}$ be a Markov process with an irreducible, recurrent state space E and with generator matrix* **G**. *Furthermore, let* **p** *be a vector of steady-state probabilities defined by Eq. (6.1). Then* **p** *is the solution to*

$$\mathbf{pG} = \mathbf{0}$$
$$\sum_{j \in E} p(j) = 1 .$$

Obviously, this equation is very similar to the analogous equation in Property (5.9). The difference is in the right-hand side of the first equation. The easy way to remember which equation to use is that the multiplier for the right-hand side is the same as the row sums. Since Markov matrices have row sums of one, the right-hand side is *one* times π; since the generator matrix has row sums of zero, the right-hand side is *zero* times **p**.

Example 6.4. A small garage operates a tractor repair service in a busy farming community. Because the garage is small, only two tractors can be kept there at any one time and the owner is the only repairman. If one tractor is being repaired and one is waiting, then any further customers who request service are turned away. The time between the arrival of customers is an exponential random variable and an average of four customers arrive each day. Repair time varies considerably and it also is an exponential random variable. Based on previous records, the owner repairs an average of five tractors per day if kept busy. The arrival and repair of tractors can be described as a Markov process with state space $\{0, 1, 2\}$ where a state represents the number of tractors in the garage. Instead of determining the Markov matrix for the imbedded chain, the generator matrix can be calculated directly. As in Markov chain problems, drawing a state diagram is helpful. State diagrams for Markov processes

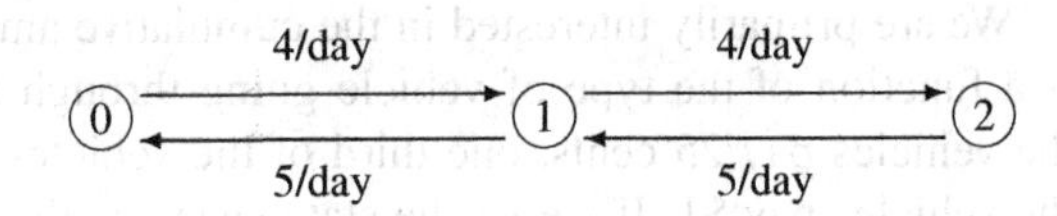

Fig. 6.3 State diagram for the Markov process of Example 6.4

show transition rates instead of transition probabilities. Fig. (6.3) shows the diagram with the transition rates. The rate of going from State 0 to State 1 and from State 1 to State 2 is four per day since that is the arrival rate of customers to the system. The rate of going from State 2 to State 1 and from State 1 to State 0 is five per day because that is the repair rate. Transition rates are only to adjacent states because both repairs and arrivals only occur one at a time. From the diagram, the generator is obtained as

$$\mathbf{G} = \begin{matrix} 1 \\ 2 \\ 3 \end{matrix} \begin{bmatrix} -4 & 4 & 0 \\ 5 & -9 & 4 \\ 0 & 5 & -5 \end{bmatrix} .$$

The steady-state probabilities are thus obtained by solving

$$\begin{aligned} -4p_0 + 5p_1 \qquad\quad &= 0 \\ 4p_0 - 9p_1 + 5p_2 &= 0 \\ 4p_1 - 5p_2 &= 0 \\ p_0 + p_1 + p_2 &= 1 , \end{aligned}$$

which yields (after deleting one of the first three equations) $p_0 = 25/61$, $p_1 = 20/61$, and $p_2 = 16/61$. As with Markov chains, there will always be one redundant equation. The steady-state equations can be interpreted to indicate that the owner is idle 41% of the time and 26% of the time the shop is full. □

An assumption was made in the above example regarding exponential random variables. The length of time that the process is in State 1 is a function of *two* exponentially distributed random variables. In effect, the sojourn time in State 1 depends on a "race" between the next arrival and the completion of the repair. The assumption is that the minimum of two exponential random variables is also exponentially distributed. Actually, that is not too difficult to prove, so we leave it as an exercise.

Example 6.5. **Poisson Process.** We are interested in counting the number of vehicles that go through a toll booth where the arriving vehicles form a Poisson process (Chap. 4) with a mean rate of 120 cars per hour. Since a Poisson process has exponential inter-arrival times, a Poisson process is also a Markov process with a state space equal to the nonnegative integers (review Fig. 4.1). The generator matrix is infinite dimensioned given by

$$\mathbf{G} = \begin{matrix} 0 \\ 1 \\ 2 \\ \vdots \end{matrix} \begin{bmatrix} -120 & 120 & 0 & 0 & \cdots \\ 0 & -120 & 120 & 0 & \cdots \\ 0 & 0 & -120 & 120 & \cdots \\ \vdots & & & \ddots & \end{bmatrix} .$$

We are primarily interested in the cumulative amount of tolls collected. The toll is a function of the type of vehicle going through the toll booth. Fifty percent of the vehicles pay 25 cents, one third of the vehicles pay 75 cents, and one sixth of the vehicles pay $1. If we let the state space be the cumulative number of quarters collected, the process of interest is a compound Poisson process (Sect. 4.3.1) which is also a Markov process. The generator for the cumulative number of quarters collected becomes

$$\mathbf{G} = \begin{matrix} 0 \\ 1 \\ 2 \\ \vdots \end{matrix} \begin{bmatrix} -120 & 60 & 0 & 40 & 20 & 0 & 0 & \cdots \\ 0 & -120 & 60 & 0 & 40 & 20 & 0 & \cdots \\ 0 & 0 & -120 & 60 & 0 & 40 & 20 & \cdots \\ \vdots & & \ddots & & & & & \end{bmatrix}.$$

□

- *Suggestion: Do Problems 6.2b, 6.3, 6.6, 6.7, 6.9, and 6.14.*

6.3 Revenues and Costs

Markov processes often represent revenue producing situations and so the steady-state probabilities are used to determine long-run average return. Suppose that whenever the Markov process is in State i, a profit is produced at a rate of $f(i)$. Thus, **f** is a vector of profit rates. The long run profit per unit time is then simply the (vector) product of the steady-state probabilities times **f**.

Property 6.2. *Let $Y = \{Y_t; t \geq 0\}$ be a Markov process with an irreducible, recurrent state space E and a profit rate vector denoted by* **f** *(i.e., $f(i)$ is the rate that profit is accumulated whenever the process is in State i). Furthermore, let* **p** *denote the steady-state probabilities as defined by Eq. (6.1). Then, the long run profit per unit time is given by*

$$\lim_{t\to\infty} \frac{1}{t} E\left[\int_0^t f(Y_s)\mathrm{d}s\right] = \sum_{j\in E} p(j)f(j)$$

independent of the initial state.

- *Suggestion: Do Problems 6.2c–6.4a.*

The left-hand side of the above equation is simply the mathematical way of writing the long-run average accumulated profit. It is sometimes important to include revenue produced at transition times so that the total profit is the sum of the accumulation of profit resulting from the sojourn times and the jumps. In other words, we would not only have the profit rate vector **f** but also a matrix that indicates a profit obtained at each jump.

Property 6.3. *Let $Y = \{Y_t; t \ge 0\}$ be a Markov process with an irreducible, recurrent state space E, a profit rate vector denoted by $\mathbf{f}$, and a matrix of jump profits denoted by $\mathbf{h}$ (i.e., $h(i,j)$ is the profit obtained whenever the process jumps from State i to State j). Furthermore, let $\mathbf{p}$ denote the steady-state probabilities as defined by Eq. (6.1). Then, the long run profit per unit time is given by*

$$\lim_{t\to\infty} \frac{1}{t} E\left[\int_0^t f(Y_s)ds + \sum_{s \le t} h(Y_{s-}, Y_s)\right] =$$

$$\sum_{i \in E} p(i) \left[f(i) + \sum_{k \in E} G(i,k)h(i,k)\right]$$

where $h(i,i) = 0$ and $\mathbf{G}$ is the generator matrix of the process.

The summation on the left-hand side of the above equation may appear different from what you may be used to seeing. The notation Y_{s-} indicates the left-hand limit of the process; that is, $Y_{s-} = \lim_{t\to s^-} Y_t$ where t approaches s from the left, i.e., $t < s$. Since Markov processes are always right-continuous, the only times for which $Y_{s-} \neq Y_s$ is when s is a jump time. (When we say that the Markov process is right-continuous, we mean that $Y_s = \lim_{t\to s^+} Y_t$ where the limit is such that t approaches s from the right, i.e., $t > s$.) Because the diagonal elements of $\mathbf{h}$ are zero, the only times that the summation includes nonzero terms are at jump times.

Example 6.6. Returning to Example 6.1 for illustration, assume that the revenue possible from each town varies. Whenever the salesman is working town 'a', his profit comes in at a rate of \$80 per day. In town 'b', his profit is \$100 per day. In town 'c', his profit is \$125 per day. There is also a cost associated with changing towns. This cost is estimated at 25 cents per mile and it is 50 miles from a to b, 65 miles from a to c, and 80 miles from b to c. The functions (using five day weeks) are

$$\mathbf{f} = (400, 500, 625)^T$$

$$\mathbf{h} = \begin{bmatrix} 0 & -12.50 & -16.25 \\ -12.50 & 0 & -20.00 \\ -16.25 & -20.00 & 0 \end{bmatrix}.$$

Therefore, using

$$\mathbf{G} = \begin{bmatrix} -\frac{1}{2} & \frac{1}{4} & \frac{1}{4} \\ \frac{3}{4} & -1 & \frac{1}{4} \\ \frac{1}{2} & \frac{1}{6} & -\frac{2}{3} \end{bmatrix}$$

and

$$p(a) = \frac{6}{11},\ p(b) = \frac{2}{11},\ p(c) = \frac{3}{11},$$

the long-run weekly average profit is

$$\frac{6}{11} \times (400 - 7.19) + \frac{2}{11} \times (500 - 14.38) + \frac{3}{11} \times (625 - 11.46) = 469.88 .$$

□

- *Suggestion: Do Problems 6.5–6.8.*

It is sometimes important to consider the time-value of the profit (or cost). In the discrete case, a discount factor of α was used. We often think in terms of a rate-of-return or an interest rate. If i were the rate of return, then the discount factor used for Markov chains is related to i according to the formula $\alpha = 1/(1+i)$; thus, the present value of one dollar obtained one period from the present is equal to α. For a continuous-time problem, we let β denote the discount rate, which in this case would be the same as the (nominal) rate-of-return except that we assume that compounding occurs continuously. Thus, the present value of one dollar obtained one period from the present equals $\mathrm{e}^{-\beta}$. When the time-value of money is important, it is difficult to include the jump time profits but the function $\mathbf{f}$ is easily utilized according to the following property.

Property 6.4. *Let $Y = \{Y_t; t \geq 0\}$ be a Markov process with a generator matrix $\mathbf{G}$, a profit rate vector $\mathbf{f}$, and a discount factor of β. Then, the present value of the total discounted profit (over an infinite planning horizon) is given by*

$$E\left[\int_0^\infty \mathrm{e}^{-\beta s} f(Y_s)\mathrm{d}s | Y_0 = i\right] = ((\beta \mathbf{I} - \mathbf{G})^{-1}\mathbf{f})(i) .$$

Example 6.7. Assume that the salesman of Example 6.1 wants to determine the present value of the total revenue using a discount rate of 5%, where the revenue rate function is given by $\mathbf{f} = (400, 500, 625)^T$. The present value of all future revenue is thus given by

$$\begin{bmatrix} \frac{11}{20} & -\frac{1}{4} & -\frac{1}{4} \\ -\frac{3}{4} & \frac{21}{20} & -\frac{1}{4} \\ -\frac{1}{2} & -\frac{1}{6} & \frac{43}{60} \end{bmatrix}^{-1} \begin{bmatrix} 400 \\ 500 \\ 625 \end{bmatrix} = \begin{bmatrix} 11.31 & 3.51 & 5.17 \\ 10.54 & 4.28 & 5.17 \\ 10.34 & 3.45 & 6.21 \end{bmatrix} \begin{bmatrix} 400 \\ 500 \\ 625 \end{bmatrix} = \begin{bmatrix} 9515 \\ 9592 \\ 9741 \end{bmatrix} .$$

Note that the total discounted revenue depends on the initial state. Thus, if at time 0 the salesman is in city a, then the present value of his total revenue is \$9515; whereas, if he started in city c, the present value of his total revenue would be \$9741.

□

- *Suggestion: Do Problem 6.4b,c.*

6.4 Time-Dependent Probabilities

A thorough treatment of the time-dependent probabilities for Markov processes is beyond the scope of this book; however, we do present a brief summary in this section for those students who may want to continue studying random processes and would like to know something of what the future might hold. For other students, this section may be skipped with no loss in continuity.

As before, the Markov process will be represented by $\{Y_t; t \geq 0\}$ with finite state space E and generator matrix $\mathbf{G}$. The basic relationship that probabilities for a Markov process must satisfy are analogous to the relationship that the transition probabilities of a Markov chain satisfy. As you recall from Sect. 5.2, the following holds for Markov chains:

$$
\begin{aligned}
P\{X_{n+m} = j|X_0 = i\} \\
&= \sum_{k\in E} \Pr\{X_n = k|X_0 = i\} \times \Pr\{X_{n+m} = j|X_n = k\} \\
&= \sum_{k\in E} \Pr\{X_n = k|X_0 = i\} \times \Pr\{X_m = j|X_0 = k\} .
\end{aligned}
$$

The second equality is a result of the stationary property for Markov chains and is called the Chapman-Kolmogorov equation for Markov chains. The Chapman-Kolmogorov equation for Markov processes can be expressed similarly as

$$
\begin{aligned}
P\{Y_{t+s} = j|Y_0 = i\} \\
&= \sum_{k\in E} \Pr\{Y_t = k|Y_0 = i\} \times \Pr\{Y_{t+s} = j|Y_t = k\} \\
&= \sum_{k\in E} \Pr\{Y_t = k|Y_0 = i\} \times \Pr\{Y_s = j|Y_0 = k\} ,
\end{aligned}
$$

for $t, s \geq 0$ and $i, j \in E$. In other words, the Chapman-Kolmogorov property indicates that the probability of going from State i to State j in $t+s$ time units is the sum of the product of probabilities of going from State i to an arbitrary State k in t time units and going from State k to State j in s time units (see Fig. 6.4).

Before giving a closed-form expression for the time-dependent probabilities, it is necessary to first discuss the exponentiation of a matrix. Recall that for scalars, the following power series holds for all values of a

$$
\mathrm{e}^a = \sum_{n=0}^{\infty} \frac{a^n}{n!} .
$$

This same relationship holds for matrices and becomes the definition of the exponentiation of a matrix; namely,

$$
\mathrm{e}^{\mathbf{A}} = \sum_{n=0}^{\infty} \frac{\mathbf{A}^n}{n!} . \tag{6.3}
$$

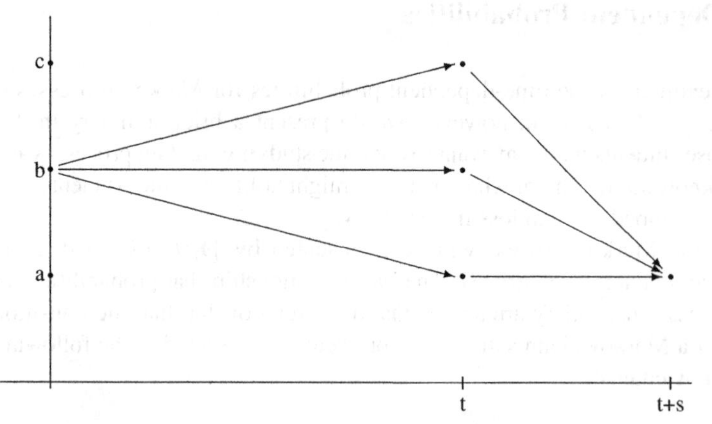

Fig. 6.4 Graphical representation for the Chapman-Kolmogorov equations for a three-state Markov process

Note that $e^{\mathbf{A}}(i,j)$ refers to the $i-j$ element of the matrix $e^{\mathbf{A}}$; thus, in general, $e^{\mathbf{A}}(i,j) \neq e^{A(i,j)}$.

Using the above definition and the Chapman-Kolmogorov equation, it is possible to derive the following property:

Property 6.5. *Let $Y = \{Y_t; t \geq 0\}$ be a Markov process with state space E and generator matrix* $\mathbf{G}$*, then for $i, j \in E$ and $t \geq 0$*

$$P\{Y_t = j | Y_0 = i\} = e^{\mathbf{G}t}(i,j) ,$$

where $\mathbf{G}t$ is the matrix formed by multiplying each element of the generator matrix by the scalar t.

Example 6.8. We are interested in an electronic device that is either "on-standby" or "in-use". The on-standby periods are exponentially distributed with a mean of 20 seconds, and the in-use periods are exponentially distributed with a mean of 12 seconds. Since the exponential assumption is satisfied, a Markov process $\{Y_t; t \geq 0\}$ with state space $E = \{0, 1\}$ can be used to model the electronic device as it alternates between on-standby and in-use. State 0 denotes on-standby, and State 1 denotes in-use. The generator matrix is given by

$$\mathbf{G} = \begin{matrix} 0 \\ 1 \end{matrix} \begin{bmatrix} -3 & 3 \\ 5 & -5 \end{bmatrix} ,$$

where the time unit is minutes. Using Property 6.1, it is easy to show that the long-run probability of being in use is $3/8$. However, we are interested in the time-dependent probabilities. If the generator can be written in diagonal form, then it is not difficult to apply Property 6.5. In other words, if it is possible to find a matrix $\mathbf{Q}$

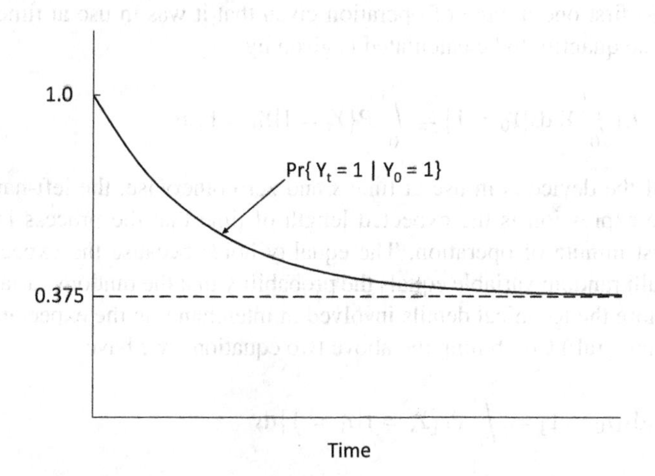

Fig. 6.5 Time-dependent probabilities for Example 6.8

such that

$$\mathbf{G} = \mathbf{Q}\mathbf{D}\mathbf{Q}^{-1}$$

where **D** is a diagonal matrix, then the following holds

$$e^{\mathbf{G}t} = \mathbf{Q}e^{\mathbf{D}t}\mathbf{Q}^{-1}\,.$$

Furthermore, $e^{\mathbf{D}t}$ is a diagonal matrix whenever **D** is diagonal, in which case $e^{\mathbf{D}t}(i,i) = e^{tD(i,i)}$. For the above generator we have

$$\begin{bmatrix} -3 & 3 \\ 5 & -5 \end{bmatrix} = \begin{bmatrix} 1 & 3 \\ 1 & -5 \end{bmatrix} \cdot \begin{bmatrix} 0 & 0 \\ 0 & -8 \end{bmatrix} \cdot \begin{bmatrix} 0.625 & 0.375 \\ 0.125 & -0.125 \end{bmatrix};$$

therefore,

$$e^{\mathbf{G}t} = \begin{bmatrix} 1 & 3 \\ 1 & -5 \end{bmatrix} \cdot \begin{bmatrix} 1 & 0 \\ 0 & e^{-8t} \end{bmatrix} \cdot \begin{bmatrix} 0.625 & 0.375 \\ 0.125 & -0.125 \end{bmatrix}$$

$$= \begin{bmatrix} 0.625 + 0.375e^{-8t} & 0.375 - 0.375e^{-8t} \\ 0.625 - 0.625e^{-8t} & 0.375 + 0.625e^{-8t} \end{bmatrix}.$$

The probability that the electronic device is in use at time t given that it started in use at time 0 is thus given by

$$\Pr\{Y_t = 1 | Y_0 = 1\} = 0.375 + 0.625e^{-8t}\,.$$

(See Fig. 6.5 for a graphical representation of the time-dependent probabilities.)

We would like to know the expected amount of time that the electronic device is in use during its first one minute of operation given that it was in use at time 0. Mathematically, the quantity to be calculated is given by

$$E[\int_0^1 Y_s \,\mathrm{d}s | Y_0 = 1] = \int_0^1 P\{Y_s = 1 | Y_0 = 1\}\mathrm{d}s\,.$$

Because Y_s is 1 if the device is in use at time s and zero otherwise, the left-hand-side of the above expression is the expected length of time that the process is in use during its first minute of operation. The equality holds because the expected value of a Bernoulli random variable equals the probability that the random variable equals 1. (We ignore the technical details involved in interchanging the expectation operator and the integral.) Combining the above two equations, we have

$$\begin{aligned} E[\int_0^1 Y_s \mathrm{d}s | Y_0 = 1] &= \int_0^1 \Pr\{Y_s = 1 | Y_0 = 1\}\mathrm{d}s \\ &= \int_0^1 0.375 + 0.625\mathrm{e}^{-8s}\mathrm{d}s \\ &= 0.375 + \frac{0.625}{8}(1 - \mathrm{e}^{-8}) = 0.453\,\text{minutes}\,. \end{aligned}$$

It is often useful to determine the total cost (or profit) of an operation modeled by a Markov process. In order to express the expected cost over a time interval, let $C_{(0,t)}$ be a random variable denoting the total cost incurred by the Markov process over the interval $(0,t)$. Then, for $i \in E$,

$$E[C_{(0,t)} | Y_0 = i] = \sum_{k \in E} f(k) \int_0^t \Pr\{Y_s = k | Y_0 = i\}\mathrm{d}s\,,$$

where $f(k)$ is the cost rate incurred by the process while in State k.

For example, suppose that the electronic device costs the company 5 cents every minute it is on standby and it costs 35 cents every minute the device is in use. We are interested in estimating the total cost of operation for the first 2 minutes of utilizing the device given that the device was on standby when the process started; thus

$$\begin{aligned} E[C_{(0,2)} | Y_0 = 0] &= 0.05 \int_0^2 \Pr\{Y_s = 0 | Y_0 = 0\}\mathrm{d}s \\ &\quad + 0.35 \int_0^2 \Pr\{Y_s = 1 | Y_0 = 0\}\mathrm{d}s \\ &= 0.05 \times (0.625 \times 2 + \frac{0.375}{8}(1 - \mathrm{e}^{-8\times 2})) \\ &\quad + 0.35 \times (0.375 \times 2 - \frac{0.375}{8}(1 - \mathrm{e}^{-8\times 2})) = 0.798\,. \end{aligned}$$

□

• *Suggestion: Do Problems 6.10–6.13.*

Appendix

We re-visit the previous example to demonstrate an Excel simulation of a Markov process. Because the process of the example only has two states, the simulation is easier than for a process with a larger state space, which we leave to the reader as homework problem 6.14.

Example 6.9. Consider the Markov process of Example 6.8. We prepare the spreadsheet by writing down the generator matrix and give the steady-state probabilities as follows:

	A	B	C
1	`State`		`G matrix`
2	`standby`	`-3`	`3`
3	`in-use`	`5`	`-5`
4			
5			`Steady-state`
6		`=C3/(B2+C3)`	`=B2/(B2+C3)`

(Verify for yourself that the formulas used in Cells B6:C6 satisfy Property 6.1.) To setup the simulation, type the following:

	A	B	C	D
8	`Time`	`Current State`	`Random Number`	`Interarrive Time`
9	`0`	`in-use`	`=RAND()`	`=LN(C9)/$C$3`

For Cell A10, type `=A9+D9`; for Cell B10, type

```
=IF(B9="standby","in-use","standby")
```

for Cell C10, type `=RAND()`; and for Cell D10, type

```
=IF(B10="standby",LN(C10)/$B$2,LN(C10)/$C$3)
```

where a formula like `=LN(C9)/$C$3` yields an exponential random variate, see Table 2.13. Now copy Cells A10:D10 down through row 30000 to complete the simulation.

An estimate of the steady-state probability of the system being in the "in-use" state is obtained by dividing the total time the process spends in that state by the total time that the process operates. Thus, in Cell F8, type `Estimatation` and in Cell F9 type

```
=SUMIF(B:B,"in-use",D:D)/MAX(A:A).
```

Notice that the `SUMIF` function can use different columns for the search criterion and for the summation values. Since Column A contains increasing values, the maximum value in the column will be the total time for the simulation. By using only the column designation in the `SUMIF` and `MAX` functions, the estimate does not depend on knowing ahead of time how many rows are being copied down; however, the final inter-arrive time should be deleted since it is not reflected in the total time. □

Because the process of Example 6.9 had only two states, the state transition were relatively simple. If there were more states, then a second column of random numbers would be required, and a more complicated `IF` statement would be needed as used at the end of Example 5.16. As a general rule, two random numbers are needed for each time step of a Markov process: one random number to determine the size of the time step and another random number for the state transition.

Problems

6.1. The following matrix is a generator for a Markov process. Complete its entries.

$$\mathbf{G} = \begin{bmatrix} \_ & 3 & 0 & 7 \\ 5 & -12 & 4 & \_ \\ 1 & \_ & -6 & 5 \\ 3 & 2 & 9 & \_ \end{bmatrix}$$

6.2. Let Y be a Markov process with state space $\{a,b,c,d\}$ and an imbedded Markov chain having a Markov matrix given by

$$\mathbf{P} = \begin{bmatrix} 0.0 & 0.1 & 0.2 & 0.7 \\ 0.0 & 0.0 & 0.4 & 0.6 \\ 0.8 & 0.1 & 0.0 & 0.1 \\ 1.0 & 0.0 & 0.0 & 0.0 \end{bmatrix}.$$

The mean sojourn times in states a,b,c, and d are 2, 5, 0.5, and 1, respectively.
(a) Give the generator matrix for this Markov process.
(b) What is $\lim_{t\to\infty} \Pr\{Y_t = a\}$?
(c) Let $r = (10,25,30,50)^T$ be a reward vector and determine $\lim_{t\to\infty} E[\int_0^t r(Y_s)ds]/t$.

6.3. Let T and S be exponentially distributed random variables with means $1/a$ and $1/b$, respectively. Define the random variable $U = \min\{T, S\}$. Justify the relationship $P\{U > u\} = P\{T > u\} \times P\{S > u\}$ and derive the CDF for U.

6.4. A revenue producing system can be in one of four states: high income, medium income, low income, costs. The movement of the system among the states is according to a Markov process, Y, with state space $E = \{h,m,l,c\}$ and with generator matrix given by

$$\mathbf{G} = \begin{matrix} h \\ m \\ l \\ c \end{matrix} \begin{bmatrix} -0.2 & 0.1 & 0.1 & 0.0 \\ 0.0 & -0.4 & 0.3 & 0.1 \\ 0.0 & 0.0 & -0.5 & 0.5 \\ 1.5 & 0.0 & 0.0 & -1.5 \end{bmatrix}.$$

While the system is in state h,m,l, or c it produces a profit at a rate of \$500, \$250, \$100, or –\$600 per time unit. The company would like to reduce the time spent in

the fourth state and has determined that by doubling the cost (i.e., from \$600 to \$1,200) incurred while in that state the mean time spent in the state can be cut in half. Is the additional expense worthwhile?
(a) Use the long-run average profit for the criterion.
(b) Use a total discounted-cost criterion assuming a discount rate of 10%.
(c) Use a total discounted-cost criterion assuming the company uses a 25% annual rate-of-return, and the time step for this problem is assumed to be in weeks.

6.5. Let Y be a Markov process with state space $\{a,b,c,d\}$ and generator matrix given by

$$\mathbf{G} = \begin{bmatrix} -5 & 4 & 0 & 1 \\ 6 & -10 & 4 & 0 \\ 0 & 0 & -1 & 1 \\ 2 & 0 & 0 & -2 \end{bmatrix}.$$

Costs are incurred at a rate of \$100, \$300, \$500, and \$1,000 per time unit while the process is in states a,b,c, and d, respectively. Furthermore, each time a jump is made from state d to state a an additional cost of \$5,000 is incurred. (All other jumps do not incur additional costs.) For a maintenance cost of \$400 per time unit, all cost rates can be cut in half and the "jump" cost can be eliminated. Based on long-run averages, is the maintenance cost worthwhile?

6.6. A small gas station has one pump and room for a total of three cars (one at the pump and two waiting). The time between car arrivals to the station is an exponential random variable with the average arrival rate of 10 cars per hour. The time each car spends in front of the pump is an exponential random variable with a mean of five minutes (i.e., a mean rate of 12 per hour). If there are three cars in the station and another car arrives, the newly arrived car keeps going and never enters the station.
(a) Model the gas station as a Markov process, Y, where Y_t denotes the number of cars in the station at time t. Give its generator matrix.
(b) What is the long-run probability that the station is empty?
(c) What is the long-run expected number of cars in the station?

6.7. A small convenience store has a room for only 5 people inside. Cars arrive to the store randomly, with the inter-arrival times between cars being an exponential random variable with a mean of 10 cars arriving each hour. The number of people within each car is a random variable, N, where $P\{N=1\}=0.1$, $P\{N=2\}=0.7$, and $P\{N=3\}=0.2$. People from the cars come into the store and stay in the store an exponential length of time. The mean length of stay in the store is 10 minutes and each person acts independent of all other people, leaving the store singly and waiting in their cars for the others. If a car arrives and the store is too full for everyone in the car to enter the store, the car will leave and nobody from that car will enter the store. Model the store as a Markov process, Y, where Y_t denotes the number of individuals in the store at time t. Give its generator matrix.

6.8. A certain piece of electronic equipment has two components. The time until failure for component A is described by an exponential distribution function with

a mean time of 100 hours. Component B has a mean life until failure of 200 hours and is also described by an exponential distribution. When one component fails, the equipment is turned off and maintenance is performed. The time to fix the component is exponentially distributed with a mean time of 5 hours if it was A that failed and 4 hours if it was B that failed. Let Y be a Markov process with state space $E = \{w, a, b\}$, where State w denotes that the equipment is working, a denotes that component A has failed, and b denotes that component B has failed.
(a) Give the generator for Y.
(b) What is the long-run probability that the equipment is working?
(c) An outside contractor does the repair work on the components when a failure occurs and charges \$100 per hour for time plus travel expenses, which is an additional \$500 for each visit. The company has determined that they can hire and train their own repair person. If they have their own employee for the repair work, it will cost the company \$40 per hour while the machine is running as well as when it is down. Ignoring the initial training cost and the possibility that an employee who is hired for repair work can do other things while the machine is running, is it economically worthwhile to hire and train their own person?

6.9. An electronic component works as follows: Electric impulses arrive to the component with exponentially distributed inter-arrival times such that the mean arrival rate of impulses is 90 per hour. An impulse is "stored" until the third impulse arrives, then the component "fires" and enters a "recovery" phase. If an impulse arrives while the component is in the recovery phase, it is ignored. The length of time that the component remains in the recovery phase is an exponential random variable with a mean time of one minute. After the recovery phase is over, the cycle is repeated; that is, the third arriving impulse will instantaneously fire the component.
(a) Give the generator matrix for a Markov process model of the dynamics of this electronic component.
(b) What is the long-run probability that the component is in the recovery phase?
(c) How many times would you expect the component to fire each hour?

6.10. Consider the electronic device that is either in the "on standby" or "in use" state as described in Example 6.8. Find the following quantities.
(a) The expected cost incurred during the time interval between the third and fourth minutes, given that at time zero the device was "in use".
(b) The expected cost incurred during the time interval between the third and fourth minutes, given that at time zero the device was "in use" with probability 0.8 and "on standby" with probability 0.2.

6.11. Consider a Markov process with state space $E = \{a, b\}$ and with initial conditions given by the vector $\boldsymbol{\mu}$. Let the mean rates for states a and b be denoted by λ_a and λ_b, and let $\Lambda = \lambda_a + \lambda_b$. Develop an expression for the time dependent probability of being in state a. Hints: The generator matrix is always diagonalizable using

$$\mathbf{D} = \begin{bmatrix} 0 & 0 \\ 0 & -\Lambda \end{bmatrix},$$

and the inverse of a 2×2 matrix is given by

$$\begin{bmatrix} a & b \\ c & d \end{bmatrix}^{-1} = \frac{1}{ad-bc} \begin{bmatrix} d & -b \\ -c & a \end{bmatrix} .$$

6.12. Let μ be a vector of initial probabilities for a Markov process Y, and let $\mathbf{f}$ denote a cost rate vector associated with Y. Write a general expression for the expected cost per unit time incurred by the Markov process during the time interval $(t, t+s)$.

6.13. Let $\mathbf{f}$ denote a profit rate vector associated with a Markov process Y, and let β denote a discount rate. Write a general expression for the total profit returned by the process during the interval $(t, t+s)$ given that the process was in state i at time 0. (See Property 4.8).

6.14. Simulate the Markov process defined by the following generator matrix:

$$\mathbf{G} = \begin{matrix} a \\ b \\ c \end{matrix} \begin{bmatrix} -1.0 & 0.3 & 0.7 \\ 0.1 & -0.2 & 0.1 \\ 0.3 & 0.2 & -0.5 \end{bmatrix} .$$

(a) Estimate the steady-state probabilities.
(b) Estimate the probability that the process will be in State b at time 10 given that the process started in State a.

Chapter 7
Queueing Processes

Many phenomena for which mathematical descriptions are desired involve waiting lines either of people or material. A queue is a waiting line, and queueing processes are those stochastic processes arising from waiting line phenomena. For example, the modeling of the arrival process of grain trucks to an elevator, the utilization of data processing services at a computer center, and the flow of jobs at a job shop facility all involve waiting lines. Although queues are ubiquitous, they are usually ignored when deterministic models are developed to describe systems. Furthermore, the random fluctuations inherent in queueing processes often cause systems to act in a counter intuitive fashion. Therefore, the study of queues is extremely important for the development of system models and an understanding of system behavior.

In this chapter we present modeling techniques employed for queueing systems governed by the exponential process. The final section of the chapter deals with some approximation techniques useful for implementing these models within complex systems when the exponential assumptions are not satisfied. The next chapter deals with processes that arise when several queueing system operate within one system; that is, Chap. 8 deals with queueing networks. The final chapter (Chap. 13) in the textbook presents some advanced analytical techniques useful for modeling non-exponential queueing systems.

7.1 Basic Definitions and Notation

A queueing process involves the arrival of customers to a service facility and the servicing of those customers. All customers that have arrived but are not yet being served are said to be in the *queue*. The queueing *system* includes all customers in the queue and all customers in service (see Fig. 7.1).

Several useful conventions have evolved over the last 20-40 years that help in specifying the assumptions used in a particular analysis. D.G. Kendall [3] is usually given credit for initiating the basic notation of today, and it was standardized in 1971 (*Queueing Standardization Conference Report*, May 11, 1971). Kendall's notation is

R.M. Feldman, C. Valdez-Flores, *Applied Probability and Stochastic Processes*, 2nd ed., DOI 10.1007/978-3-642-05158-6_7, © Springer-Verlag Berlin Heidelberg 2010

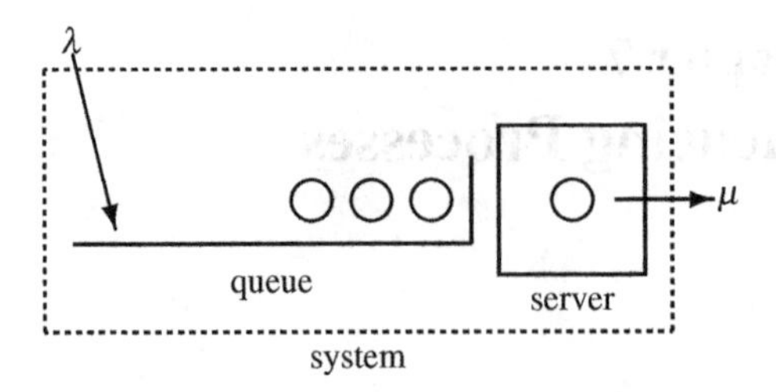

Fig. 7.1 Representation of a queueing system with a mean arrival rate of λ, and mean service rate of μ, four customers in the system, and three in the queue

a shorthand to indicate quickly the assumptions used in a particular queueing model. For example, a formula developed for a G/D/1/∞/FIFO queue would be a formula that could be used for any general arrival process, only a deterministic service time, one server, an unlimited system capacity, and a discipline that serves on a "first-in first-out" basis. The general notation has the following form.

$$\left(\begin{array}{c}\text{arrival}\\ \text{process}\end{array} \Big/ \begin{array}{c}\text{service}\\ \text{process}\end{array} \Big/ \begin{array}{c}\text{number}\\ \text{of servers}\end{array} \Big/ \begin{array}{c}\text{maximum}\\ \text{possible}\\ \text{in system}\end{array} \Big/ \begin{array}{c}\text{queue}\\ \text{discipline}\end{array} \right)$$

Table (7.1) gives the common abbreviations used with this notation. Whenever an infinite capacity and FIFO discipline are used, the last two descriptors can be left off; thus, an M/M/1 queue would refer to exponential inter-arrival and service times, one server, unlimited capacity, and a FIFO discipline.

Table 7.1 Queueing symbols used with Kendall's notation

Symbols	Explanation
M	Exponential (Markovian) inter-arrival or service time
D	Deterministic inter-arrival or service time
E_k	Erlang type k inter-arrival or service time
G	General inter-arrival or service time
$1, 2, \cdots, \infty$	Number of parallel servers or capacity
FIFO	First in, first out queue discipline
LIFO	Last in, first out queue discipline
SIRO	Service in random order
PRI	Priority queue discipline
GD	General queue discipline

Our purpose in this chapter is to give an introduction to queueing processes and introduce the types of problems commonly encountered while studying queues. To maintain the introductory level of this material, arrival processes to the queueing systems will be assumed to be Poisson processes (see Chap. 4), and service times will be exponential. Thus, this chapter is mainly concerned with investigating M/M/c/K systems for various values of c and K. The last section of the chapter will present a simple approximation that can be used for simple systems.

7.2 Single Server Systems

The simplest queueing systems to analyze are those involving a Poisson arrival process and a single exponential server. Such systems are not only relatively easy to study, but they will serve to demonstrate some general queueing analysis techniques that can be extended to more complicated system. We shall start by considering a system that has unlimited space for arriving customers and then move to systems with limited space.

7.2.1 Infinite Capacity Single-Server Systems

We begin with an M/M/1 system, or equivalently, an M/M/1∞/FIFO system. The M/M/1 system assumes customers arrive according to a Poisson process with mean rate λ and are served by a single server whose time for service is random with an exponential distribution of mean $1/\mu$. If the server is idle and a customer arrives, then that customer enters the server immediately. If the server is busy and a customer arrives, then the arriving customer enters the queue which has infinite capacity. When service for a customer is completed, the customer leaves and the customer that had been in the queue the longest instantaneously enters the service facility and service begins again. Thus, the flow of customers through the system is a Markov process with state space $\{0,1,\cdots\}$. The Markov process is denoted by $\{N_t; t \geq 0\}$ where N_t denotes the number of customers in the system at time t. The steady-state probabilities are

$$p_n = \lim_{t\to\infty} Pr\{N_t = n\} .$$

We let N be a random variable with probability mass function $\{p_0, p_1, \cdots\}$. The random variable N thus represents the number of customers in the system at steady-state, and p_n represents the long-run probability that there are n customers in the system. (You might also note that another way to view p_n is as the long-run fraction of time that the system contains n customers.) Sometimes we will be interested in the number of customers that are in the queue and thus waiting for service; therefore, let the random variable N_q denote the steady-state number in the queue. In other words, if the system is idle, $N_q = N$; when the system is busy, $N_q = N - 1$.

Our immediate goal is to derive an expression for $p_n, n = 0,1,\cdots$, in terms of the mean arrival and service rates. This derivation usually involves two steps: (1) obtain a system of equations defining the probabilities and (2) solve the system of equations. After some experience, you should find Step (1) relatively easy; it is usually Step (2) that is difficult. In other words, the system of equations defining p_n is not hard to obtain, but it is sometimes hard to solve.

An intuitive approach for obtaining the system of equations is to draw a state diagram and then use a rate balance approach, which is a system of equations formed by setting "rate in" equal to "rate out" for each node or state of the queueing system. Figure 7.2 shows the state diagram for the M/M/1 system. Referring to this figure,

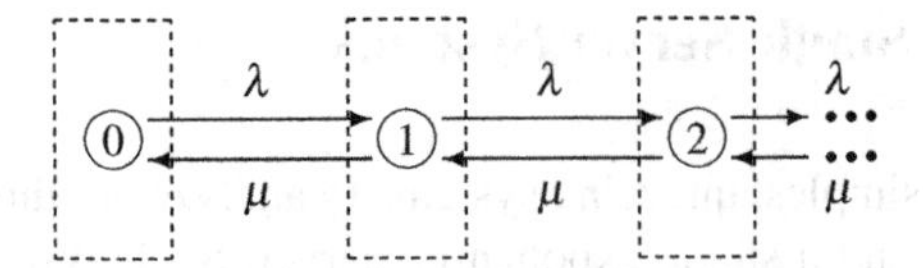

Fig. 7.2 State diagram for an M/M/1 queueing system illustrating the rate balance approach

it is seen that the rate into the box around Node 0 is μp_1; the rate out of the box around Node 0 is λp_0; thus, "rate in" = "rate out" yields

$$\mu p_1 = \lambda p_0 .$$

The rate into the box around Node 1 is $\lambda p_0 + \mu p_2$; the rate out of the box around Node 1 is $(\mu + \lambda) p_1$; thus

$$\lambda p_0 + \mu p_2 = (\mu + \lambda) p_1 .$$

Continuing in a similar fashion and rearranging, we obtain the system

$$p_1 = \frac{\lambda}{\mu} p_0 \text{ and} \tag{7.1}$$

$$p_{n+1} = \frac{\lambda + \mu}{\mu} p_n - \frac{\lambda}{\mu} p_{n-1} \text{ for } n = 1, 2, \cdots .$$

A more rigorous approach for obtaining the system of Eqs. (7.1) is to first give the generator matrix (from Chap. 6) for the M/M/1 system. Since the inter-arrival and service times are exponential, the queueing system is a Markov process and thus Property 6.1 can be used. The rate at which the process goes from State n to State $n+1$ is λ, and the rate of going from State n to State $n-1$ is μ; therefore, the generator is the infinite dimensioned matrix given as

$$\mathbf{G} = \begin{bmatrix} -\lambda & \lambda & & \\ \mu & -(\mu+\lambda) & \lambda & \\ & \mu & -(\mu+\lambda) & \lambda \\ & & \ddots & \ddots & \ddots \end{bmatrix} . \tag{7.2}$$

The system of equations formed by $\mathbf{pG} = \mathbf{0}$ then yields Eqs. (7.1) again.

The system of Eqs. (7.1) can be solved by successively forward substituting solutions and expressing all variables in terms of p_0. Since we already have $p_1 = (\lambda/\mu) p_0$, we look at p_2 and then p_3:

$$\begin{aligned} p_2 &= \frac{\lambda + \mu}{\mu} p_1 - \frac{\lambda}{\mu} p_0 \\ &= \frac{\lambda + \mu}{\mu} \left(\frac{\lambda}{\mu} p_0\right) - \frac{\lambda}{\mu} \frac{\mu}{\mu} p_0 = \frac{\lambda^2}{\mu^2} p_0 , \end{aligned} \tag{7.3}$$

$$p_3 = \frac{\lambda+\mu}{\mu} p_2 - \frac{\lambda}{\mu} p_1$$
$$= \frac{\lambda+\mu}{\mu}\left(\frac{\lambda^2}{\mu^2} p_0\right) - \frac{\lambda}{\mu}\frac{\mu}{\mu}\left(\frac{\lambda}{\mu} p_0\right) = \frac{\lambda^3}{\mu^3} p_0 .$$

At this point, a pattern begins to emerge and we can assert that

$$p_n = \frac{\lambda^n}{\mu^n} p_0 \quad \text{for } n \geq 0 . \tag{7.4}$$

The assertion is proven by mathematical induction; that is, using the induction hypothesis together with the general equation in Eq. (7.1) yields

$$p_{n+1} = \frac{\lambda+\mu}{\mu}\left(\frac{\lambda^n}{\mu^n} p_0\right) - \frac{\lambda}{\mu}\frac{\mu}{\mu}\left(\frac{\lambda^{n-1}}{\mu^{n-1}} p_0\right)$$
$$= \frac{\lambda^{n+1}}{\mu^{n+1}} p_0 ,$$

and thus Eq. (7.4) is shown to hold. The ratio λ/μ is called the *traffic intensity* for the queueing system and is denoted by ρ for the M/M/1 system. (More generally, ρ is usually defined as the arrival rate divided by the maximum system service rate.)

We now have p_n for all n in terms of p_0 so the long-run probabilities become known as soon as p_0 can be obtained. If you review the material on Markov processes, you should see that we have taken advantage of Property 6.1, except we have not yet used the second equation given in the property. Thus, an expression for p_0 can be determined by using the norming equation, namely

$$1 = \sum_{n=0}^{\infty} p_n = p_0 \sum_{n=0}^{\infty} \frac{\lambda^n}{\mu^n} = p_0 \sum_{n=0}^{\infty} \rho^n \tag{7.5}$$
$$= \frac{p_0}{1-\rho} .$$

The equality in the above expression made use of the geometric progression[1] so it is only valid for $\rho < 1$. If $\rho \geq 1$, the average number of customers and time spent in the system increase without bound and the system becomes unstable. In some respects this is a surprising result. Based on deterministic intuitive, a person might be tempted to design a system such that the service rate is equal to the arrival rate, thus creating a "balanced" system. This is false logic for random interarrival or service times since in that case the system will never reach steady-state.

The above value for p_0 can be combined with Eq. (7.4) to obtain for the M/M/1 system

$$p_n = (1-\rho)\rho^n \quad \text{for } n = 0, 1, \cdots , \tag{7.6}$$

where $\rho = \lambda/\mu$ and $\rho < 1$.

---

[1] The geometric progression is $\sum_{n=0}^{\infty} r^n = 1/(1-r)$ for $|r| < 1$.

The steps followed in deriving Eq. (7.6) are the pattern for many other Markov queueing systems. Once the derivation of the M/M/1 system is known, all other queueing system derivations in this text will be easy. It is, therefore, good to review these steps so that they become familiar.

1. Form the Markov generator matrix, **G** (Eq. 7.2).
2. Obtain a system of equations by solving $\mathbf{pG} = \mathbf{0}$ (Eq. 7.1).
3. Solve the system of equations in terms of p_0 by successive forward substitution and induction if possible (Eq. 7.4).
4. Use the norming equation to find p_0 (Eq. 7.5).

Once the procedure becomes familiar, the only difficult step will be the third step. It is not always possible to find a closed-form solution to the system of equations, and often techniques other than successive forward substitution must be used. However, these techniques are beyond the scope of this text and will not be presented.

Example 7.1. An operator of a small grain elevator has a single unloading dock. Arrivals of trucks during the busy season form a Poisson process with a mean arrival rate of four per hour. Because of varying loads (and desire of the drivers to talk) the length of time each truck spends in front of the unloading dock is approximated by an exponential random variable with a mean time of 14 minutes. Assuming that the parking spaces are unlimited, the M/M/1 queueing system describes the waiting lines that form. Accordingly, we have

$$\lambda = 4/\text{hr}, \;\; \mu = \frac{60}{14}/\text{hr}, \;\; \rho = 0.9333 \, .$$

The probability of the unloading dock being idle is

$$p_0 = 1 - \rho = 0.0667 \, .$$

The probability that there are exactly three trucks waiting is

$$Pr\{N_q = 3\} = Pr\{N = 4\} = p_4 = 0.9333^4 \times 0.0667 = 0.05 \, .$$

Finally, the probability that four or more trucks are in the system is

$$Pr\{N \geq 4\} = \sum_{n=4}^{\infty} p_n = (1-\rho) \sum_{n=4}^{\infty} \rho^n = \rho^4 = 0.759 \, .$$

(In the above expression, the second equality is obtained by using Eq. (7.6) to substitute out p_n. The third equality comes by observing that ρ^4 is a multiplicative factor in each term of the series so that it can be "moved" outside the summation sign, making the resulting summation a geometric progression.) □

Several measures of effectiveness are useful as descriptors of queueing systems. The most common measures are the expected number of customers in the system, denoted by L, and the expected number in the queue, denoted by L_q. These expected values are obtained by utilizing Eq. (7.6) and the derivative of the geometric progression[2], yielding an expression for the expected number in an M/M/1 system as:

$$\begin{aligned} L = E[N] &= \sum_{n=0}^{\infty} n p_n \\ &= \sum_{n=1}^{\infty} n p_n = \sum_{n=1}^{\infty} n \rho^n (1-\rho) \\ &= (1-\rho)\rho \sum_{n=1}^{\infty} n \rho^{n-1} = \frac{\rho}{1-\rho}\,. \end{aligned} \tag{7.7}$$

The expected number of customers waiting within an M/M/1 queueing system is obtained similarly:

$$\begin{aligned} L_q &= 0 \times (p_0 + p_1) + \sum_{n=1}^{\infty} n p_{n+1} \\ &= \sum_{n=1}^{\infty} n \rho^{n+1} (1-\rho) \\ &= (1-\rho)\rho^2 \sum_{n=1}^{\infty} n \rho^{n-1} = \frac{\rho^2}{1-\rho}\,. \end{aligned} \tag{7.8}$$

When describing a random variable, it is always dangerous to simply use its mean value as the descriptor. For this reason, we also give the variance of the number in the system and queue:

$$V[N] = \frac{\rho}{(1-\rho)^2}\,, \tag{7.9}$$

$$V[N_q] = \frac{\rho^2(1+\rho-\rho^2)}{(1-\rho)^2}\,. \tag{7.10}$$

Waiting times are another important measure of a queueing system. Fortunately for our computational effort, there is an easy relationship between the mean number waiting and average length of time a customer waits. Little [4] showed in 1961 that for almost *all steady-state queueing systems* there is a simple relationship between the mean number in the system, the mean waiting times, and the arrival rates.

Property 7.1. ***Little's Law.*** *Consider a queueing system for which steady-state occurs. Let $L = E[N]$ denote the mean long-run number in the system, $W = E[T]$ denote the mean long-run waiting time within the system, and λ_e the mean arrival rate of jobs* into *the system. Also let $L_q = E[N_q]$ and $W_q = E[T_q]$ denote the analogous quantities restricted to the queue. Then*

[2] Taking the derivative of both sides of the geometric progression yields $\sum_{n=1}^{\infty} n r^{n-1} = 1/(1-r)^2$ for $|r| < 1$.

$$L = \lambda_e W$$
$$L_q = \lambda_e W_q \,.$$

Notice that λ_e refers to the *effective* mean arrival rate into the system; whereas, λ refers to the mean arrival rate to the system. In other words, λ includes those customers who come to the system but for some reason, like a finite capacity system that is full, they do not enter; λ_e only counts those customers who make it to the server. For the M/M/1 system, the effective arrival rate is the same as the arrival rate (i.e., $\lambda_e = \lambda$); thus

$$W = E[T] = \frac{1}{\mu - \lambda}\,, \qquad (7.11)$$
$$W_q = E[T_q] = \frac{\rho}{\mu - \lambda}\,,$$

where T is the random variable denoting the time a customer (in steady-state) spends in the system and T_q is the random variable for the time spent in the queue.

When the arrival process is Poisson, there is a generalization of Little's formula that holds for variances.

Property 7.2. *Consider a queueing system for which steady-state occurs and with a Poisson arrival stream of customers entering the system. Let N denote the number in the system, T denote the customer waiting time within the system, and λ_e the mean arrival rate of jobs* into *the system. Also let N_q and T_q denote the analogous quantities restricted to the queue. Then the following hold:*

$$V[N] - E[N] = \lambda_e^2 V[T]$$
$$V[N_q] - E[N_q] = \lambda_e^2 V[T_q]\,.$$

Little's Law (Property 7.1) is a very powerful result because of its generality. The version applied to variances (Property 7.2) is not quite as powerful since it is restricted to Poisson arrivals. Applying Property 7.2 to the M/M/1 system we obtain

$$V[T] = \frac{1}{(\mu - \lambda)^2} = \frac{1}{\mu^2(1-\rho)^2} \qquad (7.12)$$
$$V[T_q] = \frac{2\rho - \rho^2}{(\mu - \lambda)^2} = \frac{\rho(2-\rho)}{\mu^2(1-\rho)^2}\,.$$

In Example 7.1, the mean number of trucks in the system is 14 (Eq. 7.7) with a standard deviation of 14.5 (Eq. 7.9), the mean number of trucks in the queue is approximately 13.1 (Eq. 7.8) with standard deviation of 14.4 (Eq. 7.10), the mean time each truck spends in the system is 3.5 hours (Eq. 7.11) with a standard deviation of 3.5 hours (Eq. 7.12), and the mean time each truck waits in line until his turn at the dock is 3 hours and 16 minutes (Eq. 7.11) with a standard deviation of 3 hours and 29 minutes (Eq. 7.12).

Another easy and obvious relationship for queueing systems due to the definition of the system and the queue is that the mean waiting time in the system must equal the mean time in the queue plus the mean service time; thus, we have the following.

Property 7.3. *Consider a queueing system for which steady-state occurs. Let W denote the mean long-run waiting time within the system, W_q the mean long-run waiting time in the queue, and μ the mean service rate; then*

$$W = W_q + \frac{1}{\mu}.$$

Notice that this property is general like Little's Law; namely, it holds for non-exponential and multi-server systems as well as for finite and infinite capacity systems.

Example 7.2. A large car dealer has a policy of providing cars for its customers that have car problems. When a customer brings the car in for repair, that customer has use of a dealer's car. The dealer estimates that the dealer cost for providing the service is \$10 per day for as long as the customer's car is in the shop. (Thus, if the customer's car was in the shop for 1.5 days, the dealer's cost would be \$15.) Arrivals to the shop of customers with car problems form a Poisson process with a mean rate of one every other day. There is one mechanic dedicated to those customer's cars. The time that the mechanic spends on a car can be described by an exponential random variable with a mean of 1.6 days. We would like to know the expected cost per day of this policy to the car dealer. Assuming infinite capacity, we have the assumptions of the M/M/1 queueing system satisfied, with $\lambda = 0.5$/day and $\mu = 0.625$/day, yielding a $\rho = 0.8$. (Note the mean rate is the *reciprocal* of the mean time.) Using the M/M/1 equations, we have $L = 4$ and $W = 8$ days. Thus, whenever a customer comes in with car problems, it will cost the dealer \$80. Since a customer comes in every other day (on the average) the total cost to the dealer for this policy is \$40 per day. In other words, cost is equal to $\$10 \times W \times \lambda$. But by Little's formula, this is equivalent to $\$10 \times L$. The cost structure illustrated with this example is a very common occurrence for queueing systems. In other words if c is the cost per item per time unit that the item spends in the system, the expected system cost per time unit is cL. □

- *Suggestion: Do Problems 7.1–7.4.*

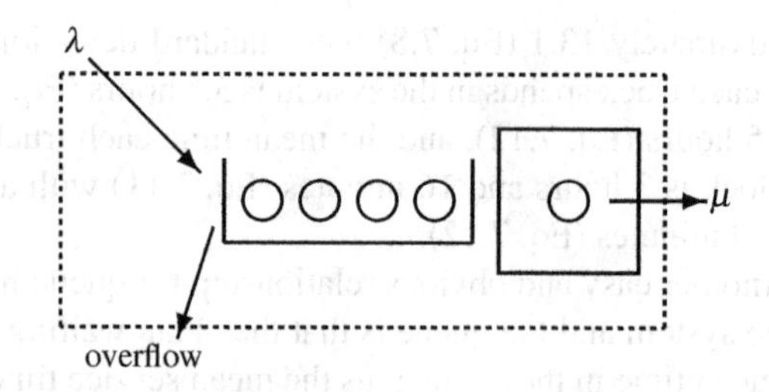

Fig. 7.3 Representation of a full M/M/1/5 queueing system

7.2.2 Finite Capacity Single Server Systems

The assumption of infinite capacity is often not suitable. When a finite system capacity is necessary, the state probabilities and measures of effectiveness presented in Eqs. (7.6–7.12) are inappropriate to use; thus, new probabilities and measures of effectiveness must be developed for the M/M/1/K system (see Fig. 7.3).

As will be seen, the equations for the state probabilities are identical except for the norming equation. If K is the maximum number of customers possible in the system, the generator matrix is of dimension $(K+1)\times(K+1)$ and has the form

$$\mathbf{G} = \begin{bmatrix} -\lambda & \lambda & & & \\ \mu & -(\lambda+\mu) & \lambda & & \\ & \ddots & \ddots & \ddots & \\ & & \mu & -(\lambda+\mu) & \lambda \\ & & & \mu & -\mu \end{bmatrix}.$$

The system $\mathbf{pG} = \mathbf{0}$ yields

$$\begin{aligned} \mu p_1 &= \lambda p_0 \\ \mu p_{n+1} &= (\lambda+\mu)p_n - \lambda p_{n-1} \quad \text{for } n = 1,\cdots,K-1 \\ \mu p_K &= \lambda p_{K-1}\,. \end{aligned} \tag{7.13}$$

Using successive substitution, we again have

$$p_n = \rho^n p_0 \text{ for } n = 0,1,\cdots,K\,,$$

where $\rho = \lambda/\mu$. The last equation in (7.13) is ignored since there is always a redundant equation in a finite irreducible Markov system. The norming equation is now used to obtain p_0 as follows:

$$1 = \sum_{n=0}^{K} p_n = p_0 \sum_{n=0}^{K} \rho^n = \begin{cases} p_0 \frac{1-\rho^{K+1}}{1-\rho} & \text{for } \rho \neq 1, \\ p_0(K+1) & \text{for } \rho = 1\,. \end{cases}$$

Since the above sum is finite, we used the finite geometric progression[3] so that ρ may be larger than one. Therefore, for an M/M/1/K system,

$$p_n = \begin{cases} \rho^n \frac{1-\rho}{1-\rho^{K+1}} & \text{for } \rho \neq 1, \\ \frac{1}{K+1} & \text{for } \rho = 1, \end{cases} \tag{7.14}$$

for $n = 0, 1, \cdots, K$.

The mean for the number in the system and queue are

$$L = \sum_{n=0}^{K} n p_n = \sum_{n=1}^{K} n p_n = p_0 \rho \sum_{n=1}^{K} n \rho^{n-1} \tag{7.15}$$

$$= \begin{cases} \rho \frac{1+K\rho^{K+1}-(K+1)\rho^K}{(1-\rho)(1-\rho^{K+1})} & \text{for } \rho \neq 1 \\ \frac{K}{2} & \text{for } \rho = 1 \end{cases}$$

$$L_q = \sum_{n=1}^{K-1} n p_{n+1} = p_0 \rho^2 \sum_{n=1}^{K-1} n \rho^{n-1} \tag{7.16}$$

$$= \begin{cases} L - \frac{\rho(1-\rho^K)}{1-\rho^{K+1}} & \text{for } \rho \neq 1 \\ \frac{K(K-1)}{2(K+1)} & \text{for } \rho = 1 \,. \end{cases}$$

The variances for these quantities for the M/M/1/K system are

$$V[N] = \begin{cases} [\rho/(1-\rho^{K+1})(1-\rho)^2] \\ \quad \times [1+\rho-(K+1)^2\rho^K \\ \quad +(2K^2+2K-1)\rho^{K+1} - K^2\rho^{K+2}] - L^2 & \text{for } \rho \neq 1, \\ K(K+2)/12 & \text{for } \rho = 1 \,. \end{cases}$$

$$V[N_q] = V[N] - p_0(L + L_q) \tag{7.17}$$

The probability that an arriving customer enters the system is the probability that the system is not full. Therefore, to utilize Little's formula, we set $\lambda_e = \lambda(1 - p_K)$ for the effective arrival rate to obtain the waiting time equations as follows:

$$W = \frac{L}{\lambda(1-p_K)} \,,$$

[3] The finite geometric progression is $\sum_{n=0}^{k-1} r^n = (1-r^k)/(1-r)$ if $r \neq 1$.

$$W_q = W - \frac{1}{\mu}.$$

Notice that the expression for W_q is simply a restatement of Property 7.3.

The temptation is to use the formulas given by Property 7.2 to obtain expressions for the variances of the waiting times. However, the "Little-like" relationships for variances are based on the assumption that the system sees Poisson arrivals. The finite capacity limitation prohibits Poisson arrivals *into* the system so the variance generalization of Little's Law cannot be used.

Example 7.3. A corporation must maintain a large fleet of tractors. They have one repairman that works on the tractors as they break down on a first-come first-serve basis. The arrival of tractors to the shop needing repair work is approximated by a Poisson distribution with a mean rate of three per week. The length of time needed for repair varies according to an exponential distribution with a mean repair time of $1/2$ week per tractor. The current corporate policy is to utilize an outside repair shop whenever more than two tractors are in the company shop so that, at most, one tractor is allowed to wait. Each week that a tractor spends in the shop costs the company \$100. To utilize the outside shop costs \$500 per tractor. (The \$500 includes lost time.) We wish to review corporate policy and determine the optimum cutoff point for the outside shop; that is, we shall determine the maximum number allowed in the company shop before sending tractors to the outside repair facility. The total operating costs per week are

$$\text{Cost} = 100L + 500\lambda p_K .$$

For the current policy, which is an M/M/1/2 system, we have $L = 1.26$ and $p_2 = 0.474$; thus, the cost is

$$\text{Cost}_{K=2} = 100 \times 1.26 + 500 \times 3 \times 0.474 = \$837/wk .$$

If the company allows three in the system, then $L = 1.98$ and $p_3 = 0.415$ which yields

$$\text{Cost}_{K=3} = 100 \times 1.98 + 500 \times 3 \times 0.415 = \$820/wk .$$

If a maximum of four are allowed in the shop, then

$$\text{Cost}_{K=4} = 100 \times 2.76 + 500 \times 3 \times 0.384 = \$852/wk .$$

Therefore, the recommendation is to send a tractor to an outside shop only when more than three are in the system. □

- *Suggestion: Do Problems 7.6 and 7.7.*

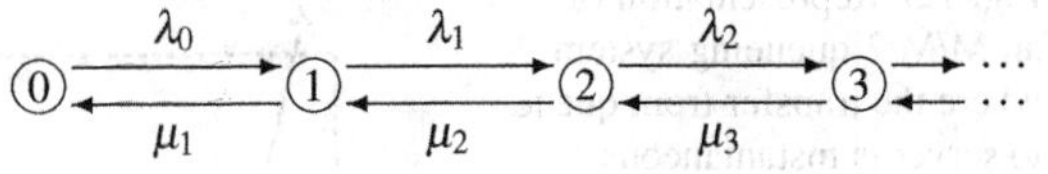

Fig. 7.4 State diagram for a birth-death process

7.3 Multiple Server Queues

A birth-death process is a special type of Markov process which is applicable to many types of Markov queueing systems. The birth-death process is a process in which changes of state are only to adjacent states (Fig. 7.4). The generator matrix for a general birth-death process is given by

$$\mathbf{G} = \begin{bmatrix} -\lambda_0 & \lambda_0 & & & \\ \mu_1 & -(\mu_1+\lambda_1) & \lambda_1 & & \\ & \mu_2 & -(\mu_2+\lambda_2) & \lambda_2 & \\ & & \ddots & \ddots & \ddots \end{bmatrix},$$

where λ_n and μ_n are the birth rate (arrival rate) and death rate (service rate), respectively, when the process is in state n.

As before, the long-run probabilities are obtained by first forming the system of equations defined by $\mathbf{pG} = \mathbf{0}$. (Or, equivalently, the system of equations may be obtained using the rate-balance approach applied to the state diagram of Fig. 7.4.) The resulting system is

$$\begin{aligned} \lambda_0 p_0 &= \mu_1 p_1 \\ (\lambda_1+\mu_1)p_1 &= \lambda_0 p_0 + \mu_2 p_2 \\ (\lambda_2+\mu_2)p_2 &= \lambda_1 p_1 + \mu_3 p_3 \,. \\ &\vdots \end{aligned}$$

This system is solved in terms of p_0 by successive substitution, and we have

$$\begin{aligned} p_1 &= p_0 \frac{\lambda_0}{\mu_1} \\ p_2 &= p_0 \frac{\lambda_0 \lambda 1}{\mu_1 \mu 2} \\ &\vdots \\ p_n &= p_0 \frac{\lambda_0 \times \cdots \times \lambda_{n-1}}{\mu_1 \times \cdots \times \mu_n} \\ &\vdots \end{aligned} \tag{7.18}$$

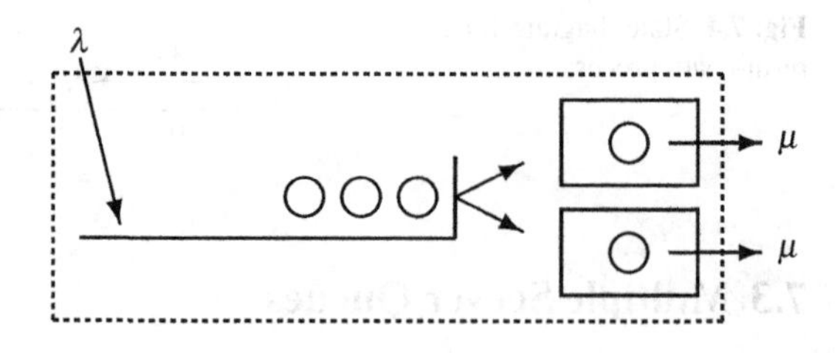

Fig. 7.5 Representation of an M/M/2 queueing system where the transfer from queue to server is instantaneous

where

$$p_0 = \left[1 + \sum_{n=1}^{\infty} \prod_{k=0}^{n-1} \frac{\lambda_k}{\mu_{k+1}}\right]^{-1} .$$

The birth-death process can be used for many types of queueing systems. Equation (7.6), for the M/M/1 system, arises from Eq. (7.18) by letting the birth rates be the constant λ and the death rates be the constant μ. The M/M/1/K system equations can be obtained by letting $\lambda_n = 0$ for all $n > K$.

The M/M/c queueing system is a birth-death process (see Fig. 7.5) with the following values for the birth rates and death rates:

$$\lambda_n = \lambda \quad \text{for } n = 0, 1, \cdots , \tag{7.19}$$

$$\mu_n = \begin{cases} n\mu \text{ for } n = 1, \cdots, c-1 \\ c\mu \text{ for } n = c, c+1, \cdots . \end{cases}$$

The reason that $\mu_n = n\mu$ for $n = 1, \cdots, c-1$ is that when there are less than c customers in the system, each customer in the system is being served and thus the service rate would be equal to the number of customers since unoccupied servers remain idle (i.e., free servers do not help busy servers). If there are more than c customers in the system, then exactly c servers are busy and thus the service rate must be $c\mu$. Substituting Eq. (7.19) into Eqs. (7.18) for the M/M/c system, yields

$$p_n = \begin{cases} p_0 r^n / n! & \text{for } n = 0, 1, \cdots, c-1 \\ p_0 r^n / (c^{n-c} c!) & \text{for } n = c, c+1, \cdots , \end{cases} \tag{7.20}$$

$$p_0 = \left[\frac{c r^c}{c!(c-r)} + \sum_{n=0}^{c-1} \frac{r^n}{n!}\right]^{-1} ,$$

where $r = \lambda/\mu$ and $\rho = r/c < 1$.

The measures of effectiveness for the M/M/c queueing system involve slightly more manipulations, but it can be shown that

$$L_q = \sum_{n=c}^{\infty} (n-c) p_n$$
$$= \frac{p_0 r^c \rho}{c!(1-\rho)^2} .$$

Little's formula is then applied to obtain

$$W_q = \frac{L_q}{\lambda},$$

and by Property 7.3

$$W = W_q + \frac{1}{\mu},$$

and finally applying Little's formula again

$$L = L_q + r.$$

The reason that we let $r = \lambda/\mu$ is because most textbooks and technical papers usually reserve ρ to be the arrival rate divided by the maximum service rate, namely, $\rho = \lambda/(c\mu)$. It can then be shown that ρ gives the server utilization; that is, ρ is the fraction of time that an arbitrarily chosen server is busy.

For completeness, we also give the formula needed to obtain the variances for the M/M/c system as

$$E[(N_q(N_q-1)] = \frac{2p_0 r^c \rho^2}{c!(1-\rho)^3} \tag{7.21}$$

$$E[T_q^2] = \frac{2p_0 r^c}{\mu^2 c^2 c!(1-\rho)^3} \tag{7.22}$$

$$V[T] = V[T_q] + \frac{1}{\mu^2} \tag{7.23}$$

$$V[N] = \lambda^2 V[T] + L. \tag{7.24}$$

Example 7.4. The corporation from the previous example has implemented the policy of never allowing more than three tractors in their repair shop. For \$600 per week, they can hire a second repairman. Is it worthwhile to do so if the expected cost is used as the criterion? To answer this question, the old cost for the M/M/1/3 system (refer back to page 212) is compared to the proposed cost for an M/M/2/3 system. The birth-death equations (7.18) are used with

$$\lambda_n = \begin{cases} \lambda \text{ for } n = 0,1,2 \\ 0 \text{ for } n = 3,4,\cdots \end{cases}$$

$$\mu_n = \begin{cases} \mu \text{ for } n = 1 \\ 2\mu \text{ for } n = 2 \text{ and } 3, \end{cases}$$

where $\lambda = 3/\text{week}$ and $\mu = 2/\text{week}$. This gives

$$
\begin{aligned}
p_1 &= 1.5p_0 \\
p_2 &= 1.125p_0 \\
p_3 &= 0.84375p_0 \\
p_0 &= \frac{1}{1+1.5+1.125+0.84375} = 0.224\,.
\end{aligned}
$$

The expected number in the system is

$$
\begin{aligned}
L &= 0.224 \times (1 \times 1.5 + 2 \times 1.125 + 3 \times 0.84375) \\
&= 1.407\,.
\end{aligned}
$$

The cost of the proposed system is

$$
\begin{aligned}
\text{Cost}_{c=2} &= 100 \times 1.407 + 500 \times 3 \times 0.189 + 600 \\
&= \$824.20/\text{week}\,.
\end{aligned}
$$

Therefore, it is not worthwhile to hire a second man since this cost is greater than the $820 calculated in the previous example. □

This section is closed with a common example that illustrates the versatility of the birth-death equation. Arrivals to a system usually come from an infinite (or very large) population so that an individual arrival does not affect the overall arrival rate. However, in some circumstances the arrivals come from a finite population, and thus, the arrival rate cannot be assumed constant.

Example 7.5. A small corporation has three old machines that continually breakdown. Each machine breaks down on the average of once a week. The corporation has one repairman that takes, on the average, one half of a week to repair a machine. (See Fig. 7.6 for a schematic of this "machine-repair" queueing system.) Assuming breakdowns and repairs are exponential random variables, the birth-death equations can be used as follows:

$$
\begin{aligned}
\lambda_n &= \begin{cases} (3-n)\lambda & \text{for } n = 0,1,2 \\ 0 & \text{for } n = 3,\cdots \end{cases} \\
\mu_n &= \mu \quad \text{for } n = 1,2,3
\end{aligned}
$$

with $\lambda = 1$ per week and $\mu = 2$ per week. This yields

$$
\begin{aligned}
p_1 &= 1.5p_0 \\
p_2 &= 1.5p_0 \\
p_3 &= 0.75p_0 \\
p_0 &= \frac{1}{1+1.5+1.5+0.75} = 0.21 \text{ and} \\
L &= 0.21 \times (1 \times 1.5 + 2 \times 1.5 + 3 \times 0.75) = 1.42\,.
\end{aligned}
$$

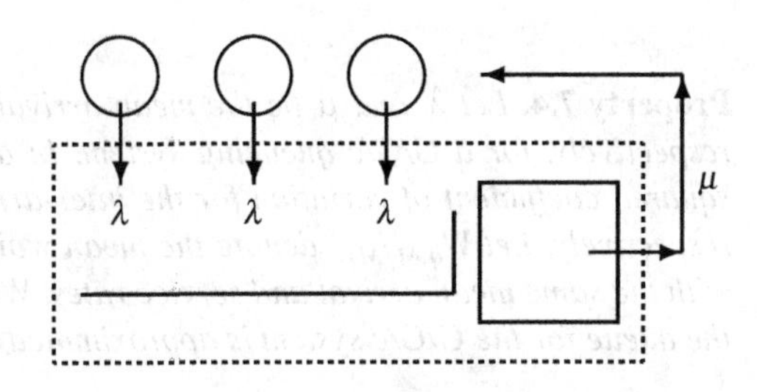

Fig. 7.6 Representation of the machine-repair queueing system of Example 7.5

Let us further assume that for every hour that a machine is tied up in the repair shop, the corporation losses \$25. The cost of this system per hour due to the unavailability of the machines is calculated as

$$\begin{aligned}\text{Cost} &= 25 \times L = 25 \times 1.42 \\ &= \$35.50/\text{ hr}\,.\end{aligned}$$

□

- *Suggestion: Do Problems 7.5 and 7.8–7.14.*

7.4 Approximations

The models developed in the previous sections depended strongly on the use of the exponential distribution. Unfortunately, the exponential assumption is not appropriate for many practical systems for which queueing models are desired. However, without the exponential assumption, exact results are much more difficult. In response to the need for numerical results for nonexponential queueing systems, there has been a significant amount of research dealing with approximations in queueing theory. In this section we report on some of these approximations. We recommend that the interested student read the survey paper by Ward Whitt [6] for a thorough review.

Assume that we are interested in modeling a G/G/c queueing system. An approximation obtained from a diffusion approximation developed by Kingman (see [1, Chap. 3]) uses the squared coefficient of variation; therefore, we define c_a^2 to be the variance of the interarrival times divided by the square of the mean of the interarrival times and c_s^2 to be the variance of the service times divided by the square of the mean of the service times.

To obtain an approximation for the mean waiting time, W, spent in a G/G/c system, we first determine the mean time spent in the queue and then add the mean service time to the queue time; namely, we use the fact that $W = W_q + 1/\mu$. The value for W_q can be obtained according to the following property.

Property 7.4. *Let λ and μ be the mean arrival rate and mean service rate, respectively, for a G/G/c queueing system. In addition, let c_a^2 and c_s^2 be the squared coefficient of variation for the inter-arrival times and service times, respectively. Let $W_{q,M/M/c}$ denote the mean waiting time for a M/M/c queue with the same mean arrival and service rates. When $\lambda < c\mu$, the time spent in the queue for the G/G/c system is approximated by*

$$W_q \approx \left(\frac{c_a^2 + c_s^2}{2}\right) W_{q,M/M/c} \,. \tag{7.25}$$

Notice that the equation holds as an equality if the interarrival and service times are indeed exponential, since the exponential has a squared coefficient of variation equal to one. Equation (7.25) is also exact for an M/G/1 queueing system, and it is known to be an excellent approximation for the M/G/c system. In general, the approximation works best for $c_a \geq 1$ with c_a and c_s being close to the same value.

Example 7.6. Suppose we wish to model the unloading dock at a manufacturing facility. At the dock, there is only one crew who does the unloading. The questions of interest are the average waiting time for arriving trucks and the average number of trucks at the dock at any point in time. Unfortunately, we do not know the underlying probability laws governing the arrivals of trucks or the service times; therefore, data are collected over a representative time period to obtain the necessary statistical estimates. The data yield a mean of 31.3 minutes and a standard deviation of 35.7 minutes for the interarrival times. For the service times, the results of the data give 20.4 minutes for the mean and 10.2 minutes for the standard deviation. Thus, $\lambda =$ 1.917 per hour, $c_a^2 = 1.30$, $\mu = 2.941$ per hour, and $c_s^2 = 0.24$ with a traffic intensity of $\rho = 0.652$. (It would be appropriate to collect more data and perform a statistical "goodness-of-fit" test to help determine whether or not these data may come from an exponential distribution. Assuming a reasonable number of data points, it becomes intuitively clear that at least the service times are not exponentially distributed since an exponential distribution has the property that its mean is equal to its standard deviation.)

In order to obtain W, we first determine W_q. From Eq. (7.11), we get that $W_{q,M/M/1} = 0.652/(2.941 - 1.917)$ hr $= 40$ min. Now using Eq. (7.25), the queue time for the trucks is $W_q \approx 40 \times (1.30 + 0.24)/2 = 30.8$ min, yielding a waiting time in the system of

$$W \approx 30.8 + 60/2.941 = 51.2 \text{ min} \,.$$

(Notice that since the mean service rate is 2.941 per hour, the reciprocal is the mean service time which must be multiplied by 60 to convert it to minutes.) It now follows from Little's Law (Property 7.1) that the mean number in the system is $L \approx 1.636$. (Notice, before using Little's formula, care must be taken to insure that the units are consistent, i.e., the quantity W must be expressed in terms of hours, because λ is in terms of hours.) □

A simple adjustment to Eq. 7.25 to account for multiple servers has been proposed by Sakasegawa [5] and used by others (e.g., [2] and [1]) as an extension to Property 7.4.

Property 7.5. *Let λ and μ be the mean arrival rate and mean service rate, respectively, for a G/G/c queueing system. In addition, let c_a^2 and c_s^2 be the squared coefficient of variation for the inter-arrival times and service times, respectively. Then the time spent in the queue for the G/G/c system is approximated by*

$$W_q \approx \left(\frac{c_a^2 + c_s^2}{2}\right) \frac{\rho^{\sqrt{2c+2}-1}}{c(1-\rho)} \frac{1}{\mu},$$

for $\rho < 1$ where $\rho = \lambda/(c\mu)$.

- *Suggestion: Do Problem 7.15.*

Appendix

The simulation of a single-server queueing system is relatively easy; however, it is not easily extended to multiple-server systems. For multiple-servers, it is best to use a event driven simulation as discussed in Chap. 9. The following material is taken from Curry and Feldman [1] to illustrate single-server simulations.

Consider a G/G/1 queueing system in which each job is numbered sequentially as it arrives. Let the service time of the nth job be denoted by the random variable S_n, the delay time (time spent in the queue) by the random variable D_n, and the inter-arrival time between the n-1st and n^{th} job by the random variable A_n. The delay time of the nth job must equal the delay time of the previous job, plus the previous job's service time, minus the inter-arrival time; however, if inter-arrival time is larger than the previous job's delay time plus service time, then the queueing delay will be zero. In other words, the following must hold

$$D_n = \max\{0,\, D_{n-1} + S_{n-1} - A_n\}\,. \tag{7.26}$$

Thus, to simulate the G/G/1 system, we need only to generate random variates for A_n and S_n for $n = 1, \cdots, n_{\max}$.

We shall simulate a G/G/1 queueing system with a mean arrival rate of 4 customers per hour and a mean service rate of 5 per hour. The inter-arrival times has a large variation having a squared coefficient of variation equal to $c_a^2 = 4$. The service times are less variable having service times distributed according to an Erlang Type-4 distribution. We shall use the gamma distribution (Eq. 1.18) to model the interarrival times. It has been the authors experience that the gamma distributed random variates with shape parameters less than one are not always reliable; however, they are fairly easy to simulate. (Can you give a good reason why normal random variates are not used for the inter-arrival times?) Since the Erlang is a special case of the gamma distribution (see p. 19), we shall also use the gamma for services. To begin the simulation, type the following in the first three rows of an Excel spreadsheet. If you do not remember how to generate the gamma random variates, see Table 2.13.)

	A	B	C
1	InterArrive	Service	Delay
2	0	`=GAMMAINV(RAND(),4,3)`	0
3	`=GAMMAINV(RAND(),0.25,60)`	`=GAMMAINV(RAND(),4,3)`	`=MAX(0,C2+B2-A3)`

Notice that the references in the C3 cell are relative references and that two of the references are to the previous row, but the third reference is to the same row. Now copy the third row down for 30,000 rows and obtain an average of the values in the C column. This average is an estimate for the mean cycle time. Hitting the F9 key will give some idea of the variability of this estimate.

- *Suggestion: Do Problem 7.16.*

Problems

7.1. Cars arrive to a toll booth 24 hours per day according to a Poisson process with a mean rate of 15 per hour.
(a) What is the expected number of cars that will arrive to the booth between 1:00 p.m. and 1:30 p.m.?
(b) What is the expected length of time between two consecutively arriving cars?
(c) It is now 1:12 p.m. and a car has just arrived. What is the expected number of cars that will arrive between now and 1:30 p.m.?
(d) It is now 1:12 p.m. and a car has just arrived. What is the probability that two more cars will arrive between now and 1:30 p.m.?
(e) It is now 1:12 p.m. and the last car to arrive came at 1:05 p.m. What is the probability that no additional cars will arrive before 1:30 p.m.?
(f) It is now 1:12 p.m. and the last car to arrive came at 1:05 p.m. What is the expected length of time between the last car to arrive and the next car to arrive?

7.2. A large hotel has placed a single fax machine in an office for customer services. The arrival of customers needing to use the fax follows a Poisson process with a mean rate of eight per hour. The time each person spends using the fax is highly variable and is approximated by an exponential distribution with a mean time of 5 minutes.
(a) What is the probability that the fax office will be empty?
(b) What is the probability that nobody will be waiting to use the fax?
(c) What is the average time that a customer must wait in line to use the fax?
(d) What is the probability that an arriving customer will see two people waiting in line?

7.3. A drill press in a job shop has parts arriving to be drilled according to a Poisson process with mean rate 15 per hour. The average length of time it takes to complete each part is a random variable with an exponential distribution function whose mean is 3 minutes.
(a) What is the probability that the drill press is busy?
(b) What is the average number of parts waiting to be drilled?
(c) What is the probability that at least one part is waiting to be drilled?
(d) What is the average length of time that a part spends in the drill press room?
(e) It costs the company 8 cents for each minute that each part spends in the drilling room? For an additional expenditure of $10 per hour, the company can decrease the average length of time for the drilling operation to 2 minutes. Is the additional expenditure worthwhile?

7.4. Derive results for the M/M/2 system using the methodology developed in Sect. 7.2 (i.e., ignore the general birth-death derivations of Sect. 7.3). Denote the mean arrival rate by λ and the mean service rate for each server by μ.
(a) Give the generator matrix for the system.
(b) Solve the system of equations given by $\mathbf{pG} = \mathbf{0}$ by using successive substitution to obtain $p_n = 2\rho^n p_0$ for $n = 1, 2, \cdots$, and $p_0 = (1-\rho)/(1+\rho)$, where $\rho = \lambda/(2\mu)$.
(c) Show that $L = 2\rho/(1-\rho^2)$ and $L_q = 2\rho^3/(1-\rho^2)$.

7.5. Derive results for the M/M/3 system using the birth-death equations in Sect. 7.3. Denote the mean arrival rate by λ, the mean service rate for each server by μ, and the traffic intensity by $\rho = \lambda/(3\mu)$
(a) Show that $p_1 = 3\rho p_0$, $p_n = 4.5\rho^n p_0$ for $n = 2, 3, \cdots$, and $p_0 = (1-\rho)/(1+2\rho+1.5\rho^2)$.
(b) Show that $L_q = 9\rho^4/((1-\rho)(2+4\rho+3\rho^2))$.

7.6. A small gasoline service station next to an interstate highway is open 24 hours per day and has one pump and room for two other cars. Furthermore, we assume that the conditions for an M/M/1/3 queueing system are satisfied. The mean arrival rate of cars is 8 per hour and the mean service time at the pump is 6 minutes. The expected profit received from each car is $5.00. For an extra $60 per day, the owner of the station can increase the capacity for waiting cars by one (thus, becoming an M/M/1/4 system). Is the extra $60 worthwhile?

7.7. A repair center within a manufacturing plant is open 24 hours a day and there is always one person present. The arrival of items needing to be fixed at the repair center is according to a Poisson process with a mean rate of 6 per day. The length of time it takes for the items to be repaired is highly variable and follows an exponential distribution with a mean time of 5 hours. The current management policy is to allow a maximum of three jobs in the repair center. If three jobs are in the center and a fourth job arrives, then the job is sent to an outside contractor who will return the job 24 hours later. For each day that an item is in the repair center, it costs the company $30. When an item is sent to the outside contractor, it costs the company $30 for the lost time, plus $75 for the repair.
(a) It has been suggested that management change the policy to allow four jobs in

the center; thus jobs would be sent to the outside contractor only when four are present. Is this a better policy?
(b) What would be the optimum cut-off policy? In other words, at what level would it be best to send the overflow jobs to the outside contractor?
(c) In order to staff and maintain the repair center 24-hours per day, it costs $400 per day. Is that a wise economic policy or would it be better to shut down the repair center and use only the outside contractor?
(d) We assume the above questions were answered using a minimum long-run expected cost criterion. Discuss the appropriateness of other considerations besides the long-run expected cost.

7.8. A small computer store has two clerks to help customers (but infinite capacity to hold customers). Customers arrive to the store according to a Poisson process with a mean rate of 5 per hour. Fifty percent of the arrivals want to buy hardware and 50% want to buy software. The current policy of the store is that one clerk is designated to handle only software customers, and one clerk is designated to handle only hardware customers; thus, the store actually acts as two independent M/M/1 systems. Whether the customer wants hardware or software, the time spent with one of the store's clerks is exponentially distributed with a mean of 20 minutes. The owner of the store is considering changing the operating policy of the store and having the clerks help with both software and hardware; thus, there would never be a clerk idle when two or more customers are in the store. The disadvantage is that the clerks would be less efficient since they would have to deal with some things they were unfamiliar with. It is estimated that the change would increase the mean service time to 21 minutes.
(a) If the goal is to minimize the expected waiting time of a customer, which policy is best?
(b) If the goal is to minimize the expected number of customers in the store, which policy is best?

7.9. In a certain manufacturing plant, the final operation is a painting operation. The painting center is always staffed by two workers operating in parallel, although because of the physical setup they cannot help each other. Thus the painting center acts as an M/M/2 system where arrivals occur according to a Poisson process with a mean arrival rate of 100 per day. Each worker takes an average of 27 minutes to paint each item. There has been recent concern about excess work in process so management is considering two alternatives to reduce the average inventory in the painting center. The first alternative is to expand the painting center and hire a third worker. (The assumption is that the third worker, after a training period, will also average 27 minutes per part.) The second alternative is to install a robot that can paint automatically. However, because of the variability of the parts to be painted, the painting time would still be exponentially distributed, but with the robot the mean time would be 10 minutes per part.
(a) Which alternative reduces the inventory the most?
(b) The cost of inventory (including the part that is being worked on) is estimated to be $0.50 per part per hour. The cost per worker (salary and overhead) is estimated to

be \$40,000 per year, and the cost of installing and maintaining a robot is estimated to be \$100,000 per year. Which alternative, if any, is justifiable using a long-term expected cost criterion?

7.10. In the gasoline service station of Problem 6.6, consider the alternative of adding an extra pump for \$90 per day. In other words, is it worthwhile to convert the M/M/1/3 system to an M/M/2/3 system?

7.11. A parking facility for a shopping center is large enough so that we can consider its capacity infinite. Cars arrive to the parking facility according to a Poisson process with a mean arrival rate of λ. Each car stays in the facility an exponentially distributed length of time, independent from all other cars, with a mean time of $1/\mu$. Thus, the parking facility can be viewed as an M/M/∞ queueing system.
(a) What is the probability that the facility contains n cars?
(b) What is the long-run expected number of cars in the facility?
(c) What is the long-run variance of the number of cars in the facility?
(d) What is the long-run expected queue length for the M/M/∞ system?
(e) What is the mean expected time spent in the system?

7.12. A company offers a correspondence course for students not passing high school algebra. People sign up to take the course according to a Poisson process with a mean of two per week. Students taking the course progress at their own rate, independent of how many other students are also taking the correspondence course. The actual length of time that a student remains in the course is an exponential random variable with a mean of 15 weeks. What is the long-run expected number of students in the course at an arbitrary point in time?

7.13. A company has assigned one worker to be responsible for the repair of a group of five machines. The machines break down independent of each other according to an exponential random variable. The mean length of working time for a machine is 4 days. The time it takes the worker to repair a machine is exponentially distributed with a mean of two days.
(a) What is the probability that none of the machines are working?
(b) What is the expected number of machines working?
(c) When a machine fails, what is the expected length of time until it will be working again?

7.14. A power plant operating 24 hours each day has four turbine-generators it uses to generate power. All turbines are identical and are capable of generating 3 megawatts of power. The company needs 6 megawatts of power, so that when all turbines are in a working condition, it keeps one turbine on "warm-standby", one turbine on "cold-standby", and two turbines operating. If one turbine is down, then two are operating and one is on "warm-standby". If two turbines are down, both working turbines are operating. If only one turbine is working, then the company must purchase 3 megawatts of power from another source. And, if all turbines are down, the company must purchase 6 megawatts. If a turbine is in the operating mode, its time until failure is 3 weeks. If a turbine is in "warm-standby", its time until failure is 9 weeks. And, if a turbine is in "cold-standby", it cannot fail. (We assume all switch-overs from warm standby to working or cold standby to warm standby are instantaneous.) The company has two workers that can serve to repair a failed turbine and it takes a worker one half a week, on the average, to repair a failed turbine. Assuming all times are exponentially distributed, determine the expected megawatt hours that must be purchased each year.

7.15. A store manager with training in queueing theory wants to take quick action on the first day at work. One of the biggest complaints that have been heard is the length of the waiting time and the length of the line. The manager asked one of the employees to record arrival times of customers to the cashier arriving roughly between 8:00 AM and noon. The following arrival times were collected: 8:05, 8:07, 8:17, 8:18, 8:19, 8:25, 8:27, 8:32, 8:35, 8:40, 8:45, 8:47, 8:48, 8:48, 9:00, 9:02, 9:14, 9:15, 9:17, 9:23, 9:27, 9:29, 9:35, 9:37, 9:45, 9:55, 10:01, 10:12, 10:15, 10:30, 10:32, 10:39, 10:47, 10:50, 11:05, 10:07, 11:25, 11:27, 11:31, 11:33, 11:43, 11:49, 12:05.

Another employee measured the service times of the cashier. The service times had a mean of 3.5 minutes and a standard deviation of 5.0 minutes. There is only one cashier that services the customers.
(a) Estimate the expected length of time that a customer has to wait before service and the expected number of customers in front of the cashier using Property 7.4 or 7.5.
(b) Suppose that the manager's knowledge of queueing theory was marginal and that an M/M/1 queueing model was wrongly used assuming that the arrival rate was Poisson with the mean estimated from the data and that the service times were exponential with mean 3.5 minutes. Determine the approximate difference between the estimates obtained in part (a) and the estimates that would be obtained using the (incorrect) Markovian assumptions.
(c) The store manager has two alternatives to reduce the waiting time at the cashier. One alternative is to buy a bar reader machine that would reduce the standard deviation of the service time to 1.7 minutes and pay a monthly service fee of $200. The other alternative is to hire a second cashier to work with a currently available cashier machine. The new cashier would work at the same rate as the other and will cost the store $350 per month. Assuming that it costs $0.25 per minute per customer waiting to pay, the manager wants to know what is the best strategy. (Assume this is a problem only four hours per day, five days a week. Furthermore, assume that the steady-state solution is a good approximation to the queueing behavior that occurs during those four hours.)

7.16. Continue the simulation of the appendix for several years. For each year, calculate the average time spent in the queue per customer and the number of customers who arrive during the year. Give 95% confidence intervals for mean waiting time per customer and the annual arrivals. Each data point in the random sample used for the estimates is composed of the annual data; thus, if the simulation is run for 25 years, there will be 25 data points for each estimate. Compare the simulated results with analytical estimates for W_q and the number of arrival per year.

References

1. Curry, G.L., and Feldman, R.M. (2009). *Manufacturing Systems Modeling and Analysis*, Springer-Verlag, Berlin.
2. Hopp, W.J., and Spearman, M.L. (1996). *Factory Physics: Foundations of Manufacturing Management*, Irwin, Chicago.
3. Kendall, D.G. (1953). Stochastic Processes Occuring in the Theory of Queues and their Analysis by the Method of Imbedded Markov Chains. *Annals of Mathematical Statistics*, **24**:338–354.
4. Little, J.D.C. (1961). A Proof for the Queuing Formula $L = \lambda W$. *Operations Research*, **9**:383–387.
5. Sakasegawa, H. (1977). An Approximation Formula $L_q = \alpha\beta^{\rho}/(1-\rho)$, *Annals of the Institute for Statistical Mathematics*, **29**:67–75.
6. Whitt, W. (1993). Approximations for the GI/G/m Queue, *Production and Operations Management*, **2**:114–161.

Chapter 8
Queueing Networks

The application of queueing theory often involves several queueing processes connected to one another as departures from one process become arrivals to another process. For example, if a bank were to be modeled, some customers from the "New Accounts" desk might go next to a teller line or to the safety deposit room. Or if the Panama Canal were to be simulated, each of the three locks would have to be modeled as a queueing process with customers (i.e., ships) flowing from one lock to another or from the entrance of the canal to a lock. Many manufacturing systems and supply chains can be modeled as a complex queueing network in which output from one process flows to other queueing processes, sometimes with probabilistic routing and sometimes with deterministic routing.

In this chapter, we present an introduction to queueing networks and some approximations useful for their analyses. We present both open networks and closed networks and discuss some of the computational issues involved. In the previous chapter, the term "customers" was used to describe the generic entities that entered a queue. In this chapter, the term for the entities that flow through a network will be called "jobs". Thus, if we were to model a bank, the term "job" would refer to customers; in the modeling of the Panama Canal, the term "job" may refer to ships; and in the modeling of supply chains, the term "job" may refer to goods.

8.1 Jackson Networks

Modeling some network systems may be relatively easy if the network has some special structure. In particular, if the system has Poisson arrivals, and each queue within the network has unlimited capacity and exponential servers, the system is called a Jackson network, and some of its measures of effectiveness can be computed using the methodology developed in the previous chapter. In particular, the following formally defines those networks that are most easily analyzed.

Definition 8.1. A network of queues is called a *Jackson network* if the following conditions are satisfied:

R.M. Feldman, C. Valdez-Flores, *Applied Probability and Stochastic Processes*, 2nd ed., DOI 10.1007/978-3-642-05158-6_8, © Springer-Verlag Berlin Heidelberg 2010

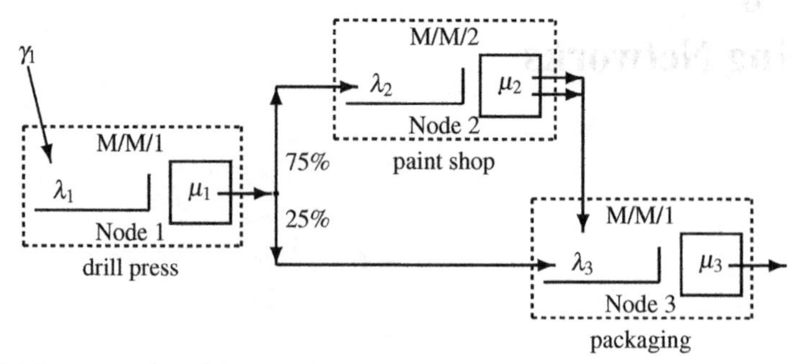

Fig. 8.1 Representation of the queueing network of Example 8.1

1. All outside arrivals to each queueing system in the network must be according to a Poisson process.
2. All service times must be exponentially distributed.
3. All queues must have unlimited capacity.
4. When a job leaves one queueing system, the probability that it will go to another queueing system is independent of that job's past history and is independent of the location of any other job.

□

The effect of these conditions is that the M/M/c formulas, developed in the previous chapter, can be used for the analysis of networks. As can be seen, the first three conditions are from the standard M/M/c assumptions. The fourth condition is not quite as straight forward, but it can be viewed as a Markov assumption; that is, its effect is that the movement of jobs through the network can be described as a Markov chain.

It is possible to have either an open Jackson network or a closed Jackson network. An open network will have jobs arriving from the "outside" according to Poisson processes and eventually jobs leaving the network. A closed network will have no arrivals and no departures. The machine-repair problem of Example 7.5 can be restructured to be a closed queueing network. The first node would be the queueing system containing working machines. When a machine breaks down, it departs from this first queueing system and arrives into the second system which consists of all machines needing repair. When a machine is repaired, it returns to the first system.

8.1.1 Open Jackson Networks

We begin this section with an example of an open Jackson network and then discuss how to analyze the network. It turns out that the analysis is a two-step process: (1) the net inflow rates are determined and (2) each node is analyzed separately as if the nodes were independent.

Example 8.1. Jobs arrive to a job shop according to a Poisson process with a mean rate of 8 jobs per hour. Upon arrival, the jobs first go through a drill press, then 75% of them go to the paint shop, and then go to a final packaging step. Those jobs that do not go to the paint shop, go directly to the final packaging step. There is one drill press, two painters, and one packer. The mean time spent on the drill press is 6 minutes, the mean time spent during the painting operation is 15 minutes, and the mean time for packaging is 5 minutes; all times are exponentially distributed. (See Fig. 8.1 for a schematic representation of the network.) Thus, this is a Jackson network, and the movement of jobs within the network can be described by the sub-Markov matrix (called the *routing matrix*):

$$\mathbf{P} = \begin{bmatrix} 0 & 0.75 & 0.25 \\ 0 & 0 & 1 \\ 0 & 0 & 0 \end{bmatrix},$$

where the drill press is Node 1, the paint shop is Node 2, and packaging is Node 3. □

The basic idea for the analysis of Jackson networks is that a net arrival rate is determined for each node, and then each queueing system (node) is treated as if it were an independent system. To illustrate, we return to the above job shop example, and let λ_i denote the arrival rate into Node i; thus $\lambda_1 = 8$, $\lambda_2 = 0.75 \times 8 = 6$, and $\lambda_3 = 0.25 \times 8 + 6 = 8$. The other data of the example indicate that $\mu_1 = 10$, $\mu_2 = 4$, and $\mu_3 = 12$; thus, the traffic intensities are $\rho_1 = 0.8$, $\rho_2 = 0.75$ (note that Node 2 has two servers), and $\rho_3 = 0.667$. With these parameters, we apply the previously developed M/M/1 (Eqs. 7.6 and 7.7) and M/M/2 (see Problem 7.4) formulas. The average number of jobs at each node is now easily calculated as $L_1 = 4$, $L_2 = 3.43$, and $L_3 = 2$, and the probabilities that there are no jobs in the various nodes are given as $p_0^1 = 0.2$, $p_0^2 = 0.143$, and $p_0^3 = 0.333$.

The only potentially difficult part of the analysis is determining the input rate to each queueing node. As in the above example, we let λ_i denote the total input to Node i. First, observe that for a steady-state system, the input must equal the output, so the total output from Node i also equals λ_i. Secondly, observe that the input to Node i must equal the input from the outside, plus any output from other nodes that are routed to Node i. Thus, we have the general relation

$$\lambda_i = \gamma_i + \sum_k \lambda_k P(k,i) , \tag{8.1}$$

where γ_i is the arrival rate to Node i from outside the network, and $P(k,i)$ is the probability that output from Node k is routed to Node i. There is one such equation for each node in the network, and the resulting system of equations can be solved to determine the net input rates. In general, the following property holds.

Property 8.1. *Let the routing matrix* $\mathbf{P}$ *describe the flow of jobs within a Jackson network, and let* γ_i *denote the mean arrival rate of jobs going directly into Node i from outside the network with* $\boldsymbol{\gamma}$ *being the vector of these rates. Then*

$$\boldsymbol{\lambda} = \boldsymbol{\gamma}(\mathbf{I}-\mathbf{P})^{-1}$$

where the components of the vector $\boldsymbol{\lambda}$ *give the arrival rates into the various nodes; i.e.,* λ_i *is the net rate into Node i.*

After the net rate into each node is known, the network can be decomposed and each node treated as if it were a set of independent queueing systems with Poisson input. (If there is any feedback within the queueing network, the input streams to the nodes will not only be dependent on each other, but also there will be input processes that are not Poisson. The ingenious contribution of Jackson [3] was to prove that if the analyst ignores that fact that some of the internal arrival streams are not Poisson, the correct state probabilities will be obtained.)

Property 8.2. *Consider a Jackson network containing m nodes with Node i having* c_i *(exponential) servers. Let* N_i *denote a random variable indicating the number of jobs at Node i (the number in the queue plus the number in the server(s)). Then,*

$$\Pr\{N_1 = n_1, \cdots, N_m = n_m\} = \Pr\{N_1 = n_1\} \times \cdots \times \Pr\{N_m = n_m\}\,,$$

and the probabilities $\Pr\{N_i = n_i\}$ *for* $n_i = 0, 1, \cdots$ *can be calculated using the appropriate M/M/*c_i *formula (or the birth-death equations of 7.20).*

Example 8.2. Now let us change the previous example slightly. Assume that after the painting operation, 10% of the jobs are discovered to be defective and must be returned to the drill press operation, and after the packaging operation, 5% of the jobs are discovered to be defective and must also be returned to the beginning of the drill press operation (see Fig. 8.2). Now the routing matrix describing the movement of jobs within the network is

$$\mathbf{P} = \begin{bmatrix} 0 & 0.75 & 0.25 \\ 0.10 & 0 & 0.90 \\ 0.05 & 0 & 0 \end{bmatrix}.$$

The equations defining the net arrival rates are

$$\begin{aligned} \lambda_1 &= 8 + 0.1\lambda_2 + 0.05\lambda_3 \\ \lambda_2 &= 0.75\lambda_1 \\ \lambda_3 &= 0.25\lambda_1 + 0.9\lambda_2\,. \end{aligned}$$

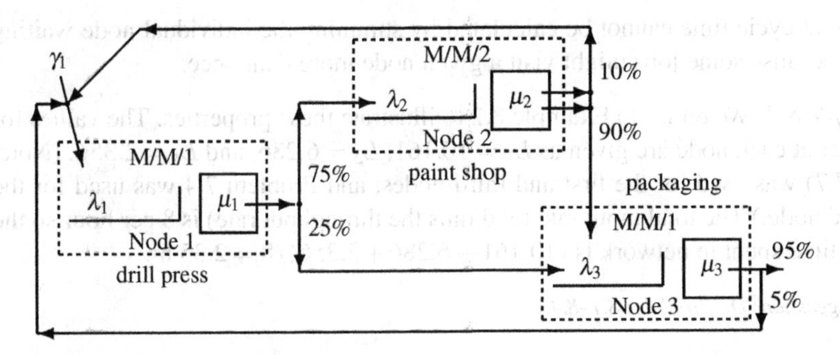

Fig. 8.2 Representation of the queueing network of Example 8.2

The vector $\boldsymbol{\gamma}$ is (8,0,0) and solving the above system yields $\lambda_1 = 9.104$, $\lambda_2 = 6.828$, and $\lambda_3 = 8.421$, which implies traffic intensities of $\rho_1 = 0.9104$, $\rho_2 = 0.8535$, and $\rho_3 = 0.702$. Again we use the M/M/1 and M/M/2 formulas and obtain the average number in the network to be 18.8. Thus, the inspection and resulting feedback doubles the mean number of jobs in the system. (Note that the number of jobs in the system is the sum of the number of jobs at the individual nodes.) The probability that there are no jobs in the system is 0.002. (Note that the probability that there are no jobs in the system equals the product of the probabilities that there are no jobs at each of the individual nodes.) □

Three measures that are often of interest in a network analysis are (1) mean system throughput, (2) mean number of jobs within the system (sometimes called work-in-process or system WIP), and (3) mean time an arbitrary job spends in the system (sometimes called system cycle time). Because we are only considering steady-state conditions, the mean system throughput equals the mean arrival rate to the system (i.e., the sum of the individual γ_i terms). The mean system size is obtained by summing the mean number at each node, and then the cycle time is obtained through the application of Little's Law. These are summarized in the following property under the assumption that each node is operating under steady-state conditions.

Property 8.3. *Consider a Jackson network containing m nodes. Denote by N_{net} the total number of jobs within the network, T_{net} the time that a job spends in the network, and $\boldsymbol{\gamma}$ the vector of external arrival rates. Thus we have that*

$$E[N_{\text{net}}] = \sum_{i=1}^{m} L_i\,, \quad \textit{and}$$

$$E[T_{\text{net}}] = \frac{E[N_{\text{net}}]}{\sum_{i=1}^{m} \gamma(i)}\,,$$

where the individual L_i terms are given by the appropriate M/M/c formula.

Note that cycle time cannot be calculated by summing the individual node waiting times because some jobs might visit a given node more than once.

Example 8.3. We return to Example 8.2 to illustrate these properties. The values for number at each node are given as $L_1 = 10.161$, $L_2 = 6.286$, and $L_3 = 2.356$. (Note, Eq. (7.7) was used for the first and third nodes, and Problem 7.4 was used for the second node.) The total input rate (and thus the throughput rate) is 8 per hour so the mean time spent in network is $(10.161 + 6.286 + 2.356)/8 = 2.35$ hr. □

- *Suggestion: Do Problems 8.1–8.4.*

8.1.2 Closed Jackson Networks

A common production control strategy is called a CONWIP strategy. This is a "pull" strategy in which a constant WIP level is maintained within the production facility. The idea is that a new job is not started until a finished job leaves; thus, the job that leaves the system "pulls" the next job into the system. Such a system can be modeled using a closed queueing network as illustrated in the following example.

Example 8.4. Consider a manufacturing facility in which all jobs start in the machine room. The machine room has one machine that has an exponential processing time with a mean time of 80 minutes. After finishing on the machine, the job leaves the machine room and 70% of the jobs go to the electrical components room with the other 30% going directly to the finishing room. There is one machine installing electrical components and the time of this installation process is random taking an exponentially distributed length of time with a mean of 111.4 minutes. Defects are often discovered during the electrical component installation and approximately 20% of the jobs are returned to the machine room where the processing must be reworked. (The distribution of processing time in the machine room and electrical components room is the same whether a job is new or is being reworked.) After the electrical components have been successfully installed, the jobs proceed to the finishing room. There is one finisher who takes an exponentially distributed time to complete the job and ready it for shipment. The average finishing time is 64 minutes.

Company policy is to maintain a WIP level of 5 units so that a new job is started as soon as a job leaves the finishing center. The movement of jobs described above is illustrated in Fig. 8.3 and yield a routing matrix of

$$\mathbf{P} = \begin{bmatrix} 0 & 0.7 & 0.3 \\ 0.2 & 0 & 0.8 \\ 1 & 0 & 0 \end{bmatrix}.$$

Note that we model this process as if a job travels from the third state to the first state even though physically it is a new job that enters the first state whenever a job leaves the third state. □

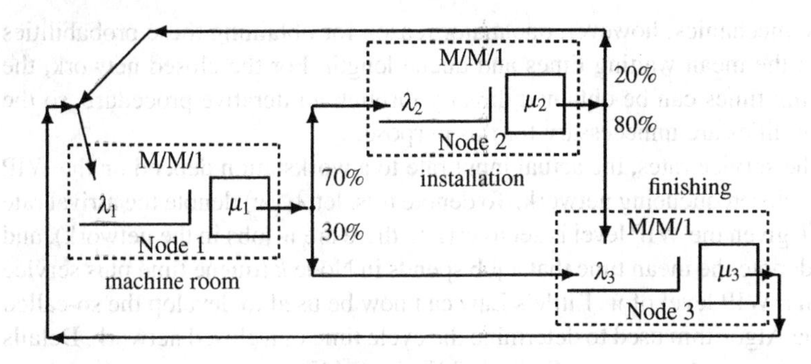

Fig. 8.3 Representation of the closed queueing network of Example 8.4

The analysis of the system of Example 8.4 starts with determining arrival rates; however, there are two difficulties with the use of Property 8.1: (1) the matrix $(\mathbf{I}-\mathbf{P})$ is singular (i.e., it has no inverse) and (2) since there no external arrivals, the vector γ is all zeros. It is only possible to obtain *relative* arrival rates. To differentiate between the relative arrival rates and the actual arrival rates, the relative rates will be denoted by the vector $\mathbf{r}$.

To determine the relative arrival rates, we arbitrarily set the first rate to 1.0 and then the remaining rates are obtained using equations of the form $r_k = \sum_i r_i p_{i,k}$ for each k. These equations are the same as in Eq. (8.1) with each γ_i term equal to zero. This results in the following property.

Property 8.4. *Let* $\mathbf{P}$ *denote the routing matrix associated with a closed queueing network containing m nodes. Form the matrix* $\mathbf{Q}$ *by setting it equal to* $\mathbf{P}$ *after deleting the first row and first column; that is,* $q_{i,j} = p_{i+1,j+1}$ *for* $i, j = 1, \cdots, m-1$. *Then the vector of relative arrival rates,* $\mathbf{r}$, *is given by*

$$(r_2, \cdots, r_m) = (p_{1,2}, \cdots, p_{1,m})\,(\mathbf{I}-\mathbf{Q})^{-1}$$

and $r_1 = 1$.

Notice that the row vector to the left of the $(\mathbf{I}-\mathbf{Q})^{-1}$ matrix is the first row of the routing matrix minus the first element.

The application of Property 8.4 to Example 8.4 results in

$$(r_2, r_3) = (0.7, 0.3)\begin{bmatrix} 1 & -0.8 \\ 0 & 1 \end{bmatrix}^{-1} = (0.7, 0.86)\,; \qquad (8.2)$$

thus, $\mathbf{r} = (1, 0.7, 0.86)$.

Property 8.2 giving the probabilities of the number of jobs at each node holds for both open and closed networks; however, the probabilities that are obtained are relative and must be normed. There are several queueing books (for example, [2]) that cover these mechanics; however, one major reason for obtaining these probabilities is to obtain the mean waiting times and queue length. For the closed network, the mean waiting times can be obtained directly through an iterative procedure, so the state probabilities are unnecessary for this purpose.

Given the service rates, the actual input rate to a workstation depend on the WIP level of the closed queueing network. To denote this, let $\lambda_k(w)$ denote the arrival rate into Node k given the WIP level is set to w (i.e., there are w jobs in the network), and let $W_k(w)$ denote the mean time that a job spends in Node k (queue time plus service time) given a WIP level of w. Little's Law can now be used to develop the so-called Mean Value Algorithm used to determine the cycle time of a closed network. Details of its derivation can be found in Curry and Feldman [1].

Property 8.5. ***Mean Value Analysis.*** *Consider a closed network with m nodes containing $w_{\max}$ jobs. Each node has a single exponential server with mean service rate $1/\mu_i$, for $i = 1, \cdots, m$, and relative arrival rates given by the m-dimensioned vector $\mathbf{r}$ determined from Property 8.4. The following algorithm can be used to obtain the mean waiting times for each node.*

1. *Set $W_k(1) = 1/\mu_k$ for $k = 1, \cdots, m$ and set $w = 2$.*
2. *Determine $W_k(w)$ for $k = 1, \cdots, m$ by*

$$W_k(w) = \frac{1}{\mu_k}\left[1 + \frac{(w-1)\, r_k W_k(w-1)}{\sum_{j=1}^{m} r_j W_j(w-1)}\right].$$

3. *If $w = w_{\max}$, stop; otherwise, increment w by 1 and return to Step 2.*

Once the mean waiting times are derived, the mean arrival rate into each node is obtained by

$$\lambda_k(w) = \frac{w\, r_k}{\sum_{j=1}^{n} r_j W_j(w)}. \tag{8.3}$$

Example 8.5. We continue Example 8.4 and determine waiting times and system throughput. The relative arrival rates are $\mathbf{r} = (1, 0.7, 0.86)$ from Eq. (8.2). These rates allow the use of Property 8.5 from which the values of Table 8.1 are derived.

Table 8.1 Mean cycle time results (in min.) for Example 8.5

Iteration	$W_1(w)$	$W_2(w)$	$W_3(w)$	$\sum r_j W_j(w)$
$w = 1$	80.0	111.4	64.0	213.0
$w = 2$	110.0	152.2	80.5	285.8
$w = 3$	141.6	194.4	95.0	359.4
$w = 4$	174.6	238.0	107.7	433.7
$w = 5$	208.8	282.5	118.6	508.6

Once the mean waiting time for each node is obtained, the arrival rate can be derived through Eq. (8.3). Then with the waiting time and arrival rate determined, the number of jobs at each node can be determined using Little's Law (Property 7.1). The traffic intensity, $\rho_k = \lambda_k/\mu_k$, (or server utilization) at each node is also an important measure commonly used to help with identifying bottlenecks. These factors are presented in Table 8.2. In calculating these quantities, it is important to take care that the time units of the various quantities are consistent (i.e., be careful not to mix minutes and hours within the same equation).

Table 8.2 Workstation characteristics for Example 8.5 at a CONWIP level of 5

Measure	Node 1	Node 2	Node 3
$W_k(5)$	208.8 min	282.5 min	118.6 min
$\lambda_k(5)$	0.59/hr	0.41/hr	0.51/hr
$L_k(5)$	2.1	1.9	1
$\rho_k(5)$	0.786	0.767	0.541

□

The Mean Value Analysis Algorithm is relatively easy to implement in Excel and this is discussed in the Appendix. The algorithm can also be extended to more general situations like multi-server exponential nodes and non-exponential servers. The interested reader should refer to Chap. 8 of [1].

- *Suggestion: Do Problem 8.5.*

8.2 Network Approximations

Many systems that involve networks are not Jackson networks but can be modeled through approximations. A very common scenario is that instead of each job being routed through the network according to a Markov chain, there are classes of jobs where each job within a class has a fixed route. For example, consider a job shop with two machines. Further suppose that jobs arrive at a rate of 10 per hour, where arriving jobs always start on machine one, proceed to machine two, then return to machine one for a final processing step (see the left-hand schematic in Fig. 8.4). Such a route is deterministic, and thus does not satisfy the Markov assumption (Condition 4) of the Jackson network definition. However, a Jackson network can be used to approximate a network with deterministic routing by fixing the routing probabilities so that the mean flow rates in the Jackson network are identical to the flow rates in the network with deterministic routes. In this job shop example, we could assume that all jobs start at machine one, then there is a 50% probability that a job leaving machine one exits the system or proceeds to machine two; thus we have the routing matrix

$$\mathbf{P} = \begin{bmatrix} 0 & 0.5 \\ 1 & 0 \end{bmatrix},$$

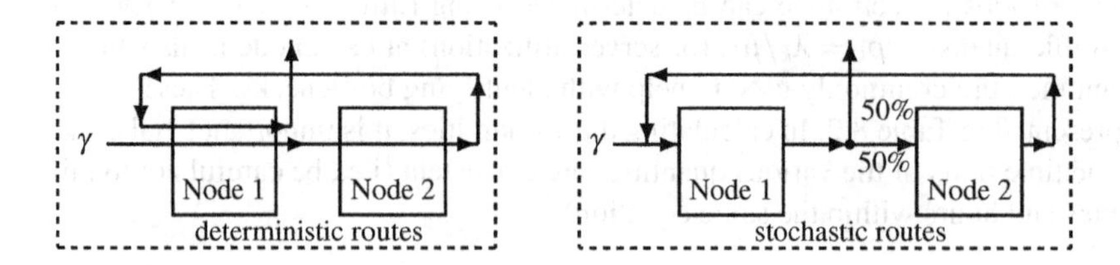

Fig. 8.4 Comparison between a network with deterministic routing and with stochastic routing

governing the flow of jobs within the two-node network (see the right-hand schematic in Fig. 8.4). Notice that in both networks, an input rate of $\gamma = 10$/hr results in a net flow of 20 per hour into Node 1 and 10 per hour into Node 2. The difficulty with the Jackson network approximation is that the variances of the two networks will not be the same.

A Jackson network model will always produce a high variability in flow times; thus, if flow times are deterministic as in the above example, a Jackson network model would be inappropriate. However, consider a network system with several distinct classes. Even if each class has deterministic flow times, the actual flow times through the network would be random because of the variability in classes. Thus, when a system contains several different classes, it may be reasonable to model the aggregate as a Jackson network.

8.2.1 *Deterministic Routing with Poisson Input*

The purpose of this subsection is to show how to model a network in which there are several different classes of jobs, where all jobs within one class have a fixed route through the network. We shall continue to assume an unlimited capacity at each node. In addition, we assume that the input stream to each node is Poisson. This is clearly an approximation because as soon as a server is non-exponential, the departure stream from that server cannot be Poisson even if the input were Poisson. However, in this subsection, we shall assume that there are several classes of jobs so that the input stream to each node is the superposition of several individual streams yielding a net stream with Poisson characteristics. (It has been shown that when many independent arrival processes are added together, the resultant arrival process tends to be Poisson even if the individual processes are not. Poisson streams also arise in systems with a large number of feedback loops as in semiconductor manufacturing.) The two main issues we need to deal with are how to obtain each node's arrival rate and how to obtain the mean service times.

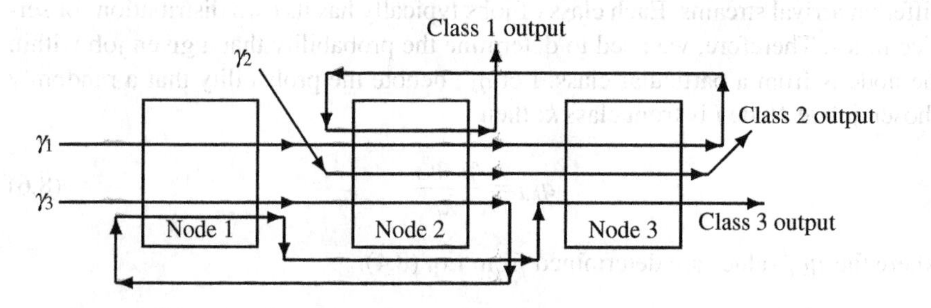

Fig. 8.5 Routings within a queueing network containing three classes for Example 8.6

Calculating the Arrival Rate to a Node. We assume that an m-node queueing network has $\bar{k}$ classes of jobs. Each class of jobs has associated with it an arrival rate, γ_k, and a "routing" vector, $(n_{k,1}, n_{k,2}, \cdots, n_{k,\bar{n}_k})$, where $\bar{n}_k$ is the total number of steps within that route. For example, assume that the routing vector for the first class is (1,2,1). In this case, $\bar{n}_1 = 3$ and first class jobs start at Node 1, then go to Node 2, and then back to Node 1 for final processing.

Example 8.6. Consider the network illustrated in Fig. 8.5. It is a queueing network with three classes of jobs, each class having its own deterministic route. Class 1 jobs have the routing vector (1,2,3,2) with $\bar{n}_1 = 4$; Class 2 jobs have the routing vector (2,3) with $\bar{n}_2 = 2$; and, Class 3 jobs have the routing vector (1,2,1,3) with $\bar{n}_3 = 4$. With deterministic routes it is also easy to determine the total flows through each node: $\lambda_1 = \gamma_1 + 2\gamma_3$, $\lambda_2 = 2\gamma_1 + \gamma_2 + \gamma_3$, and $\lambda_3 = \gamma_1 + \gamma_2 + \gamma_3$. □

From the example, it is seen that the net arrival rate is obtained primarily by counting the number of times a path goes through the node. To state this explicitly, let γ_k be the arrival rate of class k jobs to the network, and let those jobs have a route defined by $(n_{k,1}, n_{k,2}, \cdots, n_{k,\bar{n}_k})$. Let $\eta_{k,i}$ be the number of times that route k passes through Node i; that is,

$$\eta_{k,i} = \sum_{\ell=1}^{\bar{n}_k} I(i, n_{k,\ell}) , \tag{8.4}$$

where $\mathbf{I}$ is the identity matrix so that $I(i,n) = 1$ if and only if $i = n$. The net arrival rate to Node i in the network is then

$$\lambda_i = \sum_{k=1}^{\bar{k}} \gamma_k \, \eta_{k,i} , \tag{8.5}$$

or in matrix notation $\boldsymbol{\lambda} = \boldsymbol{\gamma\eta}$.

Calculating the Service Rate at a Node. After the net arrival rate has been determined, each node in the network can be considered as a separate queueing system. Although the arrival process is not Poisson, it is assumed to be Poisson based on the assumption that all arrival streams are composed of a superposition of several different arrival streams. Each class of jobs typically has its own distribution for service times. Therefore, we need to determine the probability that a given job within the node is from a particular class. Let $q_{k,i}$ denote the probability that a randomly chosen job at Node i is from class k; then

$$q_{k,i} = \frac{\gamma_k \eta_{k,i}}{\lambda_i}, \tag{8.6}$$

where the $\eta_{k,i}$ values are determined from Eq. (8.4).

Example 8.7. We return to the network of Example 8.6 and Fig. 8.5. The matrix that contains the values for $\eta_{k,i}$ is given by

$$\begin{array}{c} \text{node} \\ \begin{array}{cc} \text{job} & 1 \;\; 2 \;\; 3 \end{array} \end{array}$$

$$\boldsymbol{\eta} = \begin{matrix} 1 \\ 2 \\ 3 \end{matrix} \begin{bmatrix} 1 & 2 & 1 \\ 0 & 1 & 1 \\ 2 & 1 & 1 \end{bmatrix}.$$

(Note that, in general, the $\boldsymbol{\eta}$ matrix need not be square.) Assume that $\gamma_1 = 10$/shift, $\gamma_2 = 20$/shift, and $\gamma_3 = 50$/shift. The application of Eq. (8.5) yields input rates to each node of $\lambda_1 = 110$/shift, $\lambda_2 = 90$/shift, and $\lambda_3 = 80$/shift. The matrix of probabilities containing the terms $q_{k,i}$ is given by

$$\mathbf{q} = \begin{bmatrix} 0.091 & 0.222 & 0.125 \\ 0 & 0.222 & 0.250 \\ 0.909 & 0.556 & 0.625 \end{bmatrix}$$

according to Eq. (8.6). □

The service time distribution is a mixture of the service times so that the mean and variance can be determined from Property 1.11 (p. 35) as follows.

Property 8.6. *Assume that the mean and variance of the service times at Node i for jobs of Class k are given by* $1/\mu_{k,i}$ *and* $\sigma^2_{k,i}$*, respectively. Let* S_i *be a random variable denoting the service time at Node i of a randomly selected job. Then*

$$E[S_i] = \sum_{k=1}^{\bar{k}} q_{k,i} \frac{1}{\mu_{k,i}}$$

$$E[S_i^2] = \sum_{k=1}^{\bar{k}} q_{k,i} (\sigma^2_{k,i} + (1/\mu_{k,i})^2),$$

where the $q_{k,i}$ *values are determined by Eq. (8.6).*

Flow Time Through a Network. We can now combine Property 8.6 with the approximation of Property 7.4 to obtain an estimate for the mean flow time of a job through a non-Jackson network. The two major assumptions are that there is no queue capacity limitation and there are several classes of jobs arriving to each node so that arrivals can be approximated by a Poisson process.

Property 8.7. *Consider a network containing m nodes and $\bar{k}$ distinct classes of jobs. Let $T_{\text{net},k}$ be a random variable denoting the time that a k-class job spends in the network. Define $W_{q,i,M/M/c}$ to be the mean waiting time in an M/M/c queue using an arrival rate of λ_i (Eq. 8.5), a service rate of $1/E[S_i]$ (from Property 8.6), and a value of c equal to the number of servers at Node i. Then, the mean cycle time through the network for a job of class k is given by*

$$E[T_{\text{net},k}] \approx \sum_{i=1}^{m} \left(\frac{1}{\mu_{k,i}} + \frac{E[S_i^2]}{2E[S_i]^2} W_{q,i,M/M/c} \right) \eta_{k,i} ,$$

where the $\eta_{k,i}$ values are determined from Eq. (8.4).

Example 8.8. We illustrate the use of Property 8.7 with a production center consisting of four departments responsible for the following operations: turret lathe (T), milling machine (M), drill press (D), and inspection (I). There are five different types of jobs with the routing plan given in Table 8.3 and illustrated in Fig. 8.6.

Table 8.3 Routing plan for Example 8.8

Product	Route	Production Rate
A	T – M – D – I	65 units per 8-hour shift
B	M – T – M – D – T – I	45 units per 8-hour shift
C	T – M – I	30 units per 8-hour shift
D	M – D – T – D – I	75 units per 8-hour shift
E	T – M – D – T – I	120 units per 8-hour shift

Each department has only one machine except that there are two inspectors. The mean and standard deviation of the processing times are as in Table 8.4.

Table 8.4 Service time characteristics (in minutes) for Example 8.8

	T		**M**		**D**		**I**	
Product	Mean	StdDev	Mean	StdDev	Mean	StdDev	Mean	StdDev
A	1.3	0.2	1.1	0.3	0.9	0.2	2.1	1.4
B	0.5	0.1	0.6	0.4	1.1	0.3	3.1	2.5
C	1.2	0.5	1.5	0.4	0.0	0.0	1.9	1.1
D	1.1	0.8	1.2	0.4	1.5	0.8	2.8	1.9
E	0.8	0.3	1.0	1.0	1.0	0.5	2.7	2.0

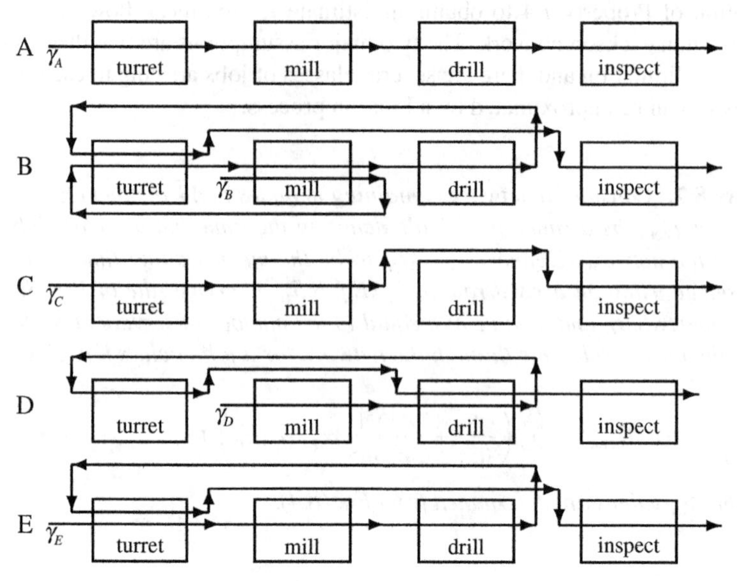

Fig. 8.6 Individual routes for the queueing network of Example 8.8

We present most of the calculations needed to obtain the mean flow times through the network by way of several tables. First, the mean production rates for the different classes are converted to a mean hourly arrival rate and these are shown in the right-hand column of Table 8.5. The numbers of visits to each node are also given in Table 8.5 using Eq. (8.4) and finally, the net arrival rates to each node are calculated by Eq. (8.5) and displayed in the final row of the table.

Table 8.5 Values for $\eta_{k,i}$ (in main body of table), λ_i, and γ_k

Classes	Nodes				γ_k (/hr)
	T	M	D	I	
A	1	1	1	1	8.125
B	2	2	1	1	5.625
C	1	1	0	1	3.750
D	1	1	2	1	9.375
E	2	1	1	1	15.000
λ_i (/hr)	62.5	47.5	47.5	41.875	

Table 8.6 shows the results of using Eq. (8.6) to determine the the probabilities needed to define the first and second moments of the service time. Although not needed, the squared coefficients of variation of service times are also given for an intuitive understanding of variability. (Remember that exponential service times have $c_s^2 = 1$ so that knowledge of the squared coefficients of variation should give a good understanding of variability.)

Table 8.6 Values for $q_{k,i}$ and the service times moments

Classes	Nodes			
	T	**M**	**D**	**I**
A	0.13	0.171	0.171	0.194
B	0.18	0.237	0.118	0.134
C	0.06	0.079	0	0.090
D	0.15	0.197	0.395	0.224
E	0.48	0.316	0.316	0.358
$E[S_i]$	0.880	1.001	1.192	2.588
$E[S_i^2]$	1.001	1.483	1.835	10.401
$c_s^2(i)$	0.293	0.480	0.291	0.553

With the mixture probabilities for the service times (i.e., the $q_{k,i}$ values), the main measures of effectiveness at each node can be calculated. The main quantities associated with the nodes are in Table 8.7.

Table 8.7 Waiting time in the queue at each node

	Nodes			
	T	**M**	**D**	**I**
arrival rate/hr	62.5	47.5	47.5	41.875
service rate/hr	68.2	59.9	50.3	23.2
number servers	1	1	1	2
$W_{q,i,M/M/c}$ (min)	9.65	3.84	20.24	11.353
$W_{q,i}$ (min)	6.24	2.84	13.07	8.82

The final step in the calculation of waiting times are shown in Table 8.8.

Table 8.8 Values for the node waiting times and network flow times

Classes	Nodes				$T_{\text{net},k}$
	T	**M**	**D**	**I**	
A	7.54	3.94	13.97	10.92	36.37
B	6.74	3.44	14.17	11.92	46.45
C	7.44	4.34	0	10.72	22.50
D	7.34	4.04	14.57	11.62	52.14
E	7.04	3.84	14.07	11.52	43.51

Notice that the formula for total waiting time within the network of Property 8.7 can be broken into individual waiting times at each node. The formula arises by determining the waiting time in the queue for an arbitrary job and then adding the service time for the specific class type. Thus, the time spent by a job of Class k in each pass through Node i is given by

$$E[T_{k,i}] \approx \frac{1}{\mu_{k,i}} + \frac{E[S_i^2]}{2E[S_i]^2} W_{q,i,M/M/c} \,. \tag{8.7}$$

Table 8.8 is constructed using Eq. (8.7) with the final row of Table 8.7. These values are then multiplied by the values in Table 8.5 to obtain the mean flow time through the network for each class of jobs (right-hand column of Table 8.8). □

- *Suggestion: Do Problem 8.6.*

8.2.2 Deterministic Routing with non-Poisson Input

To obtain some measure of the accuracy of the approximation given in Property 8.7, a simulation was written of the network contained in Example 8.8. Based on 25 replications of the simulation, the a analytical approximation appears to over estimate the waiting times an average of 8-9% per job class. One difficulty is with the Poisson assumption for the arrival streams. When the Poisson assumption is not good due to the limited number of feedback routes and a low number of different job classes, procedures are needed to approximate the squared coefficient of the arrival streams.

The basic concept for approximating the arrival stream process is a relationship between the arrival stream to a node, the service characteristics of the node, and the departure stream from the node. This is given in the following property.

Property 8.8. *The squared coefficient of variation of the inter-departure times for a queueing system with c servers can be approximated by*

$$c_d^2 \approx (1-\rho^2)\,c_a^2 + \rho^2\,\frac{c_s^2 + \sqrt{c} - 1}{\sqrt{c}}\,,$$

where ρ is the traffic intensity for the queueing system.

Because the departure process for an upstream node becomes the arrival process for the downstream node, Property 8.8 allows modeling the arrival processes as long as there is some procedure for combining streams and splitting streams.

Splitting Streams. Assume that the squared coefficient of variation is know for the arrival stream into Node k, and that there are some jobs that leave Node k to next enter Node j. The (known) squared coefficient of variation for this arrival stream will be denoted by $c_{a,k}^2$. Let p_{kj} denote that probability that a randomly chosen job at Node k will go next to Node j and let $c_{d,k,j}^2$ be a squared coefficient of variation of the inter-departures along that stream. Then

$$c_{d,k,j}^2 \approx p_{kj}\left[(1-\rho^2)\,c_{a,k}^2 + \rho^2\,\frac{c_{s,k}^2 + \sqrt{c_k} - 1}{\sqrt{c_k}}\right] + 1 - p_{kj}\,, \tag{8.8}$$

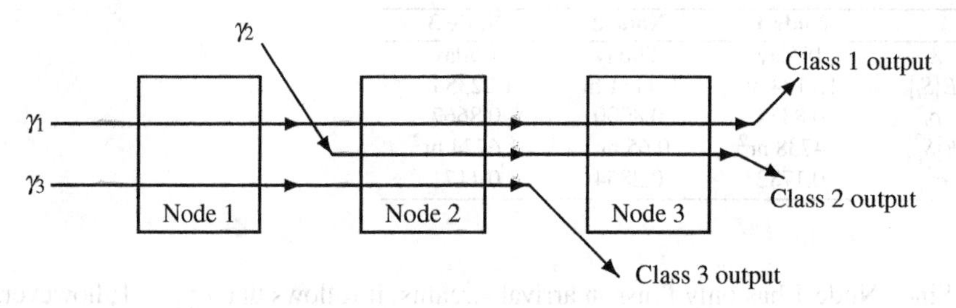

Fig. 8.7 Routings within a queueing network containing three classes for Example 8.9

where $c^2_{s,k}$ is determined by the moments obtained through Property 8.6.

Merging Streams. Assume that $\bar{i}_k$ is the total number of nodes that feed directly into Node k. From Eq. (8.8), it should be possible to determine the squared coefficient of variation along the stream from Node i to Node k, namely, $c^2_{d,i,k}$. In addition, based on the raw data, it should also be possible to determine the mean rate of departures along that stream which we denote by $\lambda_{i,k}$. The squared coefficient of variation for the net arrival stream to Node k is the weighted average of the individual squared coefficients of variation. In other words,

$$c^2_{a,k} \approx \sum_{i=1}^{\bar{i}_k} \frac{\lambda_{i,k}\, c^2_{d,i,k}}{\lambda_k} \tag{8.9}$$

where λ_k is the mean rate of arrivals to Node k (Eq. 8.5).

Example 8.9. Consider the network shown in Fig. 8.7 for a manufacturing firm that operates 24 hours per day. Each node refers to a different workstation that acts as a single-server queueing system with characteristics as shown in Table 8.9.

Table 8.9 Arrival rates (/day) and service time characteristics (hr) for Example 8.9

		Node 1		Node 2		Node 3	
Product	γ_k	Mean	StdDev	Mean	StdDev	Mean	StdDev
Class 1	9	1.25	0.4	0.8	0.3	1.2	0.5
Class 2	8	–	–	0.75	0.4	1.25	0.3
Class 3	10	1.0	0.5	0.6	0.4	–	–

Using the same operations as in Example 8.8, it is not difficult to obtain the workstation characteristics as shown in Table 8.10. After the service time characteristics for each node have been calculated, the arrival stream characteristics must be determined using Property 8.8 and Eqs. (8.8) and (8.9). Due to the structure of this network, it is possible to do this node by node; however, the technique we use here would be unworkable if there were any feedback loops in the network.

Table 8.10 Characteristics for each node of Example 8.9

	Node 1	Node 2	Node 3
λ_i	19/day	27/day	17/day
$E[S_i]$	1.1184 hr	0.7111 hr	1.2235 hr
ρ_i	0.8854	0.8000	0.8667
$E[S_i^2]$	1.4738 hr^2	0.65 hr^2	1.6724 hr^2
$c_{s,i}^2$	0.1782	0.2854	0.1171

Since Node 1 has only Poisson arrival streams, it follows that $c_{a,1}^2 = 1$; however, because the service times are not exponential, the departures from Node 1 will not be Poisson. The application of Property 8.8 yields

$$c_{d,1}^2 = (1 - 0.8854^2) \times 1 + 0.8854^2 \times 0.1782 = 0.3558\,.$$

The input stream to Node 2 is a merging of the departures from Node 1 having a squared coefficient of variation of 0.3558 and an external Poisson arrival process with squared coefficient of variation of 1.0. Taking a weighted average based on the arrival rates, the application of Eq. (8.9) and Property 8.8 yields the following for Node 2:

$$c_{a,2}^2 = \frac{19}{27}\,0.3558 + \frac{8}{27}\,1.0 = 0.5467$$
$$c_{d,2}^2 = (1 - 0.8^2) \times 0.5467 + 0.8^2 \times 0.2854 = 0.3795\,.$$

There is a branch at the output side of Node 2, since some of the departures (specifically, 10/27 of the departures) branch to the outside, and some (namely, 17/27 of the departures) go to Node 3. Thus, the arrivals to Node 3 have a squared coefficient of variation, derived from Eq. (8.8), of

$$c_{a,3}^2 = \frac{17}{27}\,0.3795 + 1 - \frac{17}{27} = 0.6093\,.$$

The approximation of Property 7.4 can now be applied to each node. The form useful for a single server queue is

$$W_{q,G/G/1} \approx \frac{c_a^2 + c_s^2}{2}\,\frac{\rho}{1-\rho}\,E[S]\,. \tag{8.10}$$

The results of using Eq. (8.10) in order to obtain the waiting times per class of jobs are in Table 8.11. The final column of the table gives the cycle time within the network. Notice that Little's Law can be applied to determine that the average WIP of Class 1 jobs in the network is 4.655, it is 2.024 for Class 2 jobs, and 3.281 for Class 3 jobs. Thus the average number of jobs within the network is 9.96. □

For a network that contains feedback loops, the approach of Example 8.9 will not work because of the circular dependencies. In such a case, the squared coefficients of variations for all nodes must be obtained simultaneously through the solution of

Table 8.11 Waiting times (in hours) for Example 8.9

	Node 1	Node 2	Node 3	Network
$W_{q,G/G/1}$	5.0914	1.1834	2.8886	
W-Class 1	6.3414	1.9834	4.0886	12.4134
W-Class 2	0	1.9334	4.1386	6.0720
W-Class 3	6.0914	1.7834	0	7.8748

a system of equations. Such procedures are beyond the scope of this textbook but are found in detail in Curry and Feldman [1, Chaps. 5 and 6].

Appendix

Excel is an excellent tool to use for the Mean Value Analysis. The first step is to calculate the relative arrival rates which requires a matrix inverse (see the Appendix of Chap. 5), and the second step is the implementation of the algorithm given in Property 8.5.

To illustrate the use of Excel, we return to Example 8.5 and place the given data (together with an identity matrix) onto the worksheet according to the following.

	A	B	C	D	E	F	G
1		Node 1	Node 2	Node 3			
2			Routing Matrix			Identity	
3	Node 1	0.0	0.7	0.3		1	0
4	Node 2	0.2	0.0	0.8		0	1
5	Node 3	1.0	0.0	0.0			
6			Mean Service Times				
7		80	111.4	64			

For determining the relative arrival rates according to Property 8.4, type and center `Relative Arrival Rates` across Cells B8:D8; type `1` in Cell B9; highlight Cells C9:D9; type

```
=MMULT(C3:D3, MINVERSE(F3:G4 - C4:D5))
```

and hit `<enter>` while holding down the `<shift>` and `<ctrl>` keys. The vector **r** appears in Cells B9:D9.

The Mean Value Analysis Algorithm is obtained as follows.

	A	B	C	D
10	WIP Level	Mean Value Algorithm		
11	1	=B7	=C7	=D7

It is important to be careful of the location of the dollar signs ($) in the following instructions: in Cell E10, type `Sum`; in Cell E11, type

```
=SUMPRODUCT($B$9:$D$9, B11:D11)
```

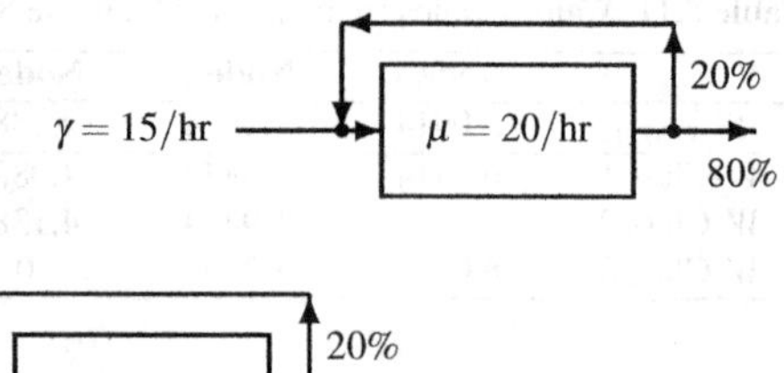

Fig. 8.8 M/M/1 queueing system with Bernoulli feedback

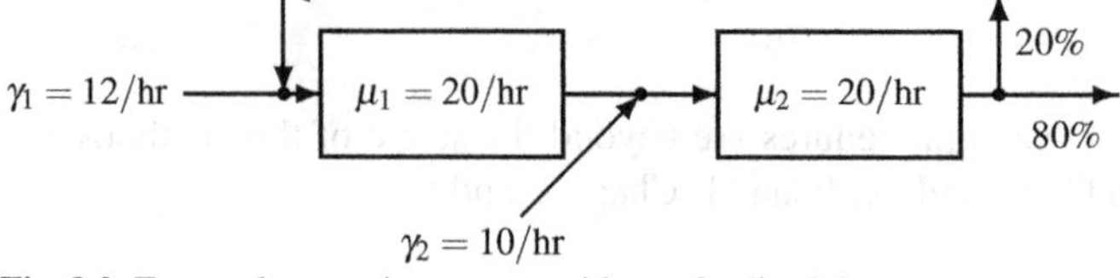

Fig. 8.9 Two-node queueing system with one feedback loop

in Cell A12, type `=A11+1`; and finally in Cell B12, type

```
=B$7*(1 + $A11*B$9*B11/$E11) .
```

Highlight Cells E11:E12 and do a "fill down" (namely, use `<ctrl>+d`) and highlight Cells B12:D12 and do a "fill right" (namely, use `<ctrl>+r`). (In case you are interested in such things, instead of doing the "fill down" to place the proper formula in Cell E12, it is also obtained using a simple "copy-paste" step or by placing the cursor in Cell E12 in hitting the `<ctrl>+<shift>+<">` keys.) The correct formulas for the Mean Value Analysis Algorithm are now in Cells A12:E12, so the mean waiting times can easily be obtained for whatever CONWIP level is desired by copying the formulas down for the proper number of rows. Since Example 8.5 was for a CONWIP level of 5, the algorithm is completed by highlighting Cells A12:E15 and doing a "fill down".

Problems

8.1. The simplest type of queueing network is an M/M/1 queueing system with Bernoulli feedback. Model the queueing system illustrated in Fig. 8.8 and answer the following questions.
(a) What is the probability that the system is empty?
(b) What is the expected number of customers in the system?
(c) What is the expected amount of time that a customer will spend in the system?
(d) What is the probability that a customer will pass through the server exactly twice?
(e) What is the expected number of times that a customer will pass through the server?

8.2. Consider the queueing network illustrated in Fig. 8.9. The first node of the network is an M/M/1 system and the second node of the network is an M/M/2 system. Answer the following questions.

(a) What is the probability that Node 1 is empty?
(b) What is the probability that the entire network is empty?
(c) What is the probability that there is exactly one job in Node 1's queue?
(d) What is the probability that there is exactly one job within the network?
(e) What is the expected number of jobs within the network?
(f) What is the expected amount of time that an arbitrarily selected job will spend in the system?
(f) There is a cost of $100 per hour that each job spends within the system. A proposal has been made that a third node could be added to the system. The third node would be an M/M/c system, and with the addition of this node, there would be *no* feedback. Each server added to the third node operates at a rate of 20 per hour and costs $200 per hour. Is the third node worthwhile? And if so, how many servers should the third node use?

8.3. Customers arrive to a bank according to a Poisson process with a mean rate of 24 customers per hour. Customers go to the receptionist 20% of the time to describe the type of transaction they want to make, 60% of the customers go directly to a teller for common bank transactions, and the rest go directly to see a bank counselor for service. After talking to the receptionist, 75% of the customers are sent to a teller, 20% to a counselor and 5% leave the bank. After completing service at a teller, 90% of the customers leave the bank while the rest go to see a bank counselor. A counselor can handle all the requests for their customers, so that when a customer is finished with a counselor, the customer leaves the bank.

There is one receptionist who takes an exponential amount of time with each customer averaging 5 minutes per customer. There are three tellers, each with an exponential service time with mean 6 minutes. There are two bank counselors that serve customers with an exponential service time with an average 12 minutes each. Assume that the Jackson network assumptions are appropriate and respond to the following.
(a) Draw a block diagram to represent the network of the bank service.
(b) Write the routing matrix for the Jackson network.
(c) For a customer who desires a teller, what is the mean amount of time spent in line waiting for the teller?
(d) What is the average number of customers in the bank?
(e) Compute the mean flow time through the network; that is, compute the average time that a customer spends inside the bank.
(f) Bank counselors often have to make telephone calls to potential customers. The bank president wants to know, on the average, the amount of time per hour that at least one counselor is available to make telephone calls.
(g) Answer parts (a) through (d) again with one change in the customer flow dynamics. Assume that after spending time with a counselor, 15% of the customers go to a teller, and 85% leave the bank. (Thus, for example, a small fraction of customers go to a teller, then talk to a counselor, and then return to a teller.)

8.4. A switchboard receptionist receives a number of requests in a day to ask trained personnel questions about different areas of computer software. Clients explain their

problem to the receptionist who routes the calls to whomever is the most knowledgeable consultant in that area. There are four consultants with different areas of expertise. Two of the consultants (A) can handle questions regarding problems on how to operate the software, one consultant (B) can answer questions related to interaction conflicts with other software packages, and the other consultant (C) can help with software-hardware questions. The receptionist tries to route the calls to the correct person, but the question sometimes cannot be answered and the consultant may suggest another consultant. When a switch is made from one consultant to another, there is a equal probability that the consultant will suggest either of the other two consultants. Calls arrive to the switchboard according to a Poisson distribution with mean of 10 calls per hour. It takes the receptionist an exponential amount of time with mean of 1.5 minutes to process a call. The receptionist routes 50% of the calls to consultants A, 25% to consultant B, 15% to consultant C, and 10% of the calls cannot be helped by a consultant. It takes each consultant an exponential amount of time to process a call. The average service time for each consultant A is 15 minutes, 18 minutes for consultant B, and 22 minutes for consultant C. Of the customers routed to each consultant 15% are sent to another consultant, 5% are sent back to the receptionist, and the rest get their answer and go out of the system. Assume that the telephone lines can hold an unlimited number of callers, and that the receptionist treats each call as if it were a new call.
(a) Compute the expected length of time that a customer has to wait before his/her call is answered by the receptionist.
(b) Compute the average number of phone calls in the system.
(c) Compute the expected time a caller spends in the network.

8.5. Consider the manufacturing facility of Example 8.4. In order to reduce the feedback from the installation of electrical components to the machine room, the engineering department has suggested a preparation step. In other words, the proposed system has all jobs starting in a preparation room where a single worker prepares each job for machining. The time needed for preparation is exponentially distributed with a mean of 15 minutes. After preparation, the job goes to the machine room. After the machine room, 70% go to the electrical components room and 30% go to finishing. With this change, only 5% (instead of 20%) of the jobs are returned to the machine room from the electrical components room. Is this extra step worthwhile?

8.6. Consider the problem given in Example 8.8. The manager of the production center has told a consultant that Product D could also be made by eliminating the first drilling operation. However, the manager estimates that if this is the case about 3% of the production will be lost to defective production. The cost of scrapping a unit of product D is $50, but the work-in-process (WIP) costs are $1/hr, $1.15/hr, $1.25/hr, $5/hr and $1/hr for products A, B, C, D and E, respectively. (For example, if we were to follow one item of part C through the manufacturing system and discover that it took 5 hours, then that part would incur a WIP carrying cost of $6.25.)
(a) In the old system (using both drilling operations for product D), what is the size of the WIP by part type?

(b) How will the proposed change to the process of Product D affect the length of the production time for the other products?
(c) Is it profitable to modify the process for Product D?

References

1. Curry, G.L., and Feldman, R.M. (2009). *Manufacturing Systems Modeling and Analysis*, Springer-Verlag, Berlin.
2. Gross, D., and Harris, C.M. (1998). *Fundamentals of Queueing Theory*, 3rd ed., John Wile & Sons, New York.
3. Jackson, J.R. (1957). Networks of Waiting Lines. *Operations Research*, **5**:518–521.

Chapter 9
Event-Driven Simulation and Output Analyses

The versatility and popularity of discrete simulation makes simulation modeling a very powerful and, sometimes, dangerous tool. Discrete simulation is powerful because it can be used to model a wide diversity of situations, and it may be dangerous because its relative simplicity may lead to misuse and misinterpretation of results. However, with good training in the use of simulation and careful statistical analysis, discrete simulation can be used for great benefit.

In this chapter, we build upon the basics that were introduced in Chap. 2. A key feature that was postponed in Chap. 2 is the requirement to deal with the timing of various concurrent activities. We introduce the major concepts through the use of simple examples that can be generated by hand.

9.1 Event-Driven Simulations

When the timing of activities is a component of a system being simulated, a method is needed to maintain and increment time in a dynamic fashion. For example, suppose we wish to simulate a queueing system with five servers. The five-server system, when all servers are full, would have six different activities (five service completions and one arrival) to maintain. In order to correctly and efficiently handle the many activities in a complex system, a simulation clock is maintained and updated as events occur. In other words, if we are simulating what happens within a manufacturing plant during a 10-hour day, the simulation study might take one second of real time to execute, but the simulation clock will show an elapsed time of 10 hours.

The movement of the simulation clock is extremely critical, and there are two common methods for updating time: (1) the simulation clock could be updated a fixed amount at each increment or (2) the clock could be updated whenever something happened to change the status of the system. The advantage of the first method is that it is conceptually simple; the disadvantage is that if the increment is too small then there are inefficiencies in "wasted" calculations, and if the increment is too large then there are mathematical difficulties in representing everything that happens

R.M. Feldman, C. Valdez-Flores, *Applied Probability and Stochastic Processes*, 2nd ed., DOI 10.1007/978-3-642-05158-6_9,

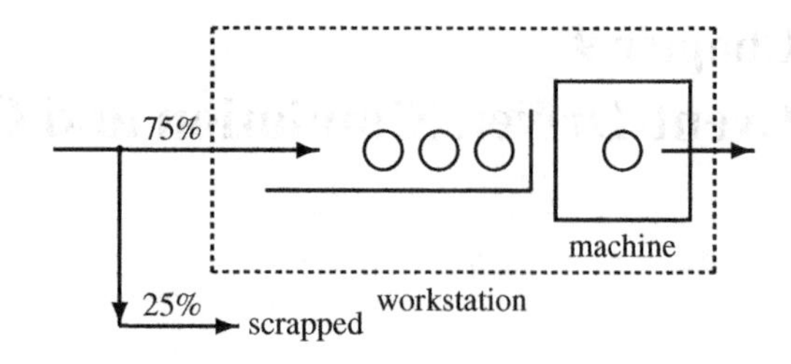

Fig. 9.1 Schematic of the manufacturing process for Example 9.1

during the increment. Therefore, most simulation procedures are based on the second method of time management, called event-driven simulations. In other words, the simulation clock is updated whenever an event occurs that might change the system. As we illustrate this concept with the following examples, observe that the simulation-by-hand is simply building tables of information. If our purpose were to develop a technique for simulation-by-hand, the following procedure would contain some clearly redundant steps; however, our purpose is to illustrate some of the basic steps that are necessary for simulating complex systems. Therefore, we suggest that you follow the basic steps that are outlined below even though shortcuts can be easily identified.

Example 9.1. Consider a simplified manufacturing process (illustrated in Fig. 9.1) where widgets are processed through a production center. One widget arrives at the facility every 12 minutes; however, because of poor quality control, 25% of these widgets are defective and must be scrapped. The remaining 75% of the widgets are sent to a work station where they are processed one at a time through a single machine. The chemical composition of the widgets varies; thus, the processing time required on the machine varies. Twenty-five percent of the widgets require 17 minutes of machining time, 50% of the widgets require 16 minutes, 12.5% of the widgets require 15 minutes, and 12.5% require 14 minutes. While a widget is on the machine, other widgets that arrive at the work station are placed in a special holding area until the widget being processed is completed and removed. The next widget in line is then placed on the machine and the process is repeated. Widgets in the holding area are said to be in a queue (waiting line) waiting for the machine to become available. To characterize this process in traditional queueing terminology, we would say that we are dealing with a single server queueing system (the machining process operates on widgets one at a time) having a deterministic arrival process, random service times, and utilizing a first-in first-out selection scheme (FIFO queueing discipline).

We wish to simulate this production center in order to answer the question, "What is the long-run average number of widgets in the facility?" For this system, two random variables must be simulated (see Table 9.1). The status of arriving widgets (defective or not defective) will be decided by tossing two fair coins: two tails indicate scrap (25% of the outcomes) and any other combination indicates a good widget. (In Chap. 2, we discussed reproducing randomness with computer generated random variates; however, for purposes of the examples in this chapter, we return to simulation-by-hand.) The second random variable, namely the machining time, will be decided by tossing three fair coins: three tails (with a chance of 1 in 8 or 12.5%)

Table 9.1 Mapping of random value to simulated outcome in Example 9.1

Random Value	Simulated Outcome
T-T	Scrap
T-H	Good
H-T	Good
H-H	Good

Random Value	Simulated Outcome
T-T-T	14 min
H-H-H	15 min
T-H-T	16 min
T-T-H	16 min
H-T-T	16 min
T-H-H	16 min
H-H-T	17 min
H-T-H	17 min

will indicate 14 minutes, three heads (again, 1 chance in 8) will indicate 15 minutes, the combinations THT, TTH, HTT, THH (with a chance of 4 in 8 or 50%) will indicate 16 minutes, and HHT, HTH (with a chance of 2 in 8 or 25%) will indicate 17 minutes of machining time.

Our simulation is actually a description of the processing of widgets within the manufacturing system. The term "entity" is used in simulations to denote the generic objects that "move" through a system and are operated upon by the system. Hence, a widget is an example of an entity. If a simulation was designed for the movement of people within a bank, the people would be called entities. If a simulation was designed to model packets of bits moving within a communication network, the packets would be called entities. Thus, a simulation deals with entities and events (events are things that happen that change the state of the system). In the widget example, there are two types of events: arrivals and service completions. (Departures from the system are also events, but in our example, departures need not be identified separately because they occur at the same time as service completions.) Events are important because it is events that move the simulation clock. As the simulation progresses, at least two tables must be maintained: a table describing the state of the manufacturing process at each point in time and a table keeping track of all scheduled future events. It is the table of future events (called the "Future Events List") that will drive the simulation.

A simulation of a large system involves many different types of lists. In this chapter, our goal is to illustrate only a few of the basic types of lists. For example, an understanding of the basic mechanics of the Future Events List is essential for a well-designed simulation model. In the next example, some additional lists and concepts will be discussed.

A simulation begins by initializing the Future Events List. For this problem, we assume that an entity (i.e., widget) arrives to the system at time 0, so the Future Events List is initialized as shown in Table 9.2. The first column in the Future Events List indicates the time that the entity is to be removed from the list and added to the system, thus moving the simulation clock to the indicated time. (In this case, we move the clock from no time to time zero.) The second column gives the entity number (entities are numbered sequentially as they are created), and the third col-

Table 9.2 Future Events List: initialization

Future Events List		
time	entity no.	type
0	1	a

umn indicates the type of event that is the cause for the clock update. In the widget example, there are two types of events: arrivals (a) and service completions (s).

With the initialization of the Future Events List, we are ready to begin the simulation. Conceptually, we remove the entity from the Future Events List and update the simulation clock accordingly. The table describing the state of the system is then updated as shown in Table 9.3. Column 1 designates the clock time as determined

Table 9.3 State of the system after time 0

(1) Clock time	(2) Type of event	(3) Time next arrival	(4) Two coin toss	(5) Good or bad item	(6) Number in system	(7) Three coin toss	(8) Time service complete	(9) Sum of time × number
0	a	12	TT	bad	0	–	–	0

from the Future Events List. If the event causing the advance of the simulation clock is an arrival, the first "order of business" is to determine the next arrival time. As soon as this next arrival time is determined, the entity that is scheduled to arrive at that time is placed in the Future Events List. In this example, there are no calculations for determining the next arrival time, but the general principle is important: as soon as an arrival occurs, the next arrival time is immediately determined and a new entity is placed on the Future Events List. After determining the next arrival, we return to describing everything that happens at the current time. Two coins are tossed to determine if the arriving widget will be scrapped or processed. Since the coin toss results in two tails, the widget is bad and is not accepted for processing (Columns 4 through 6). Since the entity does not stay in the system, the description of the system at time 0 is as depicted in Table 9.3; thus, we are ready to update the clock. We look to the Future Events List Table 9.4 in order to increment the clock. (A description of Columns 8 and 9 will be delayed until more data are generated so that the explanation will be easier.)

Table 9.4 Future Events List: after time 0

Future Events List		
time	entity no.	type
~~0~~	~~1~~	~~a~~
12	2	a

Notice that the first item on the Future Events List (as of time 0) is crossed off indicating that the entity was removed (to start the simulation). We now remove the next entity from the list in order to update (i.e., advance the simulation clock) the table (Table 9.5) giving the state of the system. The entity is added to Table 9.5 representing an arriving widget. As always, as soon as an arrival occurs, the next

Table 9.5 State of the system after time 12

(1) Clock time	(2) Type of event	(3) Time next arrival	(4) Two coin toss	(5) Good or bad item	(6) Number in system	(7) Three coin toss	(8) Time service complete	(9) Sum of time × number
0	a	12	TT	bad	0	–	–	0
12	a	24	TH	good	1	THT	28	0

arrival time is determined (Row 2, Column 3) and the new entity is placed on the Future Events List (Entity 3 in Table 9.6). Next, we toss two coins to determine

Table 9.6 Future Events List: after time 12

Future Events List		
time	entity no.	type
~~0~~	~~1~~	~~a~~
~~12~~	~~2~~	~~a~~
24	3	a
28	2	s

the acceptability of the arriving entity. Since a tail-head occurred and the widget is considered good, it enters the production facility (Row 2, Column 6) and the machining process begins. It is now necessary to determine how long the widget will be on the machine. The result of tossing three coins (Row 2, Column 7), is a 16-minute machining process. Since the current clock is 12, the 16 minutes are added to the current time and the widget is scheduled to finish the machining process and leave the facility at time 12+16=28. Since this is a future event (with respect to a clock time of 12), it is placed on the Future Events List (Table 9.6). All of the activities that occur at time 12 have been completed so it is time to return to the Future Events List (Table 9.6) to increment the simulation clock. Note that during time 12, two events were placed on the Future Events List: the arrival to occur at time 24 and the service completion that will occur at time 28.

The next (future) event within the Future Events List, as the list stands after time 12, is the arrival of Entity 3 that occurs at time 24; thus, Entity 3 is removed from the Future Events List and placed in the table giving the current state of the system (Table 9.7). Before processing Entity 3, the next arrival event is scheduled yielding Entity 4 which is immediately placed on the Future Events List (Table 9.8). Returning to Entity 3, we see that it is considered acceptable for machining after the two coin toss; however, this time the machine is busy so the widget (Entity 3) must wait until the processing of Entity 2 is finished. Note that the number in the system (Row 3, Column 6, Table 9.7) is updated to 2, but a service time is not determined

Table 9.7 State of the system after time 24

(1) Clock time	(2) Type of event	(3) Time next arrival	(4) Two coin toss	(5) Good or bad item	(6) Number in system	(7) Three coin toss	(8) Time depart system	(9) Sum of time × number
0	a	12	TT	bad	0	–	–	0
12	a	24	TH	good	1	THT	28	0
24	a	36	HH	good	2	–	–	12

until the widget is actually placed on the machine. In other words, although Entity 3 is waiting in a queue for the server to become available, we do not yet know when the entity will leave the system, so it is not yet placed on the Future Events List (Table 9.8).

Table 9.8 Future Events List: after time 24

Future Events List		
time	entity no.	type
~~0~~	~~1~~	~~a~~
~~12~~	~~2~~	~~a~~
~~24~~	~~3~~	~~a~~
28	2	s
36	4	a

The last column in the table used to describe the state of the system (as in Table 9.7) is used to keep track of the (time-weighted) average number of widgets in the system. The time-averaged number of widgets is determined by summing the number of widgets within the facility multiplied by the length of time each widget spends in the facility and then dividing by the total length of time. At time 24, one item has been in the system for 12 minutes so the last column of Table 9.7 contains a 12. To help understand this averaging process, consider Fig. 9.2. It represents a graph of the number in the system versus time for the first 106 minutes. An estimate for the average number of entities in the system is calculated by taking the area under that graph and dividing it by the total time of the graph. Therefore, if we wish to determine the average number in the system, the main quantity that must be maintained is the area under the graph. Each time the clock advances, a new rectangle of area is added to the total area, and the accumulation of the area of these rectangles is maintained in the last column. Note that the accumulation of area is accomplished by looking *backwards*. Thus, the 12 in Row 3, Column 9 is the area under the curve in Fig. 9.2 from time 0 through time 24 (or equivalently, the area of rectangle A in the figure).

To continue the simulation, we again look for the time that the next change in the system will occur (Table 9.8). Thus, the clock is advanced to time 28. The event causing the advance of the clock is a service completion, and thus the widget being processed leaves the system and the widget in the queue (Entity 3) is placed on the machine. Three coins are tossed to determine the processing time, and since they

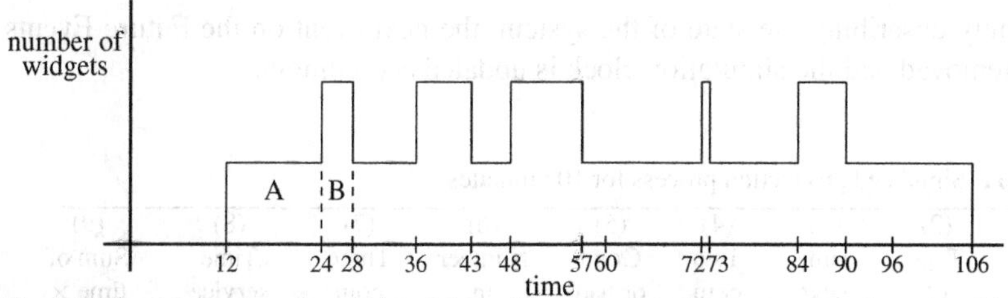

Fig. 9.2 Plot of the data from Table 9.11 to illustrate that the average number of entities in the system equals the area under the curve divided by the length (time) of the curve

come up head-head-head, we assign 15 minutes of machining time to this widget. The 15 minutes of machining time is added to the current clock time of 28 minutes, and we record in Column 8 (and in the Future Events List) that the widget will leave the facility at time 43. To obtain the number in the last column (Row 4, Column 9, Table 9.9), we need to multiply the time increment by the number in the system

Table 9.9 State of the system after time 28

(1) Clock time	(2) Type of event	(3) Time next arrival	(4) Two coin toss	(5) Good or bad item	(6) Number in system	(7) Three coin toss	(8) Time service complete	(9) Sum of time × number
0	a	12	TT	bad	0	–	–	0
12	a	24	TH	good	1	THT	28	0
24	a	36	HH	good	2	–	–	12
28	s	–	–	–	1	HHH	43	20

(4×2) and add it to the previous total. Or equivalently, we calculate the area of the rectangle B in Fig. 9.2 and add it to the previous area yielding a current total area of 20. The Future Events List at this point is shown in Table 9.10.

Table 9.10 Future Events List: after time 28

Future Events List		
time	entity no.	type
~~0~~	~~1~~	~~a~~
~~12~~	~~2~~	~~a~~
~~24~~	~~3~~	~~a~~
~~28~~	~~2~~	~~s~~
36	4	a
43	3	s

As a summary, Table 9.11 shows the updating of the state of the system for the first 106 minutes. As illustrated above, the process is to completely describe the state of the system at a particular time. During that description, one or more future events will become known, and those will be placed on the Future Events List. After completely describing the state of the system, the next event on the Future Events List is removed and the simulation clock is updated accordingly.

Table 9.11 Simulated production process for 106 minutes

(1) Clock time	(2) Type of event	(3) Time next arrival	(4) Two coin toss	(5) Good or bad item	(6) Number in system	(7) Three coin toss	(8) Time service complete	(9) Sum of time × number
0	a	12	TT	bad	0	–	–	0
12	a	24	TH	good	1	THT	28	0
24	a	36	HH	good	2	–	–	12
28	s	–	–	–	1	HHH	43	20
36	a	48	HT	good	2	–	–	28
43	s	–	–	–	1	TTT	57	42
48	a	60	HH	good	2	–	–	47
57	s	–	–	–	1	THH	73	65
60	a	72	TT	bad	1	–	–	68
72	a	84	HT	good	2	–	–	80
73	s	–	–	–	1	HTH	90	82
84	a	96	TH	good	2	–	–	93
90	s	–	–	–	1	HTT	106	105
96	a	108	TT	bad	1	–	–	111
106	s	–	–	–	0	–	–	121

The final sum in the last column equals 121, so the estimate for the average number of widgets in the facility is $121/106 = 1.14$. To understand this average, assume that the system is observed at 100 random points in time. Each time that the system is observed, the number of entities in the system is recorded. After the 100 observations are finished, we take the average of the resulting 100 numbers. This average should be close to the 1.14. (Actually, the 1.14 will have very little significance because the simulation must be run for a long time before the estimate is good.) Such an estimate is highly variable unless it involves a large number of entities. This again emphasizes the fact that a simulation is a statistical experiment, and the final numbers are only estimates of the desired information, not actual values. □

Example 9.2. Consider a bank, illustrated in Fig. 9.3, having two drive-in windows with one queue. (That is, waiting cars form one line, and the car at the head of the line will proceed to whichever window first becomes available.) There are two types of customers who come to the bank: customers with personal accounts and customers with commercial accounts. The time between personal-account customers coming to the bank for drive-in window service is 3 minutes with probability 0.5 or 5 minutes with probability 0.5. The length of time each car spends in front of the drive-in window varies and is 7 or 9 minutes with a 0.50 probability for each possibility. The time between arrivals of commercial-account customers is 8 minutes with probability 0.5 and 11 minutes with probability 0.5. The length of time that a commercial customer spends in front of the drive-in window is either 10 or 12 minutes with a 0.50 probability for each possibility. We would like to simulate

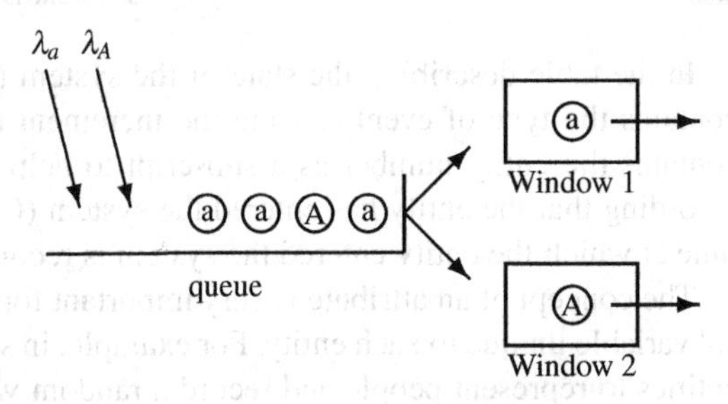

Fig. 9.3 Schematic of the bank with two windows and two customer types for Example 9.2

this process with the purpose of determining the average length of time that a customer spends in the system. We shall again use a coin toss to reproduce the relevant random events letting a tail represent the smaller number and a head represent the larger number.

There are two complications in simulating this system as compared to the previous example. First, there are two arrival processes: one arrival stream of entities representing personal-account customers, and the other arrival stream of entities representing commercial-account customers. Second, in order to collect data on the time each entity spends in the system, we need to mark *on the entity* the time at which that entity enters the system. When the entity leaves the system, we will subtract the entry time stored on the entity from the current clock time, thus obtaining the total time the entity spent in the system.

As before, we start by initializing the Future Events List; that is, each arrival stream generates its first arriving entity and places the entity on the list (Table 9.12). In this example, lower-case letters represent personal-account customers, and upper-case letters represent commercial customers.

Table 9.12 Future Events List: initialization

Future Events List		
time	entity no.	type
0	1	a
0	2	A

The simulation begins by taking the first entity off the Future Events List and recording the state of the system at the indicated time. In case of a tie (two or more events which are scheduled to come off the Future Events List at the same time), it does not matter which entity is chosen.

Table 9.13 State of the system after first entity processed

(1) Clock time	(2) Type of event	(3) Coin toss	(4) Time next arrival	(5) Number in system	(6) Coin toss	(7) Time service complete	(8) Sum of number departs	(9) Sum of times in system
0	a_1	T	3	1	H	9	0	0

In the table describing the state of the system (Table 9.13), Column 2 not only contains the type of event causing the increment to the simulation clock, but also contains the entity number as a subscript to help keep track of the entities. After recording that the entity has entered the system (Column 5), the (simulation clock) time at which the entity entered the system is recorded in an *attribute*.

The concept of an attribute is very important for simulations. An attribute is a local variable unique to each entity. For example, in some circumstance, we may want entities to represent people and record a random weight and height for each entity. In such a case, we would define two attributes in which the values of the weight and height random variables would be stored. In our drive-in window banking example, we want to record the time of entry for each entity. To do this, we define an attribute and as soon as the entity enters the system, we record the value of the simulation clock in that attribute. Now, in addition to the two previously defined lists, we add a third one called an Entities List to hold the value of the attribute for each entity. The Entities List (Table 9.14) and Future Events List (Table 9.15) are shown as they would appear after processing the first entity.

Table 9.14 Entities List: after first entity processed

Entities List	
entity number	value of attribute
1	0

Table 9.15 Future Events List: after first entity processed

Future Events List		
time	entity no.	type
~~0~~	~~1~~	~~a~~
0	2	A
3	3	a
9	1	s

We continue the simulation by processing entities one-at-a-time in a similar manner as in the previous section. The following tables are written after simulating the arriving and processing of customers for 16 minutes. As you look at Tables 9.16–9.18, there are several important observations to make. In the processing of the first two entities (Rows 1 and 2, Table 9.16), random service times are generated because there are two drive-in windows so both arriving entities immediately proceed to the window. However, there is no service time generated in the third row because that entity enters a queue; thus its service will not be known until after a service completion occurs. At time 9 Entity 1 departs, and the queued entity enters the server (i.e., gets in front of the drive-in window). Thus, a service time is generated in Row 5. The next service completion occurs at time 12. The departing entity (i.e., Entity 2) represents a commercial-account customer, but the random service time generated

Table 9.16 State of the system after time 16

(1) Clock time	(2) Type of event	(3) Coin toss	(4) Time next arrival	(5) Number in system	(6) Coin toss	(7) Time service complete	(8) Sum of number departs	(9) Sum of times in system
0	a_1	T	3	1	H	9	0	0
0	A_2	H	11	2	H	12	0	0
3	a_3	H	8	3	–	–	0	0
8	a_5	T	11	4	–	–	0	0
9	s_1	–	–	3	T	16	1	9
11	A_4	H	22	4	–	–	1	9
11	a_6	H	16	5	–	–	1	9
12	S_2	–	–	4	H	21	2	21
16	s_3	–	–	3	T	26	3	34
16	a_8	H	21	4	–	–	3	34

while the clock is at time 12 is for a personal-account customer because it is Entity 5 that is next in line. Thus, it is important to keep track of which entity came first to a queue. If there are several different types of entities within a simulation, it can become complicated keeping track of which type of entity will next perform a particular action. This is an advantage of special simulation languages (as compared to a language like C or Java); the languages designed specifically for simulation take care of much of the bookkeeping automatically.

Table 9.17 Entities List: after time 16

entity number	value of attribute
1	0
2	0
3	3
4	11
5	8
6	11
8	16

At time 9, when Entity 1 departs, we look at its attribute (in the Entities List Table) to determine when it arrived so that the total waiting time can be determined. This also occurs at time 12 and time 16. For example, consider time 16. It is Entity 3 that is departing, so when it departs we observe from Table 9.17 that Entity 3 arrived at time 3, and since the current time is 16, the total time in the system is 13. The value of 13 is added (Column 9) to the previous sum of 21 yielding a new total of 34. If the simulation were to end after time 16, our estimate for the average time in the system would be the ratio of Column 9 to Column 8, or $34/3 = 11.333$ minutes.

Notice that the Future Events List, in Table 9.18, is arranged in the order in which the events were generated. To determine which event to chose next, we search all the

Table 9.18 Future Events List: after time 16

Future Events List		
time	entity no.	type
~~0~~	~~1~~	~~a~~
~~0~~	~~2~~	~~A~~
~~3~~	~~3~~	~~a~~
~~9~~	~~1~~	~~s~~
~~11~~	~~4~~	~~A~~
~~8~~	~~5~~	~~a~~
~~11~~	~~6~~	~~a~~
~~12~~	~~2~~	~~S~~
~~16~~	~~3~~	~~s~~
22	7	A
~~16~~	~~8~~	~~a~~
21	5	s
26	4	S
21	7	a

future events and select the one with the minimum time. In a computer, the computer automatically deletes those entities that we have crossed out, and whenever an entity is placed in the list, the list is sorted so that the first entity always has the minimum time. □

- *Suggestion: Do Problems 9.1–9.6.*

9.2 Statistical Analysis of Output

Simulation modeling is a very powerful tool for analysis that can be learned by most technically inclined individuals and, thus, its use has become widespread. The typical approach is to learn the basic simulation concepts and a simulation language, but pay very little attention to the proper analysis of the output. The result is that many simulation studies are performed improperly. Simulation programs do not serve much purpose without a proper interpretation of the output. In fact, an improper understanding of the results of a simulation can be detrimental in that the analysis could lead to an incorrect decision. Therefore, now that the basic concepts of simulation have been introduced, it is necessary that the analysis of the output be discussed.

The reason that the output analysis of simulation is often given only a cursory review is that output analysis is more difficult than simple programming. Furthermore, there remain unanswered questions in the area of output analysis. In this section, we discuss enough of the concepts to equip the analyst with the minimum set of tools. We caution that there is much more to this area than we are able to cover at the level of this introductory text. Many of the concepts presented here are based on the text by Law [1] which we recommend for further reading.

Simulation output can be divided into two categories: *terminating simulations* and *steady-state simulations*. A terminating simulation is one in which there is a fixed time reference of interest. For example, assume we would like to simulate customer traffic within a bank that is open from 9:00 AM until 3:00 PM five days a week. Because the time frame if fixed and each day starts as a new day, this would be called a terminating simulation. Steady-state simulations do not have a fixed length of time; they deal with questions of long-run behavior. If we assume that the widgets from the Example 9.1 are produced 24 hours a day seven days a week, then questions dealing with the long-run behavior of the process would be of interest. A less obvious example is a manufacturing facility that is open 8 hours per day; however, if each day begins as the previous day ends, this would be an example of a steady-state simulation. Output analyses of terminating and steady-state simulations are considered separately.

9.2.1 Terminating Simulations

Consider a small service station with only one gasoline pump located along a major road leading into a large metropolitan area. It is open from 6:00 AM to 10:00 PM every day, with a rush hour occurring from 4:00 PM to 6:00 PM. (Access to the station is difficult for people going into the city; therefore, a morning rush hour does not cause congestion at the gasoline pump.) Because of space limitations, there is room for only four customers in the station; in other words, if one customer is at the pump, a maximum of three vehicles can be waiting. The owner of the station is concerned that too many customers are lost because people find the station full; therefore, a simulation study is desired to determine how many customers are lost due to the limited capacity. If the number lost is significant, the owner will consider purchasing a second pump. All interarrival times are observed to be exponentially distributed random variables, with an average of 6 arrivals per hour from 6:00 AM to 4:00 PM, 20 per hour from 4:00 PM to 6:00 PM, and 6 per hour from 6:00 PM to 10:00 PM. The time each customer spends in front of the pump is also exponentially distributed, with an average time of 4 minutes. (In other words, we are interested in an M/M/1/4 system, except that the input is a nonhomogenous Poisson process, and we are interested in transient results instead of steady-state results. However, it is possible to use the material from the previous chapter to obtain a "quick and dirty" estimate for the mean number turned away. The M/M/1/4 model with $\lambda = 20$ and $\mu = 15$ yields an overflow rate of 6.6 customers per hour. The difficulty is that steady-state is never reached, but as an estimate, this would yield an overflow of 13.2 customers per day.)

This service station example illustrates a situation suitable for a terminating simulation study. (A steady-state analysis does not make sense since the process is constantly being "restarted" every morning.) The first step in determining the feasibility of obtaining a second pump is to simulate the above situation to estimate the expected number of daily customers lost under the current configuration of one

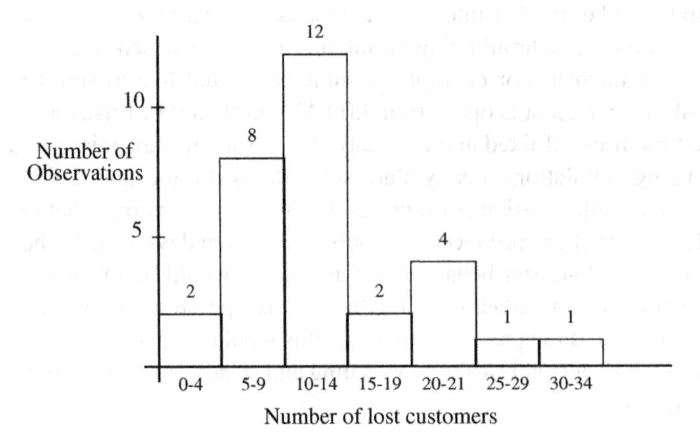

Fig. 9.4 Histogram from 30 simulations of the one pump service station for Example 9.3

pump. To illustrate this we simulated the service station and executed the program yielding the number of lost customers in a day as 25, 10, and 12, respectively, for three different runs. With variations of this magnitude, it is clear that care must be taken in estimating the mean number of customers lost. Furthermore, after an estimate is made, it is important to know the degree of confidence that we can place in the estimate. Therefore, instead of using a single value for an estimate, it is more appropriate to give a *confidence interval* (Sect. 3.3). For example, a 95% confidence interval for the expected number of customers lost each day is an interval which we can say with 95% confidence contains the true mean. That is, we expect to be wrong only 5% of the time. (The choice of using 95% confidence instead of 97.2% or 88.3% confidence is, for the most part, one of historical precedence.)

Example 9.3. To illustrate the building of confidence intervals, the simulation of the one pump service station was executed 30 times and the daily total number of lost customers was recorded in Table 9.19. The histogram of the results is in Fig. 9.4. (It

Table 9.19 Results from 30 simulations of the one pump service station

25	10	12	16	31	8
14	21	7	6	21	6
7	13	10	13	0	21
12	9	13	13	16	9
14	23	4	12	8	13

might appear that it is redundant to plot the histogram since the table gives all the information; however, we recommend that you make a practice of plotting your data in some fashion because quantitative information cannot replace a visual impression for producing an intuitive feel for the simulated process.)

The sample mean and standard deviation (Definition 3.3) for the results obtained from the 30 simulations are $\bar{x} = 12.9$ and $s = 6.7$, respectively. (Notice the use of the lower-case letters. The upper-case letters indicate a random variable; after an observation is made, a lower-case letter must be used indicating a particular value.) The critical t-value for a 95% confidence interval with 29 degrees of freedom is 2.045 (Excel function `TINV(0.05,29)`). Equation (3.4, p. 86) thus yields

$$12.9 \pm 2.045 \times \frac{6.702}{\sqrt{30}} = (10.4, 15.4)$$

as the 95% confidence interval for the mean. That is, we are 95% confident that the actual expected number of customers lost each day is between 10.4 and 15.4.

The sample variance was $s^2 = 44.89$ and the two critical values (Excel functions `CHIINV(0.975,29)` and `CHIINV(0.025,29)`) are $\chi^2_{(29,0.975)} = 16.05$ and $\chi^2_{(29,0.025)} = 45.72$. Equation (3.7, p. 88) thus yields

$$\left(\frac{29 \times 44.92}{45.72}, \frac{29 \times 44.92}{16.05} \right) = (28.49, 81.16)$$

as the 95% confidence interval for the variance. Many practitioners will use "large sample size" statistics for samples of size 25 or more; therefore, we can use the data of Table 9.19 to illustrate confidence intervals for standard deviations (Eq. 3.8, p. 88). The critical z value (Excel function `NORMSINV(0.975)`) is 1.960, which yields a 95% confidence interval of

$$\left(\frac{6.702}{1 + \frac{1.960}{\sqrt{60}}}, \frac{6.702}{1 - \frac{1.960}{\sqrt{60}}} \right) = (5.35, 8.97)$$

for the standard deviation which is very close to the square root of the confidence interval obtained for the variance. □

We need to consider the two assumptions under which the confidence interval for the mean theoretically holds. The assumption of independent observations is adequately satisfied as long as each replicate execution of the simulation program begins with a different initial random number seed (see Sect. 2.2). The normality assumption is troublesome and can cause errors in the confidence interval. This is especially true for the service station example since the histogram of Fig. 9.4 does not give the appearance of a symmetric bell-shaped curve. However, regardless of the actual underlying distribution (assuming a finite mean and variance), we can take advantage of the *central limit theorem* (Property 1.5) and assume normally distributed random variables when the sample size is large. The central limit theorem states that sample means, when properly normed, become normally distributed as the sample size increases. A commonly used rule-of-thumb is to have a sample size of at least 25 to 30 individual observations.

It is sometimes important to obtain information about the tails of the distribution. For example, the average number of customers turned away may not be as important as knowing the probability that 20 or more customers per day will be turned away. (In other words, we might want to minimize the probability of "bad" things happening, instead of concentrating on expected values.) A major problem with tail probabilities is that the sample size must be larger to estimate tail probabilities than the sample size needed to estimate means. A rule-of-thumb is to have at least 25 data points and at least 5 occurrences of the event for which the probability statement is being made. Thus, the 30 simulations contained in Table 9.19 are not enough to estimate the probability that 25 or more customers per day are turned away, but it is okay for estimating the tail probability of 20 or more customers.

Example 9.4. We continue with the data of Example 9.3 to estimate the probability that 20 or more customers will be turned away on any given day from the service station. The data of Table 9.19 indicate that 6 out of 30 runs resulted in 20 or more customers being turned away; therefore, our estimate for the probability is $\hat{p} = 0.2$ and thus the 95% confidence interval (Eq. 3.6, p. 87) is

$$0.2 \pm 1.960 \times \sqrt{\frac{0.16}{30}} = (0.06, 0.34) .$$

□

- *Suggestion: Do Problems 9.7 and 9.9.*

9.2.2 Steady-State Simulations

A large manufacturing company has hundreds of a certain type of machine that it uses in its manufacturing process. These machines are critical to the manufacturing process, so the company maintains a repair center that is open 24 hours a day, seven days a week. At any time, day or night, there is one technician who takes care of fixing machines. Each machine that is in the repair center costs the company $75 per hour in lost production due to its unavailability. It has been determined that the arrival of failed machines to the repair center can be approximated by a Poisson arrival process with a mean rate of 4 per day, i.e., independent and exponentially distributed interarrival times with a mean of $1/4$ day between arrivals. (We have assumed that the company has such a large number of these machines in use that the failure of a few of the machines does not significantly affect the arrival rate resulting from the failure of any of the remaining machines.) The time it takes to repair a machine is highly variable due to the many different things that can go wrong, and, in fact, the repair times are also described by an exponentially distributed random variable with a mean time of 4 hours. The repair center thus operates as a single server, infinite capacity, FIFO queueing system. The company would like to determine if it would be economically advantageous, in the long run, to add another technician per shift so that two machines could be under repair at the same time when the need arises. Each technician costs the company $30 per hour, including salary and overhead.

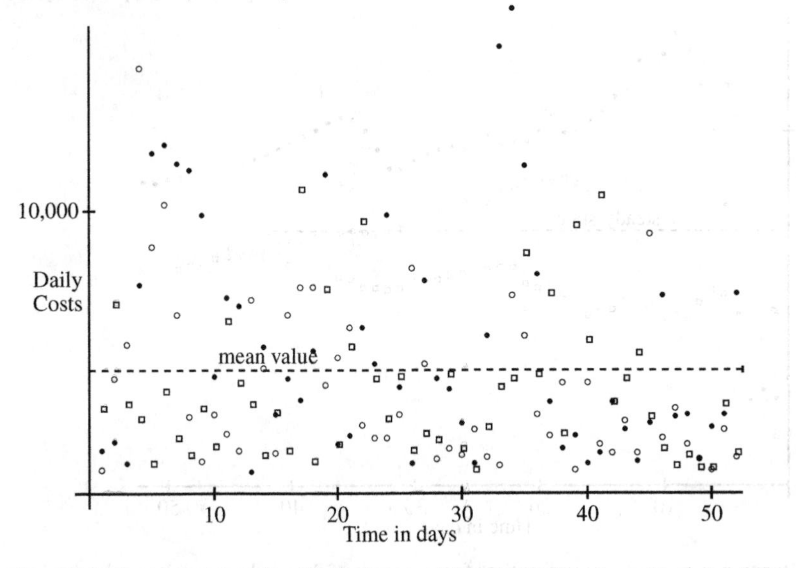

Fig. 9.5 Daily costs versus time for three runs of the machine repair problem simulation

A program was written to simulate the costs incurred with this repair facility using only one technician. The high degree of variability in daily costs is illustrated in Fig. 9.5, in which daily costs for three simulations are shown. Figure 9.6 shows the plot of the running averages of those costs. The variability between simulations for this repair facility is illustrated by comparing the three curves in Fig. 9.6.

In our example, we are interested in a long-range policy, and it is the long-run average cost per day on which the decision will be based. Consequently, we need to look at the behavior of this quantity over time. (The first question that actually needs to be asked when a steady-state analysis is contemplated is "Do steady-state values exist?" It is easy to construct situations in which no steady-state exists. For example, if a slow server is trying to handle fast arrivals, the backlog might continually increase and never reach steady-state. A rigorous answer to the "existence" question is beyond the scope of this textbook. We shall simply assume throughout this section that the processes under consideration do reach steady-state behavior.) Notice that the graphs of the average costs in Fig. 9.6 initially increase rapidly, and then they approach the straight line drawn on the graph. Extrapolating beyond the shown graph, the plot of the average costs would eventually cease to change. Its limiting value with respect to time is called its steady-state value. The straight dotted line in the figure is the theoretical steady-state value, which in this case is the cost per day incurred by the machines being in the repair room.

One of the primary purposes of a steady-state simulation is to estimate the long-run values of performance measures, including means, variances, and in some cases, probabilities. But, because values from simulations are statistical estimates, we also would like to determine the accuracy (i.e., confidence limits) of the estimates. While both obtaining an estimate and determining its accuracy present challenges, it is usu-

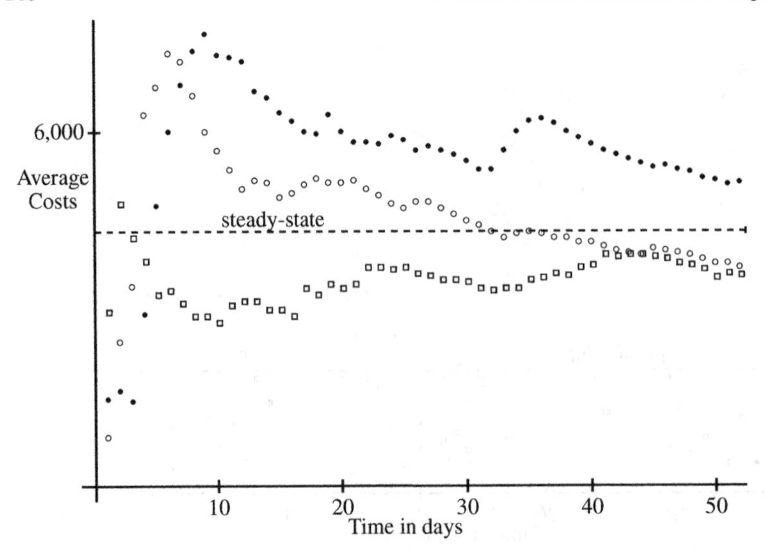

Fig. 9.6 Running average of cost versus time for three runs of the machine-repair simulation

ally easier to derive the estimates than it is to determine their confidence limits. If a steady-state value exists, the time-averaged value from any single simulation will limit to the steady-state value as the run length increases to infinity. The obvious practical difficulty is to know when to stop the simulation. Since the above simulation was stopped short of infinity, we do not expect the simulation's final average cost to be exactly equal to the true long-run value; therefore, it is necessary to determine how much confidence we have in the estimate. To illustrate the problem, assume that the simulations illustrated in Fig. 9.6 were stopped after 52 days. Taking any one of the simulations in the figure, the analyst would possibly conclude that the simulation is reasonably close to steady-state because the graphs indicate that the major fluctuations have ceased; however, none of the three examples has yet reached the asymptotic value. Of course, this particular example was chosen because we could easily determine the theoretically correct value, and with this knowledge we have the ability to measure the accuracy of our estimators. Without the theoretical values (we would not normally simulate if the theoretic values were known), we must turn to statistical methods which usually involve confidence limits.

Before discussing specific procedures for obtaining confidence intervals, let us consider some of the underlying concepts. For illustrative purposes, our context will be the determination of an estimate for mean cost. It should be remembered that constructing a confidence interval for a parameter requires an estimator *and* the variance of the estimator.

Let μ_{cost} denote the long-run mean, and let σ^2_{cost} be the long-run variance of the daily cost. Further, let $C_1, C_2, \cdots$ be the sequence of daily costs obtained by the simulation on days $1, 2, \cdots$, respectively. Suppose n such observations have been obtained by a run of the simulation, and let

$$\overline{C}(n) = \frac{1}{n}\sum_{i=1}^{n} C_i \, ; \tag{9.1}$$

thus, $\overline{C}(n)$ is the average of the first n data values. Figure 9.6 plots $\overline{C}(n)$ versus n for three different realizations, where n is the number of days. In theory,

$$\lim_{n\to\infty} \overline{C}(n) = \mu_{cost} \, , \tag{9.2}$$

which means that if we let n grow indefinitely, the estimate (9.1) converges to the true value of μ_{cost}. The intuitive meaning of Eq. 9.2 is that we can take a single simulation, run it a long time, then use the time-averaged value for an estimate of the long-run expected value[1]. In similar fashion, the long-run sample variance,

$$S^2(n) = \frac{(\sum_{i=1}^{n} C_i^2) - n(\overline{C}(n))^2}{n-1} \, , \tag{9.3}$$

converges to σ^2_{cost} as n limits to ∞. Consequently, we can control the quality of our estimates of the population mean and variance by the run lengths of the simulation.

The quality of the estimator $\overline{C}(n)$ of μ_{cost} is measured by the variance of $\overline{C}(n)$. (Actually, we need its distribution too, but as we shall see, difficulties will arise with just its variance.) You might recall that if we assume $C_1, \cdots, C_n$ are independent and identically distributed, an unbiased estimator for the variance of $\overline{C}(n)$ is $S^2(n)/n$. However, in a simulation study, the sequential random variables $C_1, \cdots, C_n$ are often highly dependent, which introduces a (possibly large) bias into $S^2(n)/n$ as an estimator for the variance of $\overline{C}(n)$. Therefore, the sample variance (Eq. 9.3) of the sequentially generated data should not be used to build a confidence interval for μ_{cost} using the standard methods (i.e., Eq. 3.4). These facts should be stressed: when n is large, $\overline{C}(n)$ is a valid estimator for μ_{cost}, $S^2(n)$ is a valid estimator for σ^2_{cost}, *but* $S^2(n)/n$ is *not* a valid estimator for the variance of $\overline{C}(n)$.

9.2.2.1 Replicates

The most straightforward method (and our recommended approach) of obtaining confidence intervals is to use several replicates of the simulation run (using different initial random number seeds), just as was done for the terminating simulation analyses. What this does is provide a way to create truly random samples by sequential estimation. To illustrate, consider the steady-state simulation repair center mentioned earlier in this section. We might decide that a "long time" is 52 days and make 30 different simulations of 52 days each to obtain a random sample, where the 30 individual data points are the 52-day averaged daily cost values. (The end-points on the graph in Fig. 9.6 yield three such data points.) A major difficulty with the replicate approach is that the results are highly dependent on the decision as to what

[1] When a process has the property that the limiting time-averaged value and the value averaged over the state space are the same, the process is said to have the ergodic property.

constitutes a long time. In theory, the steady-state results of a simulation are not dependent on the initial conditions; however, if a run is too short, the steady-state estimator will be biased due to the initial conditions.

For example, we ran the simulation program of the repair center thirty times. If we use (ridiculously short) runs of two days and use the standard equation for confidence intervals of the mean (Eq. 3.4) to estimate the daily cost due to machine failure, it would be $2,276 per day with a 95% confidence interval of ($1,822, $2,730). The true theoretical steady-state value for this problem is $4,320, so it is apparent that this estimate is too low. Using a run length of 52 days yields an estimate of $4,335 per day with a 95% confidence interval of ($3,897, $4,773). From these results, it is tempting to say that 52 days is long enough to be at steady-state. However, the property given by Eq. 9.2 indicates that if the simulation is at steady-state, there should be very little variation between runs. In other words, the maximum error represented by a confidence interval based on several simulations at steady-state should be close to zero. But the maximum error of the above interval based on 52 days is 438 or 10.1% of the mean value — not at all negligible. If we use 2,002 days to simulate a long time, 30 runs would yield a 95% confidence interval of ($4,236, $4,462), which has a maximum error of under three percent of the mean.

9.2.2.2 Start-up Time

A common approach taken to eliminate the bias of the simulation due to its initial conditions is to establish a "start-up" period and only calculate statistics after the start-up period is finished. For example, we might take 2 days for the start-up period and then calculate the average daily costs based on the costs from day 3 through day 52. Again, the difficulty with this approach is the accurate determination of the best values for the start-up time and the overall run length. If the start-up period is too short, the estimate will be biased by the initial period. If the start-up period is too long, the estimate will have more inherent variability, producing a larger confidence interval than would be expected from the length of the run. We should also mention that practical considerations do enter the picture. A large number of repetitions of long simulations may take a great deal of time. When time is at a premium, another method may be worthwhile. With these considerations in mind, we investigate an alternative method for obtaining confidence intervals for mean values.

9.2.2.3 Batch Means

Let us review again the context of our estimation problem. We make a long simulation run (preferably containing a start-up period) obtaining m data points, $C_1, \cdots, C_m$. We desire to estimate μ_{cost}, which is the long-run mean value; that is,

$$\mu_{cost} = E[C_m] ,$$

where m is assumed large. The most common estimator for μ_{cost} is the sample mean (Definition 3.3). However, in most simulations C_i and C_{i+1} are highly dependent random variables, which is the reason that estimating the standard deviation of the estimator for μ_{cost} is difficult.

As mentioned previously, steady-state results are independent of the initial conditions[2]. This fact implies that if two costs are sufficiently far apart in time, they are essentially independent; that is, C_i and C_{i+k} can be considered independent if k is large enough. This leads to one of the most popular forms of obtaining confidence intervals: the method of batch means. The concept is that the data will be grouped into batches, and then the mean of each batch will be used for the estimators. If the batches are large enough, the batch means can be considered independent because, intuitively, they are far apart. Since the batch means are considered independent, they can be used to estimate not only the overall mean, but also the standard deviation. Specifically, let us take the m data points and group them into n batches, each batch containing $\ell = m/n$ data points, and let $\overline{C}_j$ denote the mean of the jth batch; namely,

$$\begin{aligned}
\overline{C}_1 &= \frac{C_1 + \cdots + C_\ell}{\ell} \\
\overline{C}_2 &= \frac{C_{\ell+1} + \cdots + C_{\ell+\ell}}{\ell} \\
&\vdots \\
\overline{C}_n &= \frac{C_{(n-1)\ell+1} + \cdots + C_{n\ell}}{\ell}
\end{aligned} \tag{9.4}$$

where $m = n\ell$. (Figure 9.7 illustrates this concept using a simulation of the repair center example. In the figure, an initial period of two observations is ignored, then five batches of seven each are formed. Notice that the figure is for illustration purposes only; usually a single batch includes hundreds or thousands of observations.) As ℓ gets large, $\overline{C}_j$ and $\overline{C}_{j+1}$ become independent. Using the definitions for the sample mean and variance, we have the following:

$$\begin{aligned}
\overline{X}_{batch} &= \frac{\overline{C}_1 + \cdots + \overline{C}_n}{n} = \frac{C_1 + \cdots + C_m}{m}, \\
S^2_{batch} &= \frac{\overline{C}_1^2 + \cdots + \overline{C}_n^2 - n\overline{X}^2_{batch}}{n-1}.
\end{aligned} \tag{9.5}$$

The confidence interval is, thus, given by

$$\left(\overline{X}_{batch} - t_{n-1,\alpha/2}\frac{S_{batch}}{\sqrt{n}},\ \overline{X}_{batch} + t_{n-1,\alpha/2}\frac{S_{batch}}{\sqrt{n}}\right) \tag{9.6}$$

[2] To be technically correct in this paragraph, we need to discuss concepts and assumptions related to the ergodic property, but this is beyond the scope of this textbook. Here we only intend to give an intuitive feel for these principles.

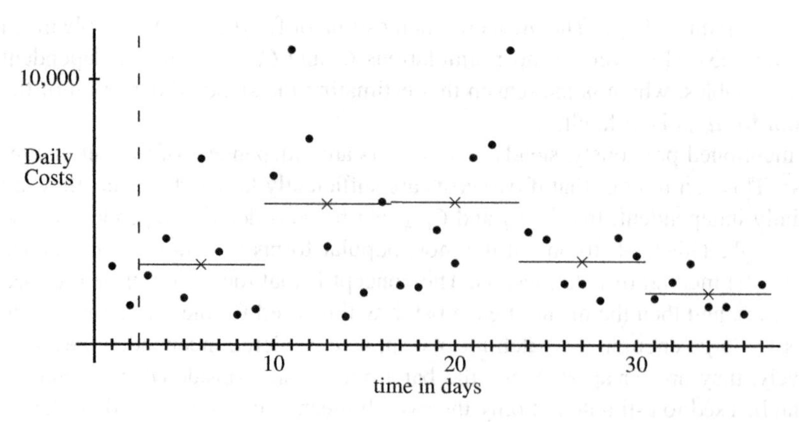

Fig. 9.7 Batch means from a simulation of the repair center example

where $t_{n-1,\alpha/2}$ has the same meaning as it does in Eq. (3.4).

The two major assumptions made in the development of the confidence interval of Eq. (9.6) are: (1) the random variables $\overline{C}_1, \cdots, \overline{C}_n$ are distributed according to a normal distribution, and (2) the random variables are independent. As long as the batch size is large, the central limit theorem again permits the assumption of normality (a rule of thumb for using the central limit theorem is to have a batch sample size of at least 30). The independence assumption is more difficult to verify. One test for independence of the sequence is to estimate the correlation coefficient (Definition 1.20) for adjacent values, namely, the correlation between $\overline{C}_i$ and $\overline{C}_{i+1}$, called the lag-1 correlation coefficient. (A zero correlation between adjacent values does not guarantee independence, but a nonzero value guarantees lack of independence.)

The most common estimator for the lag-1 correlation coefficient for the sequence $\overline{C}_1, \cdots, \overline{C}_n$ is given by

$$r = \frac{\sum_{i=1}^{n-1}[(\overline{C}_i - \overline{X}_{batch})(\overline{C}_{i+1} - \overline{X}_{batch})]}{\sum_{i=1}^{n}[(\overline{C}_i - \overline{X}_{batch})^2]}, \tag{9.7}$$

where $\overline{X}_{batch}$ is defined in Eq. (9.5). A slightly easier formula (again using Eq. 9.5) for computing this estimator is

$$r = \frac{\sum_{i=1}^{n-1}\overline{C}_i\overline{C}_{i+1} - n\overline{X}_{batch}^2 + \overline{X}_{batch}(\overline{C}_1 + \overline{C}_n - \overline{X}_{batch})}{(n-1)S_{batch}^2}. \tag{9.8}$$

(It would be worthwhile to compare this estimate for the lag-1 auto-correlation coefficient with the estimator used for the correlation coefficient between two random samples given in Eq. (3.10).)

An hypothesis test can be easily constructed to test the assumption that the correlation coefficient is zero. The estimator for the correlation coefficient of two independent normally distributed random samples can be transformed into a Student-t

distribution using the equation

$$T_{n-2} = \hat{\rho}\sqrt{\frac{n-2}{1-\hat{\rho}^2}}\,.$$

To obtain the p-value associated with testing if the correlation coefficient is zero (see discussion of p-value on p. 95), first calculate the test statistic using the above equation as

$$t_{\text{test}} = r\sqrt{\frac{n-2}{1-r^2}} \tag{9.9}$$

and then the p-value for the test that the $\rho = 0$ is given by

$$\text{p-value} = \Pr\{T_{n-2} > t_{\text{test}}\}\,.$$

(As you recall, one approach for using the p-value is that if the p-value is larger than 10%, the hypothesis is not rejected; if the p-value is less than 1%, the hypothesis is rejected; when it is between 1% and 10%, the evidence may be considered inconclusive.)

Another approach which eliminates the "uncertainty region" between 1% and 10% is to obtain the critical value for r. For this, the inverse function to Eq. (9.9) is obtained as

$$r_{\text{critical}} = \frac{t_{n-2,\alpha/2}}{\sqrt{n-2+t^2_{n-2,\alpha/2}}}\,. \tag{9.10}$$

If the absolute value of the r statistic (Eq. 9.8) is greater than r_{critical}, the hypothesis that the batches are independent can be rejected. If the absolute value of the r statistic is less than r_{critical}, it is still hard to accept with confidence that the lag-1 correlation coefficient is zero because of the high variability of the estimator.

A rule of thumb that is sometimes used is for the absolute value of the lag-1 correlation coefficient to be less than 0.05. For example, the repair center example was simulated for 2,002 days using a start-up period of 2 days. Using the last 2,000 days, the method of batch means was used to estimate a confidence interval where the size of each batch was 50 days. To determine the appropriateness of the method, the lag-1 correlation coefficient was estimated as $r = -0.04$. If the –0.04 number is an accurate estimate for ρ, then the one simulation run is all that is needed for the confidence interval. Therefore, to determine the accuracy of the $r = -0.04$ estimate, the simulation was run five more times, each time being a run of 2,002 days. The five values for r were 0.21, 0.25, –0.04, –0.13, and 0.07. The conclusion from this example is that the estimator for ρ is too variable to be reliable when only 40 batches are used; thus, more batches are needed in order to have a reliable estimator.

The difficulty with choosing a batch size is that if the batch size is large enough to obtain a small lag-1 correlation coefficient, then the number of batches will be too small to produce a reliable estimate; however, if the number of batches is large enough to produce a reliable estimate, then the batch size will be too small to maintain independence between batch means.

When the modeler has the freedom to choose the run length based on values dynamically determined within the simulation, he or she may use the textbook by Law and Kelton (mentioned at the beginning of this chapter) that contains a suggested approach for confidence intervals. The authors' procedure is to extend the run length until the absolute value of the lag-1 correlation coefficient determined from 400 batches is less than 0.4. In other words, for a given run length, the batch size necessary to produce 400 batches is used, and the lag-1 correlation coefficient is calculated based on those 400 batches. If the absolute value of the coefficient is greater than 0.4, one can infer that the batches are too small, and the run length is increased. Once the absolute value of the coefficient is less than 0.4, the run length is fixed and the batch size is increased by a factor of ten. The resulting 40 batches are then used to obtain the confidence interval. The justification for this approach is that when 400 batches produce a coefficient of less than 0.4 and then the batches are made 10 times larger, the resulting 40 batches should produce a coefficient of less than 0.05.

More commonly, the run length of the simulation cannot be dynamically determined, but must be fixed at the start of the simulation. A good "quick and dirty" estimate of confidence intervals may be made using a short start-up period, dividing the remaining run length into five to ten batches, and then using those batches to estimate the confidence interval. (A minimum of five batches is suggested in [2] and at least ten batches is suggested in [3].) The idea of using a small number of batches is flawed since it is difficult to justify using the Central Limit Theorem; in practice, however, it does seem to perform adequately.

Example 9.5. Table 9.20 (and Fig. 9.7) contains simulation output to be used in a numerical example. (We use a run length that is too small for accuracy, but it is

Table 9.20 Individual data points from a simulation run of length 37 days from the machine repair example

	Daily costs							means
week 0						2929	1441	—
week 1	2558	3958	1772	6996	3451	1094	1296	3017.9
week 2	6338	11105	7774	3710	1062	1931	5406	5332.3
week 3	2270	2182	4365	3065	7111	7570	11107	5381.4
week 4	4276	4024	2340	2345	1665	4042	3346	3148.3
week 5	1728	2379	3050	1476	1490	1201	2303	1946.7

better for illustrating the computations. Do not infer from this example that estimates should be made based on such short runs.) This example uses an initialization period of two days and a batch size of seven days that yields five batches. We calculate the average for each row in Table 9.20, which gives the five batch means as: $\overline{c}_1 = 3{,}017.9$, $\overline{c}_2 = 5{,}332.3$, $\overline{c}_3 = 5{,}381.4$, $\overline{c}_4 = 3{,}148.3$, and $\overline{c}_5 = 1{,}946.7$. The sum of the five values is 18,826.6, and the sum of their squares is 80,202,043.44. The confidence interval calculations are

$$\bar{x}_{batch} = \frac{18,826.6}{5} = 3,765.32$$

$$s^2_{batch} = \frac{80,202,043.44 - 5 \times 3,765.32^2}{4} = 2,328,467.5$$

which yields a 95% confidence interval of

$$(3765.3 - 2.776 \times \frac{1525.9}{\sqrt{5}},\ 3765.3 + 2.776 \times \frac{1525.9}{\sqrt{5}}) = (1870.9,\ 5659.7)\,.$$

Thus, we see that a run of this small size yields an estimate with a large degree of variability. With 95% confidence, we would say that the maximum error is 1,894.4. Let us consider what would have happened if we had improperly used the individual 35 daily values (i.e., a batch size of one) to obtain the confidence interval. In that case, we would have estimated the maximum error to be 878.5, which greatly understates the inherent variation of the 3,765.3 estimated value of the average daily cost. □

As a final comment, the authors feel that the use of batch means is somewhat outdated. The main advantage of batch means over using replicates is that the user does not have to delete a warm-up period for each replicate, thus saving run time. When simulations involved expensive computer time, it was important to have procedures that minimized simulation run lengths; however, today many simulation are run on desktop computers or laptops and time is not as important. Therefore, it is our recommendation to base estimates on several replicates thus insuring independent samples and not be concerned with building batch means. The main reason to include this section is for historical reasons and because many commercial simulation packages automatically calculate estimates based on batch means.

• *Suggestion: Do Problems 9.10–9.15.*

9.2.3 Comparing Systems

Not only are simulations used to analyze existing or proposed systems, but also they are used as a decision tool in comparing two or more systems. In this section, we briefly discuss analyzing simulation results when the goal is to compare two systems. For example, management may be considering adding a new workstation at a bottleneck point within a manufacturing facility. To help with that decision, a simulation of the facility with and without the new workstation could be run with the intent of comparing the throughput rate of the two options.

It is also quite common to compare several systems or to compare systems with their parameters set at multiple levels. The comparison of more than two systems should involve a careful design of the statistical experiments before any simulation runs are carried out. In the analysis of the simulations, a procedure called "analysis of variance" is used for comparing several systems. Such statistical issues are beyond the scope of this textbook; however, there are numerous textbooks that cover the statistical design and analysis for comparing multiple systems.

The context for this section is that two random samples have been obtained from a series of simulations describing two systems under consideration. If the study involves terminating simulations, the random samples come from replications. If the study involves steady-state simulations, the random sample may arise from a combination of replications and batching. The first random sample is assumed to describe system 1 and involves n_1 data points with their mean and variance being $\overline{X}_1$ and S_1^2, respectively. The second random sample (independent from the first) is assumed to describe system 2 and involves n_2 data points with their mean and variance being $\overline{X}_2$ and S_2^2.

Confidence Intervals for Comparing Means. When mean values are of interest, we generally would like to know the difference between the mean for System 1 and the mean for System 2. As usual we must assume that the random samples are normally distributed (at least for small sample size). If the two random samples are the same size, it should be possible to pair the data and use the confidence interval of Eq. (3.4). In other words, if $n_1 = n_2$, let $Y_i = X_{1,i} - X_{2,i}$ and form a confidence interval from the random sample $Y_1, \cdots, Y_n$, where $n = n_1$. The advantage of this procedure is that $X_{1,i}$ and $X_{2,i}$ need not be independent.

Example 9.6. We wish to compare two queueing systems that operate 24 hours per day. Arrivals occur to both systems according to a Poisson process with a mean rate of 50 arrivals per day. The first system is an M/M/2 system with each server having a mean service time of 48 minutes. The second system is an M/M/1 system with its server having a mean service time of 24 minutes. Based on queueing theory, we know that the second system is better in terms of mean waiting times for the customers (for equal traffic intensities, the few number of servers is always the better system); however, we would like to build our confidence in the theoretical results with simulated results.

A simulation was written for an M/M/2 queueing system and an M/M/1 queueing system, except that the *same* (random) arrival stream was used for both systems. Thus, the system are not independent, but one source of variability between the systems was eliminated. Five replications were made. Each replication had a run length of 1000 days with a two-day warmup period. The results are in Table 9.21.

Table 9.21 Waiting time data for Example 9.6

Replicate Number	M/M/2 Cycle Time	M/M/1 Cycle Time	Difference
1	2.4918	2.3396	0.1522
2	2.7489	2.3406	0.4083
3	2.7174	2.5840	0.1334
4	2.8237	2.4463	0.3774
5	2.6509	2.2341	0.4168

The sample mean and standard deviation of the differences (fourth column) are 0.2976 and 0.1422, respectively, yielding a 95% confidence interval of

$$(0.1210, 0.4742) .$$

Because zero is not within the interval, we can say that there is a statistically significant difference between the two mean waiting times. □

For large sample sizes, a confidence interval for the difference can be obtained by taking advantage of the fact that the difference in sample means can be approximated by a normal distribution. The variance for the difference is calculated by "pooling" the variations from the two samples; in other words, the estimate for the variance of the differences is given by

$$S^2_{\text{pooled}} = \frac{(n_1 - 1)S_1^2 + (n_2 - 1)S_2^2}{n_1 + n_2 - 2} .$$

A confidence interval can now be obtained by using the standard deviation as estimated from the combined data sets. In particular, the $1 - \alpha$ confidence interval for the difference in the mean values between system 1 and system 2 can be approximated by

$$(\overline{X}_1 - \overline{X}_2) \pm t_{(n_1+n_2-2, \frac{\alpha}{2})} S_{\text{pooled}} \sqrt{\frac{1}{n_1} + \frac{1}{n_2}} . \tag{9.11}$$

If the confidence interval does not include zero, we can say that we are at least $1 - \alpha$ confident that the means from system 1 and system 2 are different.

Example 9.7. Table 9.20 contains simulation results from a simulation run of 37 days using one technician for the repairs. We run the simulation of the repair center again, except we use two technicians for repairs. The daily cost values include the cost of both technicians and the cost incurred due to machines being in the repair center. For the run involving two technicians, we use a start-up period of 2 days and a batch size of 7 days and run the simulation for 401 days; thus, we have 57 batches. These 57 data points yield a mean of \$2,500 and a standard deviation of \$1,200. Our goal is to use these data to determine whether the average cost of using two technicians is different from the average cost of using one technician. The estimate for the difference in the average cost of using one technician versus two technicians is \$1,265. The pooled variance estimate is

$$S^2_{\text{pooled}} = \frac{4 \times 2328467 + 56 \times 1440000}{60} = 1{,}499{,}231 .$$

For a 95% confidence interval, we use a t-value of 2.0 (see Table C.3); therefore, the confidence interval is

$$1265 \pm 2 \times 1224.4\sqrt{0.2 + 0.0175} = (123, 2407) .$$

Because zero is not contained in the confidence interval, we can say that there is a statistical difference between the two systems, and that having two technicians will reduce the daily cost of the repair center. Similarly, we can be 95% confident that the reduction in daily costs resulting from having two technicians on duty is between \$123 and \$2,407. (If this range in the estimate seems too large, then more simulations runs must be made.) □

As a general rule, it is usually best to use the paired test (as in Example 9.6) if the simulation can developed so that the same random number stream is used for those processes that are common to both systems. Such reduction in variability is extremely helpful in identifying significant differences.

Confidence Intervals for Comparing Variances. Changes for systems are sometimes designed with the intent of decreasing variances. Comparisons involving the variances for two systems are based on the *ratio* of the variances. For example, if system 2 involves a proposed design change that claims to decrease the variance from system 1, then we would hope that the entire confidence interval centered around the ratio S_1^2/S_2^2 would be greater than one. The confidence interval involves critical values from the F-distribution, and the F-distribution has two parameters referred to as the degrees of freedom. Thus, the $1-\alpha$ confidence interval for the ratio of S_1^2 to S_2^2 is given by

$$\left(\frac{S_1^2}{S_2^2}\frac{1}{F_{(n_1,n_2,\alpha/2)}},\ \frac{S_1^2}{S_2^2}F_{(n_2,n_1,\alpha/2)}\right)\ . \tag{9.12}$$

If the confidence interval does not include 1, we can say that (with $1-\alpha$ confidence) there is a statistical difference between the variances of system 1 and system 2. To illustrate this concept, we use the numbers from Example 9.7. The critical F-values are taken from a book of statistical tables and are $F_{(5,57,0.025)} \approx 6.12$ and $F_{(57,5,0.025)} \approx 2.79$; therefore an approximate 95% confidence interval for the ratio of the variances is

$$(1.617\frac{1}{6.12}, 1.617\times 2.79) = (0.26, 4.51)\ .$$

Because one is included in the confidence interval for the ratio of the variances, we can only say that there is no statistical difference between the variances for the two systems. The variances that are being compared are the variances of mean daily costs averaged over a week. (Note that this is different than the variance for daily costs.) As with other examples based on this repair center system, the start-up periods and batch sizes should be much larger if actual estimates were to be made.

Confidence Intervals for Comparing Proportions. Sometimes design changes are considered because of their potential effect on tail probabilities. For example, after receiving complaints regarding long customer waiting times, management may try to institute changes to decrease the probability that queue times are longer than some specified level. Thus, we may want to develop a confidence interval for the difference between the tail probability for system 1 and the tail probability for system 2; denote these probabilities by $\hat{P}_1$ and $\hat{P}_2$, respectively. As long as n_1 and n_2 are large, the following can be used for the $1-\alpha$ confidence interval for the difference

$$(\hat{P}_1 - \hat{P}_2) \pm z_{\alpha/2}\sqrt{\frac{\hat{P}_1(1-\hat{P}_1)}{n_1} + \frac{\hat{P}_2(1-\hat{P}_2)}{n_2}}. \tag{9.13}$$

We again use the numbers from Example 9.7 to illustrate the use of the confidence intervals for proportions. (As before, the value of n_1 is too small to appropriately use Eq. (9.13); however, the numerical values below should be sufficient to illustrate the proper calculations.) We are concerned with a average daily cost for any given week being larger than $4,500. So our goal is to determine the amount of reduction in the probability that a week's average daily cost would be greater than $4,500. From the first simulation run involving one technician (Table 9.20), there were 2 out of 5 weeks with daily averages greater than $4,500. In the second run, there were 6 out of 57 weeks with a daily average greater than $4,500. Thus, the 95% confidence interval is given by

$$0.4 - 0.105 \pm 1.96 \times \sqrt{\frac{0.4 \times 0.6}{5} + \frac{0.105 \times 0.895}{57}} = (-0.14, 0.73).$$

Thus, there is again too much statistical variation to draw any conclusions regarding a reduction in the probability that a week's average daily cost is greater than $4,500.

• *Suggestion: Do Problems 9.8 and 9.16–9.19.*

Problems

9.1. Using three coins, simulate the production process of Example 9.1. Stop the simulation as soon as the simulated clock is greater than or equal to 100 and estimate the average number of widgets in the system. Compare your estimate with the estimate from Table 9.11.

9.2. Simulate the production process of Example 9.1, but assume that there are several machines available for the machining process. In other words, if a widget is acceptable (not destined for the scrap pile) and goes to the facility for machining, there are enough machines available so that the widget never has to wait to be machined. Thus, any number of widgets can be machined at the same time. Estimate the average number of widgets in the facility over a 100-minute period.

9.3. Simulate the production process of Example 9.1, but assume that arrivals are random. Specifically, the time between arriving widgets will equal 12 minutes half the time, 11 minutes one third of the time, and 10 minutes one sixth of the time. Use a die along with the coins to simulate this system. Estimate the number of widgets in the facility over a 100-minute period.

9.4. The manager of a service station next to a busy highway is concerned about congestion. The service station has two gas pumps and plenty of room for customers to wait if the pumps are busy. Data indicate that a car arrives every 3 minutes. Of the customers that arrive, 25% of them take 5 minutes 45 seconds in front of the pump and spend $10, 50% of them take 5 minutes 50 seconds in front of the pump and spend $10.50, and 25% of them take 5 minutes 55 seconds in front of the pump and spend $11. Simulate the arriving and servicing of cars and estimate the average receipts per hour and the average number of customers in the service station.

9.5. Consider the service station of Problem 9.4 except that the time between arrivals is not fixed. Assume that the time between arriving cars is $2\frac{1}{2}$ minutes with probability 0.25, 3 minutes with probability 0.50, and $3\frac{1}{2}$ minutes with probability 0.25. Using simulation, estimate the average receipts per hour and the average number of customers in the service station.

9.6. A computer store sells two types of computers: the MicroSpecial and MicroSuperSpecial. People arrive at the computer store randomly throughout the day, such that the time between arriving customers is exponentially distributed with a mean time of 40 minutes. The store is small and there is only one clerk who helps in selection. (The clerk handles customers one at a time in a FIFO manner.) Twenty-five percent of the customers who enter the store end up buying nothing and using exactly 15 minutes of the clerk's time. Fifty percent of the customers who enter the store end up buying a MicroSpecial (yielding a profit of $225) and taking a random amount of the clerk's time, which is approximated by a continuous uniform distribution between 30 and 40 minutes. Twenty-five percent of the customers who enter the store end up buying a MicroSuperSpecial (yielding a profit of $700) and taking a random amount of the clerk's time, which is approximated by a continuous uniform distribution between 50 and 60 minutes. The store is open only 4 hours on Saturday. The policy is to close the door after the 4 hours but continue to serve whoever is in the store when the doors close. Simulate the activities for a Saturday and answer the following questions.
(a) The store opens at 10:00 A.M. and closes the door at 2:00 P.M. Based upon your simulation experiment, at what time is the clerk free to go home?
(b) What is your estimate for the total profit on Saturday?
(c) Simulate the process a second time, and determine how much your answers change.

9.7. Consider the simulation used to estimate the average sales receipts per house from Problem 2.1. Since each house was independent from all other houses, the 25 data points formed a random sample. Use those 25 values to obtain a confidence interval for the mean receipts per house. Calculate the theoretical value and compare it with your confidence interval.

9.8. Consider the selling of microwaves in Problem 2.1. In that example, one half of the customers who entered the store did not purchase anything. The store's management feels that by making certain changes in the store's appearance and by some additional training of the sales staff, the percent of customers who enter the store and do not purchase anything can be decreased to 40%. Furthermore, it has been estimated that the cost of these changes is equivalent to \$1 per customer entering the store. Using simulation to justify your decision, do you feel that the extra expense of \$1 per customer is worth the increase in purchases?

9.9. Write a computer simulation of a non-stationary Poisson process (see the appendix of Chap. 4) with a mean rate function defined, for $t \geq 0$, by

$$\lambda(t) = 2 + \sin(2\pi t) ,$$

where the argument of the sin function is radians. Obtain 95% confidence intervals for the following based on 25 replications.
(a) The mean number of arrivals during the interval $(0,2)$.
(b) The standard deviation for the number of arrivals during the interval $(0,2)$.
(c) The probability that there will be no arrivals during the interval $(0,2)$.

9.10. The following 100 numbers are correlated data: 52.0, 27.3, 46.5, 43.0, 41.3, 44.2, 47.5, 56.8, 77.0, 50.9, 28.3, 30.9, 67.1, 50.3, 58.8, 60.0, 35.1, 40.5, 36.0, 41.1, 47.6, 63.6, 78.6, 59.6, 47.6, 44.5, 59.3, 61.0, 53.1, 26.7, 56.9, 52.4, 70.4, 58.0, 72.6, 58.1, 47.6, 37.7, 36.1, 59.3, 51.5, 40.2, 58.0, 77.9, 54.5, 57.5, 58.9, 44.6, 60.5, 57.5, 51.1, 60.2, 39.9, 37.5, 67.2, 61.5, 58.9, 58.3, 56.6, 44.9, 43.4, 44.5, 31.6, 20.1, 47.0, 58.0, 58.8, 54.6, 48.9, 58.9, 62.3, 62.8, 22.6, 66.3, 48.3, 56.8, 73.3, 47.5, 43.3, 42.5, 40.0, 54.2, 44.7, 51.7, 56.3, 59.7, 29.6, 49.0, 43.5, 33.2, 61.1, 80.1, 55.1, 76.8, 47.5, 56.5, 66.7, 65.8, 73.6, 62.6.
(a) Obtain the sample mean and sample variance of this data. (For comparison purposes, note that the true mean and variance of the underlying distribution are 50 and 200, respectively.)
(b) Estimate the lag-1 correlation coefficient and its 95% confidence interval. (For comparison purposes, note that the true lag-1 correlation coefficient is 0.2.)
(c) Using a batch size of 20, give a confidence interval for the mean.

9.11. Correlated interarrival times can be generated using a Markov chain together with two interarrival distributions. Let φ_a and φ_b be two distributions, with means and variances given by θ_a, σ_a^2, θ_b, and σ_b^2, respectively. Furthermore, let $X_0, X_1, \cdots$ be a Markov chain with state space $E = \{a,b\}$ and with transition matrix given by

$$\begin{bmatrix} p & 1-p \\ 1-p & p \end{bmatrix},$$

where $0 < p < 1$. The idea is that the Markov chain will be simulated with each time step of the chain corresponding to a new interarrival time. When the chain is in state i, interarrival times will be generated according to φ_i.

To start the simulation, we first generate a random number; if the random number is less that 0.5, set $X_0 = a$, otherwise set $X_0 = b$. The sequence of interarrival times can now be generated according to the following scheme:

1. Set $n = 0$.

2. If $X_n = a$, then generate T_n according to φ_a.
3. If $X_n = b$, then generate T_n according to φ_b.
4. Generate a random number. If the random number is less than p, let $X_{n+1} = X_n$; otherwise, set X_{n+1} to be the other state.
5. Increment n by one and return to step 2.

In this fashion, a sequence $T_0, T_1, \cdots$ of interarrival times are generated. These times are identically distributed, but not independent.
(a) Show that $E[T_0] = 0.5(\theta_a + \theta_b)$.
(b) Show that $\text{var}(T_0) = 0.5(\sigma_a^2 + \sigma_b^2 + 0.5(\theta_a - \theta_b)^2)$.
(c) Show that the lag-1 correlation coefficient is given by

$$\rho = \frac{0.5(2p-1)(\theta_a - \theta_b)^2}{\sigma_a^2 + \sigma_b^2 + 0.5(\theta_a - \theta_b)^2} .$$

9.12. Generate 100 correlated interarrival times according to the scheme described in the preceding problem. Let φ_a be exponential with mean 10, φ_b be exponential with mean 1, $p = 0.85$. Based on your sample of size 100, estimate mean, standard deviation, and lag-1 correlation coefficient. Compare your estimates with the actual values.

9.13. Generate 100 interarrival times such that the distribution of the times are normally distributed with mean 200, standard deviation 25, and lag-1 correlation 0.8. Based on your sample of size 100, estimate the mean, standard deviation, and lag-1 correlation coefficient. (Hint, refer back to Problem 1.25 and Example 2.7.)

9.14. Write a computer simulation of an M/M/5/10 queueing system using the structure of the future events list. Set the mean arrival rate to be 14 per hour and the mean service time per machine to be 25 minutes. Use your simulation to build a confidence interval for the long-run overflow rate. Is the theoretical value within your confidence interval?

9.15. Write a computer simulation, using the structure of the future events list, of the manufacturing facility described in Example 8.8. The purpose of Example 8.8 was to illustrate an analytical approximation technique. Based on your simulation, was the approximation technique good?

9.16. A small service station next to an interstate highway has one pump and room for two other cars. Furthermore, we assume that the assumptions for an M/G/1/3 queueing system are appropriate. The mean arrival rate of cars is 8 per hour and the service time distribution is approximated by a continuous uniform distribution varying from 3 to 10 minutes. The expected profit received from each car is $10.00. For an extra $75 per day, the owner of the station can increase the capacity for waiting cars by one (thus, becoming an M/G/1/4 system). Is the extra $75 worthwhile? To answer this question, write a computer simulation with or without the future events list concept.
(a) Assume that the station is open 10 hours per day.
(b) Assume that the station is open 24 hours per day.

9.17. A 12-trial test of vehicles A and B for fuel efficiency shows that vehicle A travels an average of 22.3 city miles per gallon of gasoline (mpg) with a sample standard deviation of 1.3 mpg while vehicle B uses an average of 23.5 mpg with a sample standard deviation of 1.6 mpg.
(a) Construct a 95% confidence interval on the difference of the means.
(b) Based on the results given above, is there enough evidence to conclude, at the 5% significance level, that the average mpg are different in the two vehicles? Explain?
(c) Construct a 95% confidence interval on the ratio of the variances?
(d) Are the variances statistically different at the 5% significance level?

9.18. Vehicles C and D were tested to compare the consumption of gasoline per mile traveled. The testing was performed at the same time under a set of similar conditions for both vehicles. At the end of each test, the miles traveled and the gallons of gas consumed were recorded. The following table lists the ten conditions and results.

Test	Road Conditions	Miles Traveled	Gallons Vehicle C	Gallons Vehicle D
1	Flat Surface	120	5.3	5.9
2	10% Hilly, Rainy, Summer	85	4.1	4.5
3	30% Hilly, Dry, Summer	100	4.8	4.7
4	60% Hilly, Humid, Fall	75	3.9	4.7
5	Flat Surface, Curves, Dry, Fall	80	4.5	4.4
6	15% Hilly, Dry, Winter	115	5.2	5.5
7	40% Hilly, Rainy, Winter	90	4.3	4.1
8	20% Dirt Road, Dry, Summer	110	5.9	6.3
9	Sea Level, Humid, Summer	80	4.1	4.5
10	High Mountains, Winter	95	5.1	5.6

(a) Construct a 95% confidence interval on the (paired) difference on the miles per gallon efficiency.
(b) Construct a 95% confidence interval on the average difference ignoring the pairing of the tests.
(c) Which of the above confidence intervals is better for the experiment and why?
(d) Is there a significant difference in the gasoline usage of the two vehicles?

9.19. Fifteen out of the last 100 flights departing from the city airport have been delayed due to different reasons, but mainly due to security checks. The security checks include detailed checks to randomly selected passengers. A new system that allows passengers to move ahead in the line if the expected waiting time for detailed checks exceeds the time before departure is being tested at the expense of adding another inspector that checks departure schedules and waiting times. After using the new system while monitoring 110 flights, only 5 flights were delayed.
(a) Find a 95% confidence interval on the difference of the fraction of delayed flights before and after the new security check system.
(b) Is there statistical evidence to suggest that the new system helps in reducing the flight delays?
(c) In the sample of 110 flights, what is the maximum number of flights that could be delayed for the new system to be considered statistically significantly better than the current security check system?

9.20. Simulate an M/M/1/3 queueing system with a mean arrival rate and service rate of four per hour. How many customers are turned away during the first two hours of operation? (Assume the system starts empty.)

9.21. Simulate a G/G/1/3 queueing system for two hours. Let the interarrival times and service times have a continuous uniform distribution between 0 and 30 minutes. How many customers are turned away during the first two hours of operation? (Assume the system starts empty.) Compare the answer to this problem with the previous problem.

References

1. Law, A.M. (2007). *Simulation Modeling and Analysis*, 4th ed., McGraw Hill, Boston.
2. Law, A.M., and Kelton, W.D. (1984). Confidence Intervals for Steady-State Simulations: I. A Survey of Fixed Sample Size Procedures. *Operations Research*, **32**:1221–1239.
3. Schmeiser, B. (1982). Batch Size Effects in the Analysis of Simulation Output. *Operations Research*, **30**:556–568.

Chapter 10
Inventory Theory

Applied probability can be taught as a collection of techniques useful for a wide variety of applications, or it can be taught as various application areas for which randomness plays an important role. The previous chapters have focused on particular techniques with some applications being emphasized through examples and the homework problems. We now change tactics and present two chapters dealing with specific problem domains: the first will be inventory and the second will be replacement. Inventory theory is a useful stochastic concept that deals with the uncertainties associated with demand and supply of goods and services. It may seem advantageous from a production stand point to have an infinite supply of raw material, but it may not be the most economical way of managing. Because storage rooms are limited, supply is limited, demand is stochastic, handling and storage of items have a cost, ordering often has a cost, being out of an item has a cost, and items may be perishable, the determination of an inventory policy that optimizes a cost or performance criterion is often needed. Given an objective to optimize, inventory theory uses probability principles to develop models that can answer questions such as, how much to order? Or, how often to order?

It is important to mention one caveat with respect to these inventory models. The techniques and probabilistic reasoning used to solve the problems are more important than the problems themselves. The assumptions made in developing each inventory model tend to be restrictive so that the range of possible applications is thus restricted. However, the general approach to formulating and solving the models is quite common. In actual application, approximations must often be employed to obtain solutions, and with an initial understanding of the basic modeling approach it becomes easier to develop good heuristics.

As a "fringe benefit", this chapter also contains a discussion on the optimization of discrete functions. Students of this text are familiar with the use of derivatives for optimizing continuous functions; however, the use of differences for optimizing discrete functions is often overlooked in most mathematics courses taken by engineering students; thus, a discussion of the difference operator for optimization purposes is covered in the following sections.

R.M. Feldman, C. Valdez-Flores, *Applied Probability and Stochastic Processes*, 2nd ed., DOI 10.1007/978-3-642-05158-6_10,

10.1 The News-Vendor Problem

We begin with a classical example that is simplistic, yet illustrates some of the basic reasoning used in many inventory modeling situations. Since one of the main purposes of the chapter is to help develop the modeling process, we shall motivate each new type of model by a specific example. Then, a general model will be developed that can be used for the problem type motivated by the example.

Example 10.1. A news vendor buys newspapers at the start of every day to sell on the street corner. Each newspaper costs \$0.15 and is sold for \$0.25. Any unsold newspapers will be bought back by the supplier for \$0.10. After observing demand for a few months, the news vendor has determined that daily demand follows a uniform distribution between 21 and 40 newspapers. Thus, the question of interest is, "how much inventory to purchase at the start of each day in order to maximize the expected profits?" The decision involves balancing the risk of having extra papers at the end of the day, with having too few papers and being unable to take advantage of potential sales. We shall first show how to solve the problem in general, and then apply the above numbers to illustrate with a specific example. □

The problem is to determine the inventory quantity to order, q, at the beginning of each day. The reason for the difficulty is that the daily demand, D, is not fixed and is known only through its cumulative distribution function, F (or equivalently, through its probability mass function, f). In the order quantity decision there is a trade-off between two "bad" possibilities: (1) the initial inventory order quantity may be too small (with probability $P\{D > q\}$) which results in some lost sales and (2) the initial inventory order quantity may be too large (with probability $P\{D < q\}$) resulting in having unused inventory.

Since demand is random, the actual daily receipts will vary; thus, we represent the daily profit (or loss) by a random variable C_q which depends on the initial order quantity. In other words, the news vendor problem is to find the optimum q that maximizes $E[C_q]$. The relevant cost figures per newspaper are the wholesale purchase cost c_w, the retail purchase price c_r, and the salvage value c_s. For every newspaper that is sold, a profit of $c_1 = c_r - c_w$ is realized; for every newspaper initially purchased but not sold, a *loss* of $c_2 = c_w - c_s$ is incurred. If demand on a particular day is k and k is less than or equal to q, then k papers will be sold and $q - k$ papers will be returned. If k is greater than q, then q papers will be sold and none returned. Thus, the expected profit per day is

$$E[C_q] = \sum_{k=0}^{q} [kc_1 - (q-k)c_2] f_k + qc_1 \sum_{k=q+1}^{\infty} f_k \,. \tag{10.1}$$

Because Eq. (10.1) is not continuous, we cannot maximize it by taking the derivative; therefore, if we are to use something other than brute force (determining the value of (10.1) for each possible q), we must discuss optimization procedures for discrete functions. The analogous expression for the derivative of continuous functions is the difference operator, denoted by Δ, for discrete functions (see Fig. 10.1).

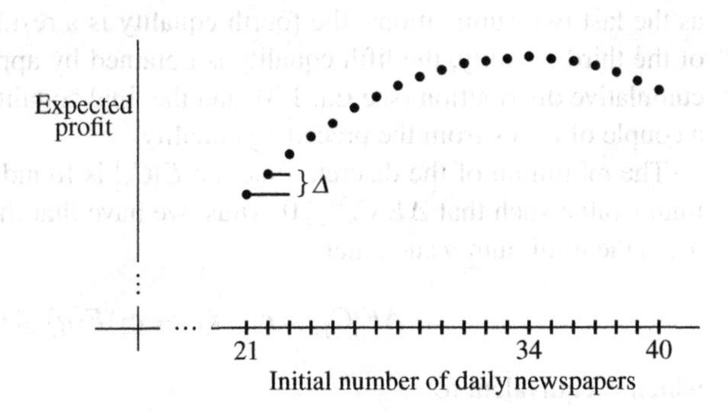

Fig. 10.1 The expected profit function versus order quantity for Eq. (10.1)

In other words, if g is a discrete function, then its difference operator is defined by $\Delta g(k) = g(k+1) - g(k)$. Assume that g is a unimodal function with its maximum at k_0; that is, for $k < k_0$ the function g is increasing, and for $k > k_0$ the function is decreasing. Or to say the same thing using the Δ operator, $\Delta g(k) > 0$ for $k < k_0$, and $\Delta g(k) < 0$ for $k \geq k_0$; thus,

$$k_0 = \min\{k : \Delta g(k) \leq 0\} .$$

The difficult part of the above equation is obtaining an expression for $\Delta g(k)$. It is not truly difficult, simply tedious. For our problem, we have the following:

$$\begin{aligned}
\Delta E[C_q] &= E[C_{q+1}] - E[C_q] \\
&= \sum_{k=0}^{q+1} [kc_1 - (q+1-k)c_2] f_k + (q+1)c_1 \sum_{k=q+2}^{\infty} f_k \\
&\quad - \sum_{k=0}^{q} [kc_1 - (q-k)c_2] f_k - qc_1 \sum_{k=q+1}^{\infty} f_k \\
&= \sum_{k=0}^{q} [kc_1 - (q+1-k)c_2] f_k + (q+1)c_1 f_{q+1} \\
&\quad + (q+1)c_1 \sum_{k=q+1}^{\infty} f_k - (q+1)c_1 f_{q+1} \\
&\quad - \sum_{k=0}^{q} [kc_1 - (q-k)c_2] f_k - qc_1 \sum_{k=q+1}^{\infty} f_k \\
&= -\sum_{k=0}^{q} c_2 f_k + c_1 \sum_{k=q+1}^{\infty} f_k \\
&= -c_2 F(q) + c_1(1 - F(q)) \\
&= c_1 - (c_1 + c_2) F(q) .
\end{aligned}$$

In the above expression, the first term is Eq. (10.1), the third equality is a result of changing the limits of the first two summation terms so that they have the same limits as the last two summations, the fourth equality is a result of simplifying the terms of the third equality, the fifth equality is obtained by applying the definition of the cumulative distribution (see Eq. 1.3), and the final equality results from rearranging a couple of terms from the preceding equality.

The minimum of the discrete function $E[C_q]$ is found by searching for the minimum value such that $\Delta E[C_q] \leq 0$. Thus, we have that the optimum order quantity, q^*, is the minimum q such that

$$\Delta E[C_q] = c_1 - (c_1 + c_2)F(q) \leq 0\,,$$

which is equivalent to

$$F(q) \geq \frac{c_1}{c_1 + c_2}\,. \tag{10.2}$$

We return to the numbers given at the start of this section and observe that the profit on each paper sold is $c_1 = \$0.10$ and the loss on each paper not sold is $c_2 = \$0.05$. Thus, we need to find the smallest q such that $F(q) \geq 0.667$. Since F is discrete-uniform between 21 and 40, we have that $F(33) = 0.65$ and $F(34) = 0.70$, which implies that $q^* = 34$; therefore, at the start of each day, 34 newspapers should be ordered.

The news vendor problem can also be solved if demand is continuous. (Of course, continuous demand hardly fits this particular example, but it may easily fit other examples.) Let us suppose that the demand random variable is continuous, so that the objective function (analogous to Eq. 10.1) becomes

$$E[C_q] = \int_{s=0}^{q} (sc_1 - (q-s)c_2)f(s)\mathrm{d}s + qc_1 \int_{s=q}^{\infty} f(s)\mathrm{d}s\,. \tag{10.3}$$

To minimize this expression with respect to the order quantity, we simply take the derivative of (10.3) with respect to q and set it equal to zero. This derivative turns out to be a little tricky since the independent variable is part of the limit of the integral so we must take advantage of Leibnitz's Rule[1] that was hopefully covered in a previous calculus course.

Property 10.1. ***Leibnitz's Rule.*** *Define the function G by*

$$G(y) = \int_{q(y)}^{p(y)} g(x,y)\mathrm{d}x \text{ for } -\infty < y < \infty$$

where q and p are functions of y and g is a function of x and y. The derivative of G is given by

[1] Gottfried Leibnitz (1646–1716) is considered the co-inventor of calculus along with Sir Isaac Newton (1643–1727). Although there was considerable controversy at the time, Leibnitz and Newton are thought to have acted independent of each other.

$$G'(y) = p'(y)g(p(y),y) - q'(y)g(q(y),y) + \int_{q(y)}^{p(y)} \frac{\partial g(x,y)}{\partial y} \mathrm{d}x \,.$$

Now using Leibnitz's Rule, the derivative of the expected cost function is given as

$$\begin{aligned} \frac{d}{dq}E[C_q] &= qc_1 f(q) - \int_{s=0}^{q} c_2 f(s)\mathrm{d}s - qc_1 f(q) + c_1 \int_{s=q}^{\infty} f(s)\mathrm{d}s \\ &= -c_2 F(q) + c_1(1 - F(q)) \,. \end{aligned}$$

Thus, the optimum order quantity, q^*, is given as that value that solves the following equation

$$F(q^*) = \frac{c_1}{c_1 + c_2} \,.$$

- *Suggestion: Do Problems 10.1–10.4.*

10.2 Single-Period Inventory

The classical single-period inventory problem is very similar to the news vendor problem. In the news vendor problem, any inventory at the end of the period was returned to the supplier. In the single-period inventory problem, the ending inventory will be held for the next period and a cost will be incurred based on the amount of inventory held over. The modeling for this is carried out in two stages: first we solve the problem when there is no setup charge for placing the order, then it is solved with the inclusion of a setup charge. Based on the chapter regarding Markov chains, we should recognize that the inventory process can be modeled as a Markov chain. In this section, however, we first consider a myopic[2] optimum; that is, an optimum that only considers one period at a time without regard to future periods.

10.2.1 No Setup Costs

The trade-off in the single-period inventory model with no setup cost will be an inventory carrying cost versus a shortage cost. Again, we motivate the problem with a hypothetical example involving a discrete random variable for demand. After the discrete model has been formulated and solved, we shall then develop the model

[2] Myopic is a synonym for what is commonly called nearsightedness. Thus, a myopic policy is one that "sees" what is close, but not what is far.

for the case when demand is continuous (or, at least, can be approximated by a continuous random variable).

Example 10.2. A camera store specializes in a particularly popular and fancy camera. They guarantee that if they are out of stock, they will special order the camera and promise delivery the next day. In fact, what the store does is purchase the camera retail from out-of-state and have it delivered through an express service. Thus, when the store is out of stock, they actually lose the sales price of the camera and the shipping charge, but they maintain their good reputation. The retail price of the camera is $680 and the special delivery charge adds another $25 to the cost. At the end of each month, there is an inventory holding cost of $20 for each camera in stock. The problem for the store is to know how many cameras to order at the beginning of each month so that costs will be minimized. Wholesale cost for the store to purchase the cameras is $400 each. (Assume that orders are delivered instantaneously, but only at the beginning of the month.) Demand follows a Poisson distribution with a mean of 7 cameras sold per month. □

To begin formulating the inventory problem, we define three costs: c_h is the cost per item held in inventory at the end of the period, c_p is the penalty cost charged for each item short (i.e., for each item for which there is demand but no stock-on-hand to satisfy it), and c_w is the wholesale cost for each item. Let x be the inventory level at the beginning of the month immediately before an order is placed, let q be the order quantity, let y be the inventory level immediately after ordering (i.e., $y = x + q$), and let $C_{x,y}$ be the monthly cost. The decision variable has been changed from the order quantity (q) to the order *up-to* quantity (y). Thus, y is the decision variable, and the cost equation[3] (see Fig. 10.2) to be minimized is

$$E[C_{x,y}] = c_w(y-x) + \sum_{k=0}^{y} c_h(y-k)f_k + \sum_{k=y+1}^{\infty} c_p(k-y)f_k \,.$$

It is customary to separate the above equation into two parts: an ordering cost and a "loss function". The loss function, $L(y)$, is a function of the order up-to quantity, y, and represents the cost of carrying inventory or being short. In other words, the loss function is defined, for $y \geq 0$, as

$$L(y) = \sum_{k=0}^{y} c_h(y-k)f_k + \sum_{k=y+1}^{\infty} c_p(k-y)f_k \,, \tag{10.4}$$

which yields an objective function of

$$E[C_{x,y}] = c_w(y-x) + L(y) \,. \tag{10.5}$$

Since the inventory involves a discrete random variable for demand, we proceed as in the previous section by working with differences instead of derivatives; namely,

---

[3] For a profit formulation of this problem, work Problem 10.7.

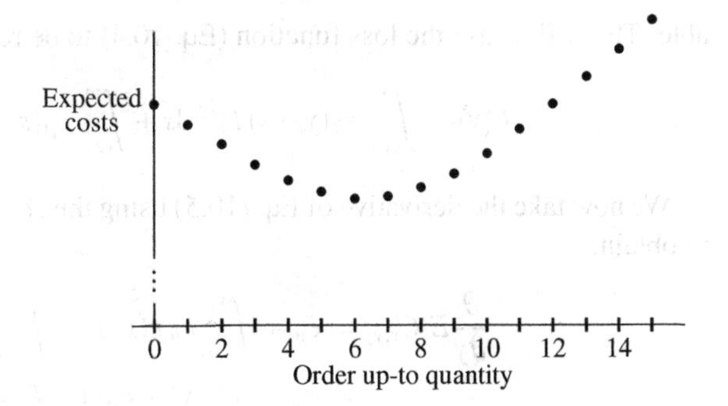

Fig. 10.2 The Expected cost function versus order up-to quantity for Eq. (10.5)

$$
\begin{aligned}
\Delta_y E[C_{x,y}] &= E[C_{x,y+1}] - E[C_{x,y}] \\
&= c_w(y+1-x) + \sum_{k=0}^{y+1} c_h(y+1-k) f_k + \sum_{k=y+2}^{\infty} c_p(k-y-1) f_k \\
&\quad - c_w(y-x) - \sum_{k=0}^{y} c_h(y-k) f_k - \sum_{k=y+1}^{\infty} c_p(k-y) f_k \\
&= c_w + c_h F(y) - c_p(1-F(y)) \,.
\end{aligned}
$$

Notice that several steps are missing between the second to last equality and the last equality; however, the steps are very similar to the analogous derivation for the news vendor problem.

Since we want to minimize cost, it follows that y^* is the smallest y such that $\Delta_y E[C_{x,y}] \geq 0$. This is equivalent to finding the smallest y such that

$$F(y) \geq \frac{c_p - c_w}{c_p + c_h} \,. \tag{10.6}$$

Notice Eq. (10.6) does not depend upon the initial inventory level, x. (Of course, the reason that it does not depend upon x is because we are only interested in myopic policies.) Thus, at the end of each period, the policy is to order enough inventory to bring the stock level up to y^*.

We now return to the example that opened this section to illustrate the use of the inventory policy. The cost constants are $c_p = 705$, $c_w = 400$, and $c_h = 20$. Thus we need to find the value y such that

$$\sum_{k=0}^{y} \frac{7^k e^{-7}}{k!} \geq 0.4207$$

which implies that $y^* = 6$; that is, at the end of each month, order enough cameras so that the beginning inventory at the start of the following month will be 6 cameras.

If demand for the cameras was in the hundreds or thousands, we might choose to approximate the Poisson random variable with a normally distributed random variable; thus, necessitating model development assuming a continuous random variable. This will cause the loss function (Eq. 10.4) to be redefined, for $y \geq 0$, as

$$L(y) = \int_{s=0}^{y} c_h(y-s)f(s)\mathrm{d}s + \int_{s=y}^{\infty} c_p(s-y)f(s)\mathrm{d}s\,. \tag{10.7}$$

We now take the derivative of Eq. (10.5) using the above continuous loss function to obtain:

$$\begin{aligned}\frac{\partial}{\partial y}E[C_{x,y}] &= c_w + \int_{s=0}^{y} c_h f(s)\mathrm{d}s - \int_{s=y}^{\infty} c_p f(s)\mathrm{d}s \\ &= c_w + c_h F(y) - c_p(1 - F(y))\,.\end{aligned}$$

The derivative is set to zero which yields an inventory policy of always ordering enough material at the end of each period to bring the inventory level at the start of the next period up to the level y^*, where y^* is defined to be that value such that

$$F(y^*) = \frac{c_p - c_w}{c_p + c_h}\,. \tag{10.8}$$

There are at least two questions that should always be asked in the context of an optimization problem: (1) "does an optimum exist?" and (2) "if it exists, is it unique?" For this particular problem, we know a unique optimum can always be found because the cost expression, $E[C_{x,y}]$ treated as a function of y is convex.

- *Suggestion: Do Problems 10.5–10.8.*

10.2.2 Setup Costs

The assumption of the previous chapter is that ordering ten items for inventory is simply ten times the cost of ordering one item for inventory; however, in many circumstances, there is a setup cost incurred whenever an order is placed.

Example 10.3. Consider a manufacturing company that assembles an electronic instrument which includes a power supply. The power supply is the most expensive part of the assembly, so we are interested in the control of the inventory for the power supply unit. Orders are placed at the end of the week and delivery is made at the start of the following week. The number of units needed each week is random and can be approximated by a continuous uniform random variable which varies between 100 and 300. The cost of purchasing q power supply units is $\$500 + 5q$. In other words, there is a \$500 setup cost and a \$5 variable cost to purchase the power supplies when they are purchased in bulk (e.g., the purchase of 100 power supply units costs \$1,000). Further assume that there is a \$0.50 cost incurred for each item remaining in inventory at the end of the week, and there is a \$25 cost for each item short during the week. (In other words, it costs \$25 for each unit special ordered during the week.) □

The cost model for this problem is identical to the previous problem except for the addition of a setup cost. We denote the setup cost by K; thus the cost model is

$$E[C_{x,y}] = \begin{cases} L(y) & \text{for } y = x\,, \\ K + c_w(y-x) + L(y) & \text{for } y > x\,. \end{cases} \tag{10.9}$$

An optimal policy based on Eq. (10.9) will lead to what is called an (s,S) inventory policy; that is, an order is placed if the ending period's inventory level is less than s, and enough inventory is ordered to bring the inventory level up to S. To understand how this model yields an (s,S) policy, consider the graphs illustrated in Fig. 10.3.

The lower graph in Fig. 10.3 is the cost function ignoring the setup cost; the upper graph is the cost including the setup cost. In other words, the lower graph gives the values of $c_w(y-x)+L(y)$ as a function of y and the upper graph is simply the same values added to the constant K. If an order up to S is placed, the top graph is relevant; therefore, the optimum order up-to quantity, denoted by S, is the minimum of the graph. If no order is placed, the lower graph is relevant. Consider an initial inventory level, (say $x = 200$ as in Fig. 10.3), that is between the two values s and S . If no order is placed, the relevant cost is the point on the lower graph (a in Fig. 10.3) that is above x; if an order up to S is placed, the relevant cost is the point on the upper graph (c in Fig. 10.3) that is above S. It is thus seen that the optimum policy will be to not order since the point (representing the expected cost) on the lower graph at x is less than the point (expected cost) on the upper graph at S. Now consider the value that is less than s (say $x = 125$ in Fig. 10.3). If no order is placed, the relevant cost is the point on the lower graph (b in Fig. 10.3) that is above the initial inventory x which again must be compared to the point (c in Fig. 10.3) on the upper graph that is above the order up-to quantity S. It is thus seen that the optimum policy will be to place an order since the point on the lower graph at x is greater than the point on the upper graph at S.

Since S is the minimum point on both the lower graph and the upper graph, we can find it by solving the "no setup cost" problem from the previous section. Thus, $S = y^*$ which is the solution from Eq. (10.9). The reorder point, s, is the value for which the no setup cost function equals the optimum set-up cost function evaluated at the order up-to quantity S. That is, the value of s is defined to be that value such that

$$c_w s + L(s) = K + c_w S + L(S)\,. \tag{10.10}$$

To illustrate the mechanics, consider the cost values given in the example at the start of this section. The value of S satisfies

$$F(S) = \frac{c_p - c_w}{c_p + c_h} = \frac{25-5}{25+0.5} = 0.7843\,.$$

In the example, the demand density is defined to be $f(t) = 1/200$ for $100 < t < 300$, or equivalently, the distribution is defined to be $F(t) = (t-100)/200$ for $100 < t < 300$. Therefore, we need to solve the equation $(S-100)/200 = 0.7843$ which yields

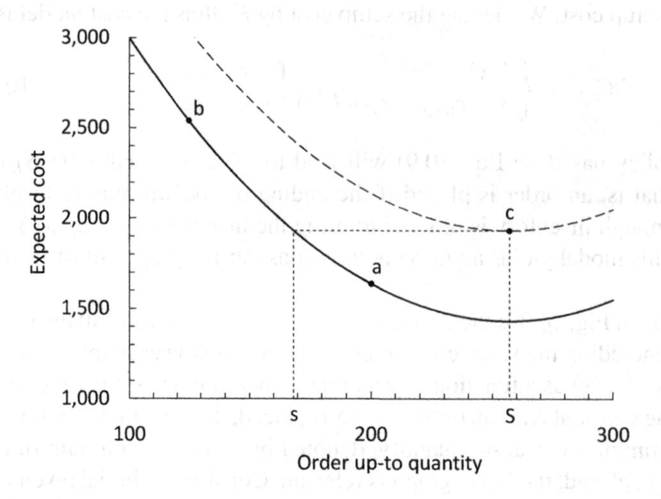

Fig. 10.3 Cost functions used for the single period inventory model with a setup cost

$S = 100 + 0.7843 \times 200 \approx 257$. To obtain s, we first determine the loss function:

$$
\begin{aligned}
L(y) &= 0.5 \int_{100}^{y} \frac{y-s}{200} \mathrm{d}s + 25 \int_{y}^{300} \frac{s-y}{200} \mathrm{d}s \\
&= \frac{51}{800} y^2 - 37.75y + 5637.5\,.
\end{aligned}
$$

The cost to order up to S is thus given as

$$
\begin{aligned}
K + c_w S + L(S) &= 500 + 5 \times 257 + L(257) \\
&= 1785 + 0.06375 \times 257^2 - 37.75 \times 257 + 5637.5 \\
&= 1931.37\,.
\end{aligned}
$$

The reorder point, s, is that value for which the inventory cost of not ordering is equal to the optimum ordering cost (Eq. 10.10). Thus, s is that value that satisfies the following

$$
\begin{aligned}
1931.37 &= c_w s + L(s) \\
&= 5s + \frac{51}{800} s^2 - 37.75s + 5637.5 \\
&= 0.06375s^2 - 32.75s + 5637.5\,.
\end{aligned}
$$

The above quadratic equation yields two solutions: one larger and another smaller than S (see Fig. 10.3). We want the smaller solution, which is $s = 168.3$. Therefore, the optimal inventory policy is to place an order whenever the ending inventory is equal to or less than 168, and when an order is placed, order enough to bring the next month's beginning inventory up to 257.

- *Suggestion: Do Problems 10.9–10.11.*

10.3 Multi-Period Inventory

It would be unusual circumstances for which a single-period inventory model is truly appropriate. However, the previous section is worthwhile not only for its pedagogical value, but also because the problem formulation and solution structure is the beginning step in the multi-period model. In this section, we discuss inventory when the demand is a discrete random variable and the planning horizon involves several (but finite) periods.

Similar to the previous sections, the probability mass function giving the amount of demand during each period is denoted by $f(\cdot)$, and we assume a planning horizon containing n_{max} periods. For notational purposes, let the demand for period n be the random variable D_n; thus, $P\{D_n = k\} = f(k)$. The ending inventory for period n is denoted by the random variable X_n. After the inventory level is observed, a decision is made regarding an order; if an order is placed, the new inventory arrives immediately so the beginning period's inventory is equal to the previous period's ending inventory plus the order quantity. If q items are ordered, there is an order cost given by $c_o(q)$. The inventory holding cost and shortage cost are given by c_h and c_p, respectively, and there are no backorders. Once an inventory ordering policy has been determined, the stochastic process $\{X_0, X_1, \cdots, X_{n_{max}}\}$ is a Markov chain.

The solution methodology we use is stochastic dynamic programming. Bellman is often considered the father of dynamic programming and he defined in [1] a concept called the "principle of optimality". Bellman's principle of optimality essentially means that an optimal policy must have the property that, at any point in time, future decisions are optimal independent of the decisions previous to the current point in time. This is fundamentally a Markov property; therefore, stochastic dynamic programming is an ideal tool for identifying optimal decisions for processes satisfying the Markov property.

At the end of a period, the ending inventory is observed and a decision is made as to what the order up-to quantity should be for the next period. Thus, the optimum decision and the cost associated with the optimum decision made at the start of period n depend on the inventory at the end of period $n-1$. Denote the optimum decision made at the beginning of period n by $y_n(x)$ and total future cost (or value) of that decision by $v_n(x)$, where x is the ending inventory level of the previous period.

Dynamic programming is usually solved by a backwards recursion. Thus, the first step towards a solution is to optimize the decision made in the final period; that is, we first determine $y_{n_{max}}$ and $v_{n_{max}}$. In other words, the first step is to solve the single-period problem:

$$v_{n_{max}}(x) = \min_{y \geq x} c_o(y-x) + L(y) , \tag{10.11}$$

and $y_{n_{max}}$ is the value of y that minimizes the expression in the right-hand-side of the above equation. The question may arise as to how we solve Eq. (10.11) if we do not know the value of the ending inventory for $n_{max} - 1$. The answer is simple: find the optimal order up-to quantity for every reasonable value of the ending inventory x_{max-1}. (Note that if c_o involves a setup cost plus a linear variable cost, then the term being minimized is $E[C_{x,y_n}]$ from Eq. 10.9.) Once $v_{n_{max}}$ is known, the remaining values of v_n can be calculated based on the typical dynamic programming recursive relationship:

$$v_{n-1}(x) = \min_{y \geq x} \{ c_o(y-x) + L(y) + E_y[v_n(X_{n-1})] \} . \tag{10.12}$$

Notice that the expected value operator in Eq. (10.12) is written with y as a subscript. This is to indicate that the expected value depends on the decision made for y, which is the beginning inventory level for period n. We illustrate these calculations through the following example.

Example 10.4. A manufacturing company needs widgets in their manufacturing process; however the exact number needed during the day is random and varies from zero to three with equal probability (i.e., f is discrete uniform). At the end of each day, an order is placed for widgets to be used during the next day, and delivery is made overnight. If not enough widgets are on hand for the day, there is a \$92 cost for each one short. If widgets must be stored for the next day, there is a \$8 storage cost due to the need to refrigerate them overnight. Because of space restrictions, the most that can be kept overnight is 5 widgets. If an order for q widgets is placed, it costs $\$50 + 20q$ (i.e., a \$50 setup cost and a \$20 variable cost). It is now Tuesday evening, and one widget is on hand for tomorrow. We would like to determine the optimal ordering policy for the next three days. □

We shall use the above example to illustrate the steps involved in using Eqs. (10.11) and (10.12). As is normal with dynamic programming problems, the amount of computations becomes tedious, but those calculations would be quick for a computer to handle as long as there is only a single state variable (there is only one state variable in this case—namely, the ending period's inventory).

We begin by calculating the single-period inventory costs. It should be clear that we would never order up to a number larger than 3 on Thursday night, because we need at most three widgets on Friday and the inventory costs beyond Friday do not affect the three-day planning horizon; however, it is not clear that we should be limited to just 3 items for either Wednesday or Thursday; therefore, we calculate the inventory ordering costs allowing for the possibility of ordering up to five widgets. The loss function calculation is according to Eq. (10.4); thus,

$$L(0) = 0 \times 0.25 + 92 \times 0.25 + 184 \times 0.25 + 276 \times 0.25 = 138$$
$$L(1) = 8 \times 0.25 + 0 \times 0.25 + 92 \times 0.25 + 184 \times 0.25 = 71$$

$$L(2) = 16 \times 0.25 + 8 \times 0.25 + 0 \times 0.25 + 92 \times 0.25 = 29$$
$$L(3) = 24 \times 0.25 + 16 \times 0.25 + 8 \times 0.25 + 0 \times 0.25 = 12$$
$$L(4) = 32 \times 0.25 + 24 \times 0.25 + 16 \times 0.25 + 8 \times 0.25 = 20$$
$$L(5) = 40 \times 0.25 + 32 \times 0.25 + 24 \times 0.25 + 16 \times 0.25 = 28\,.$$

We add the inventory ordering costs to the above loss function values and obtain the values the single-period inventory costs as shown in Table 10.1.(Since the order

Table 10.1 Values for $E[C_{x,y}]$ Used in Eq. (10.11)

	Order Up-To Quantity for Friday					
x	$y=0$	$y=1$	$y=2$	$y=3$	$y=4$	$y=5$
0	0 + 138	70+71	90+29	110 + 12	130+20	150+28
1		0+71	70+29	90+12	110+20	130+28
2			0+29	70+12	90+20	110+28
3				0+12	70+20	90+28
4					0+20	70+28
5						0+28

up-to quantity, y, must be at least as large as the previous day's ending inventory, x, the above table contains no values in the lower portion.)

Our problem is to decide the beginning inventory levels for Wednesday through Friday. Friday's decision (actually the decision made Thursday evening) is given by Eq. (10.11), which is equivalent to taking the minimum value for each row of Table 10.1. These values are given in Table 10.2.

Table 10.2 Optimum values for Thursday evening's decision

x	0	1	2	3	4	5
v_3	119	71	29	12	20	28
y_3	2	1	2	3	4	5

Note that in dynamic programming, it is necessary to determine the optimum decision for each state that might occur at the beginning of the period. The decision to be made Wednesday evening (i.e., for Thursday's beginning inventory level) is determined through the following restatement of Eq. (10.12):

$$v_2(x) = \min_{y \geq x} \{ E[C_{x,y}] + E_y[v_3(X_2)] \}\,.$$

The values for $E[C_{x,y}]$ are given in Table 10.1, but the values for $E_y[v_3(X_2)]$ must be calculated using Table 10.2 as follows:

$$E_0[v_3(X_2)] = 119$$
$$E_1[v_3(X_2)] = 71 \times 0.25 + 119 \times 0.75 = 107$$

$$E_2[v_3(X_2)] = 29 \times 0.25 + 71 \times 0.25 + 119 \times 0.50 = 84.5$$
$$E_3[v_3(X_2)] = 12 \times 0.25 + 29 \times 0.25 + 71 \times 0.25 + 119 \times 0.25 = 57.75$$
$$E_4[v_3(X_2)] = 20 \times 0.25 + 12 \times 0.25 + 29 \times 0.25 + 71 \times 0.25 = 33$$
$$E_5[v_3(X_2)] = 28 \times 0.25 + 20 \times 0.25 + 12 \times 0.25 + 29 \times 0.25 = 22.25 .$$

To understand the above equations, remember that an expected value is simply the sum over all possibilities times the probability that the possibility will occur. Consider $E_0[v_3(X_2)]$. The "0" subscript indicates that Thursday begins with no inventory. If there is no inventory for Thursday, then with probability one, there will be no inventory Thursday evening. If Thursday ends with no inventory, Friday's optimum cost (namely, $v_3(0)$ from Table 10.2) is 119. Consider $E_1[v_3(X_2)]$. If Thursday morning begins with an inventory level of 1, there is a 25% chance that Thursday evening will also have an inventory level of 1 (i.e., no demand for Thursday) in which case Friday's optimum cost ($v_3(1)$ from Table 10.2) is 71. Also, if Thursday morning begins with 1, there is a a 75% chance of no inventory being available Thursday evening in which case the optimum cost for Friday is 119. Thus, the expected value of those two cost values equals 107.

The values for Thursday's costs (Table 10.1) are added to the expected future costs (i.e., $E_y[v_3(X_2)]$) to obtain the necessary values for Eq. (10.12) as follows:

Table 10.3 Values for $E[C_{x,y}] + E_y[v_3(X_2)]$

	Order Up-To Quantity for Thursday					
x	$y=0$	$y=1$	$y=2$	$y=3$	$y=4$	$y=5$
0	138+119	141+107	119+84.5	122+57.75	150+33	178+22.25
1		71+107	99+84.5	102+57.75	130+33	158+22.25
2			29+84.5	82+57.75	110+33	138+22.25
3				12+57.75	90+33	118+22.25
4					20+33	98+22.25
5						28+22.25

Wednesday evening's decision is now made by again minimizing each row within Table 10.3. These results are in Table 10.4.

Table 10.4 Optimum values for Wednesday evening's decision

x	0	1	2	3	4	5
v_2	179.75	159.75	113.5	69.75	53	50.25
y_2	3	3	2	3	4	5

The final decision to be made is for Tuesday evening. Since we already know Tuesday's ending inventory, we only have to compute one value; namely, we need to determine

$$v_1(1) = \min_{y \geq 1} \{ E[C_{1,y}] + E_y[v_2(X_1)] \} .$$

The future costs (future from Wednesday's point of view) are calculated as

$$
\begin{aligned}
E_0[v_2(X_1)] &= 179.75 \\
E_1[v_2(X_1)] &= 159.75 \times 0.25 + 179.75 \times 0.75 = 174.75 \\
E_2[v_2(X_1)] &= 113.5 \times 0.25 + 159.75 \times 0.25 + 179.75 \times 0.50 = 158.1875 \\
E_3[v_2(X_1)] &= 69.75 \times 0.25 + 113.5 \times 0.25 + 159.75 \times 0.25 + 179.75 \times 0.25 \\
&= 130.6875 \\
E_4[v_2(X_1)] &= 53 \times 0.25 + 69.75 \times 0.25 + 113.5 \times 0.25 + 159.75 \times 0.25 = 99 \\
E_5[v_2(X_1)] &= 50.25 \times 0.25 + 53 \times 0.25 + 69.75 \times 0.25 + 113.5 \times 0.25 = 71.625\,.
\end{aligned}
$$

Thus, the relevant costs for determining Tuesday evening's decision are as follows:

Table 10.5 Values for $E[C_{x,y}] + E_y[v_2(X_1)]$

	Order Up-To Quantity for Wednesday					
x	$y=0$	$y=1$	$y=2$	$y=3$	$y=4$	$y=5$
1	—	71+174.75	99+158.1875	102+130.6875	130+99	158+71.625

Minimizing over the row gives $v_1(1) = 229$ and $y_1(1) = 4$. In other words, we order 3 additional widgets for Wednesday morning. The decision to be made Wednesday night depends on Wednesday's demand. If Wednesday evening ends with 2 or more widgets, no order is placed; if Wednesday evening ends with 0 or 1 widget, an order is placed to bring Thursday's beginning inventory up to 3 widgets. For Thursday evening, if no widgets are present, we order 2 widgets to begin Friday's schedule; however, if any widgets remain Thursday evening, no additional widgets are ordered.

- *Suggestion: Do Problems 10.12–10.17.*

Problems

10.1. The following functions are defined only on the positive integers. Use the difference operator to solve the following.
(a) $\min g(n) = 2n^2 - 9n$.
(b) $\max g(n) = 1/(3n-11)^2$.
(c) $\max g(n) = ne^{-0.3n}$.

10.2. Let $f(\cdot)$ be a probability mass function defined on the nonnegative integers, with $F(\cdot)$ its CDF and μ its mean. Define the function $g(n)$, for $n = 0, 1, \cdots$, by

$$
g(n) = n \sum_{i=n}^{\infty} (i-n) f(i)\,.
$$

(a) Using the difference operator, show that $g(\cdot)$ is maximized at the smallest integer, n, such that

$$(2n+1)(1-F(n)) + \sum_{i=0}^{n} if(i) \geq \mu \,.$$

(b) Use the answer in part (a) to obtain the optimum n when $f(\cdot)$ is the mass function for a Poisson random variable with mean 2.5.

10.3. A computer store is getting ready to order a batch of computers to sell during the coming summer. The wholesaler is giving the store a good price on the computers because they are being phased out and will not be available after the summer. Each computer will cost the store \$550 to purchase, and they will sell it for \$995. The wholesaler has also offered to buy back any unsold equipment for \$200. The manager of the computer store estimates that demand for the computer over the summer is a Poisson random variable. How many computers should be purchased by the store in preparation for the summer?
(a) The mean demand for the summer is 5.
(b) The mean demand for the summer is 50. (Note that the Normal distribution is a good approximation for the Poisson.)

10.4. Consider the news-vendor problem of Example 10.1. The newspaper company has established an incentive program. The company has grouped all news vendors in groups of 10, and at the end of each day, one person out of the ten will be randomly selected. The news vendor thus selected will receive a full refund of \$0.15 for the unsold newspapers.
(a) Give the expression for the expected profit per day analogous to Eq. (10.1).
(b) Use difference equations to find an expression analogous to Eq. (10.2) that can be used to obtain the optimum order quantity.
(c) Find the optimum order quantity.
(d) Write a computer program to simulate this process. Based on the simulation, do you believe your answer to parts (b) and (c)? (Note: the simulations suggested in these exercises can be written without using a future events list.)

10.5. Using the single-period model, determine the optimum reorder point for the camera store in Example 10.2, except assume that the demand is Poisson with a mean of 70.

10.6. A manufacturing company uses widgets for part of its manufacturing process. Widgets cost \$50 each and there is a 50 cent charge for each item held in inventory at the end of each month. If there is a shortage of widgets, the production process is seriously disrupted at a cost of \$150 per widget short. What is the optimum ordering policy for the following scenarios?
(a) The monthly demand is governed by the discrete uniform distribution varying between 3 and 7 widgets (inclusive).
(b) The monthly demand is governed by the continuous uniform distribution varying between 300 and 700 widgets.

(c) The cost structure as stated in the problem description is over simplified. The penalty for being short is divided into two pieces. There is a shortage cost due to a disruption to the manufacturing process of $1,000 independent to how many widgets are short. In addition, there is a $60 cost per widget short. Derive a formula for the optimal ordering policy assuming that the monthly demand is a continuous uniform distribution.
(d) Apply the answer to part (c) assuming that demand varies between 300 and 700 widgets.
(e) Write a computer program or Excel macro to simulate this process. Based on the simulation, do you believe your answer to parts (c) and (d)?

10.7. The single-period inventory model without setup costs can be formulated so as to maximize profits instead of minimizing costs as was done in Eq. (10.5). Show that the two formulations are equivalent.

10.8. The loss function for a single-period inventory is given by Eq. (10.7). The loss function includes a holding cost and a penalty cost. Assume that the goods are perishable and that any goods on hand at the beginning of the period must be sold during that period or they spoil and must be discarded at a cost of c_s per item. (That is, items can be held over for one period, but not for two periods.) Assume that all unsold goods (even those to be discarded) add to the inventory cost. How would Eq. (10.7) change?
(a) Use a FIFO policy; that is, assume that all items in inventory at the beginning of the period are sold first.
(b) Use a LIFO policy; that is, assume that all items in inventory at the beginning of the period are sold last so that customers get fresh goods if they are available.

10.9. Assume that there is a $170 handling cost each time that an order is placed from the camera store of Example 10.2. Using a Poisson demand process with a mean of 70, determine the store's optimal ordering policy.

10.10. For the manufacturing problem described in Problem 10.6, assume that there is an ordering cost of $750 incurred whenever an order is placed. Determine the optimal ordering policy when demand is according to a continuous uniform distribution between 300 and 700.
(a) Assume that the penalty cost is $150 per widget short.
(b) Assume that the penalty cost is $60 per widget short plus an additional $1,000 due to the process interruption.

10.11. In Example 10.3, the wholesaler of the power supplies has decided to offer a discount when large quantities are ordered. The current unit cost, excluding set-up, is $5. The wholesaler has decided that orders of 100 or more will yield a reduction of cost to $4.50 per unit. Determine if the availability of the discount changes the optimal ordering policy.

10.12. The optimal ordering policy for Example 10.4 was obtained for a planning horizon of three days. Determine the optimal ordering policy for a planning horizon of four days.

10.13. A small company produces printers on a weekly basis which are sold for \$1,700 each. If no printers are produced, there are no costs incurred. If one printer is produced, it costs \$1,400. If two printers are produced, it costs \$2,000. If three printers are produced, it costs \$2,500. The normal maximum production capacity for the company is three printers; however, by paying overtime, it is possible to produce one extra printer for an additional \$750. For each printer in inventory at the end of the week, there is a cost of \$15. Weekly demand can be approximated by a binomial random variable with $n = 4$ and $p = 0.4$. Any demand not meet from existing inventory or from the week's production is lost. The current policy is to never inventory more than four units at the end of any week. It is now Friday, and one printer remains unsold for the week.
(a) What should be next week's production if a planning period of three weeks is used.
(b) What would be the cost savings, if any, if the current policy of allowing a maximum of four units stored at the end of the week was changed to five units?
(c) Part (a) can be solved using stochastic dynamic programming as discussed in Sect. 10.3. Assume that you do not know dynamic programming and answer Part (a) through a simulation analysis. Compare the amount of work required by the two approaches. After reaching a conclusion with dynamic programming and simulation, comment on your level of confidence in the two answers.

10.14. Derive a decision rule for a two-period, no setup cost, inventory model. The basic cost structure should be the same as in Sect. 10.2.1 with continuous demand. Let the decision variables be y_1 and y_2, referring to the order up-to quantities and periods one and two respectively.
(a) Show that y_1 can be obtained by minimizing the following:

$$E[C_{x,y_1}] = c_w(y_1 - x) + \int_{s=0}^{y_1} (c_h - c_w)(y_1 - s) f(s) ds + \int_{s=y_1}^{\infty} c_p(s - y_1) f(s) ds + c_w y_2^* + L(y_2^*) ,$$

where y_2^* is the optimum order up-to quantity for period 2.
(b) Obtain expressions that can be used to find y_1^* and y_2^*.
(c) Find the optimum values for the order up-to quantities using the data from Example 10.2, except with a mean demand of 70.
(d) Write a simulation for the situation of Example 10.2, and implement a two-period decision rule. Using your simulation, answer the following question: "How much difference is there between using a two-period rule and a single-period rule?"

10.15. The multi-period inventory problem can be modeled as a Markov chain. Consider an inventory policy with an order up-to quantity given by y, where backorders are not allowed, and with a demand per period being given by the pmf $f(\cdot)$. Furthermore, let the costs c_w, c_h, and c_p be the wholesale, holding, and shortage costs, respectively. Define a Markov chain with state space $E = \{0, 1, \cdots, y\}$, where X_n denotes the on-hand inventory at the end of each period, immediately before an order

for additional inventory is placed. Finally, assume that orders are delivered instantaneously.
(a) Give the Markov matrix for the Markov chain.
(b) Give the Markov matrix using the numbers from Example 10.2 (including $y = 6$).
(c) Determine the steady-state probabilities for the Markov chain and its associated long-run cost.
(d) What is the optimal ordering policy?

10.16. Develop a Markov chain formulation for the multi-period inventory problem with set-up costs.
(a) Give the Markov matrix for the Markov chain.
(b) Give the Markov matrix using the numbers from Example 10.2 with the addition of a \$30 ordering cost. Use $s = 2$ and $S = 7$.
(c) Determine the steady-state probabilities for the Markov chain and its associated long-run cost.
(d) What is the optimal ordering policy?

10.17. A manufacturer of specialty items has the following inventory policy. Three items are kept in inventory if there are no orders being processed. Whenever an order is received, the company immediately sends a request to begin the manufacturing of another item. If there is an item in inventory, the customer is shipped the item in inventory. If no items are in inventory, the customer's request is backordered and a discount of \$600 is applied to the purchase price. The time to manufacture one item is exponentially distributed with a mean of two weeks. The items cost \$12,000 to manufacture, and any item in inventory at the end of a week costs the company \$75. The company averages 24 orders per year, and the arrival of orders appears to be according to a Poisson process. The purpose of this exercise is to formulate the inventory problem as a queueing process. For this purpose view the ordering process as an M/M/∞ system. In which case, notice that p_0 for the queueing system is equivalent to the company having an inventory level of three.
(a) What is the value of $r = \lambda / \mu$ for the M/M/∞ system?
(b) What is the expected number of items being manufactured within the facility? (Notice that this is L for the queueing system.)
(c) What is the probability that when an order is received, it must be backordered?
(d) Derive an expression for the long-run average weekly cost for this inventory problem.
(e) The current inventory policy is to maintain three items in stock when there are no outstanding orders. Is this optimum?

References

1. Bellman, R. (1957). *Dynamic Programming*, Princeton University Press, Princeton, NJ.

Chapter 11
Replacement Theory

Almost all systems deteriorate with age or usage, unless some corrective action is taken to maintain them. Repair and replacement policies are often implemented to reduce system deterioration and failure risk. Increasingly complex systems have generated interest in the research of replacement theory. Originally most of the models were developed for many industrial and military systems; however, more recent applications of these models have extended to areas such as health, ecology, and the environment.

Determining repair and replacement policies for stochastically deteriorating systems involves a problem of decision-making under uncertainty, because it is impossible to predict with certainty the timing and extent of the deterioration process. In repair and replacement models, these uncertainties are usually assumed to have known probability distribution functions so that the deterioration process is stochastically predictable. In this chapter we present some of the basic models dealing with replacement and maintenance issues. The concern is with the trade-off between maximizing the life of the equipment and minimizing increased costs due to early replacements and unforeseen failures.

In Sect. 11.1, we present the classical age-replacement maintenance policy, and in Sect. 11.2 an extension of the age replacement policy that includes repairs that do not change the failure distribution is developed. Finally in Sect. 11.3, the block replacement maintenance policy is presented.

11.1 Age Replacement

An age replacement policy is one in which the equipment is replaced as soon as it fails or when it reaches a prespecified age, whichever occurs first. As with the inventory chapter, we shall consider both a discrete case and a continuous case. The basic concept for these models is the consideration for two types of actions that may be taken at any point in time: (1) replace the equipment with (probabilistically) identical equipment, or (2) leave the equipment alone.

R.M. Feldman, C. Valdez-Flores, *Applied Probability and Stochastic Processes*, 2nd ed., 305
DOI 10.1007/978-3-642-05158-6_11, © Springer-Verlag Berlin Heidelberg 2010

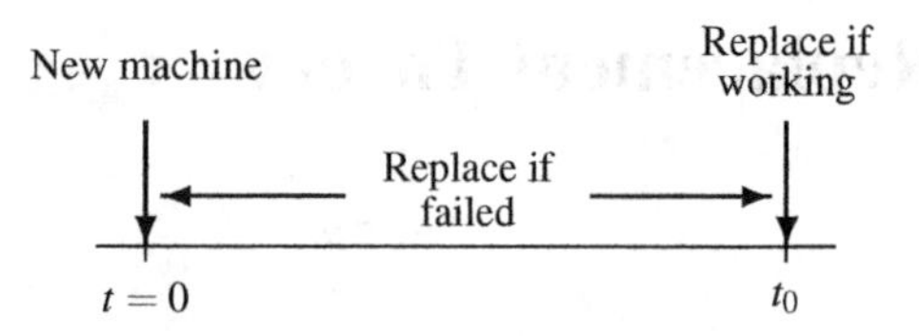

Fig. 11.1 Schematic representation for an age-replacement policy

11.1.1 Discrete Life Times

We begin by assuming that the random variable describing the time until failure for the equipment is a discrete random variable. We also assume that the equipment being replaced is only one component. In order to provide a proper framework for the replacement problem, an example is used for motivation.

Example 11.1. Consider a specially designed switch used to initiate production processing each day. At the start of each day, the switch is used to begin processing; however, it is not fully reliable, and it periodically must be replaced. Experience indicates that the switch will fail somewhere between the 20th day and the 30th day according to the following probability mass function:

day	20	21	22	23	24	25	26	27	28	29	30
prob.	0.02	0.03	0.05	0.1	0.3	0.2	0.15	0.08	0.04	0.025	0.005

The switch costs \$300 to replace; however, an additional cost of \$200 in lost production is incurred if the switch is replaced during the morning at failure time. Since production only occurs during the day, it may be worthwhile to replace the switch as a preventative measure during some evening before it fails. The disadvantage of the preventative action is that we do not get full utilization out of the switch. Thus our problem is to determine a replacement strategy that optimally makes the trade-off between avoiding the additional replacement cost and fully utilizing the life of the switch. □

A typical replacement strategy is to determine a replacement age, call it t_0, such that if the switch fails before t_0, it is replaced; if the switch works on day t_0, replace it that evening. (A schematic of this replacement policy is given in Fig. 11.1.) With this strategy, we can now formulate the optimization problem with an objective of minimizing the expected long-run average daily cost. To state the problem generally, let T denote the random variable representing the failure time with its CDF being $F(t)$. Furthermore, let c_r denote the cost of replacement, c_f represent the additional cost associated with failure, and $z(t_0)$ denote the long-run average daily cost associated with the replacement age of t_0. Since some switches will fail and some will be replaced before failure, different switches will have different replacement costs. Let C_{t_0} be the random variable indicating the replacement costs for a switch while using a replacement age of t_0. With this notation, the objective function is given as

$$z(t_0) = \frac{E[C_{t_0}]}{E[\min(T,t_0)]} \,.$$

To obtain the numerator, we make two observations. First, the replacement cost, c_r, is always incurred, and second, the failure cost, c_f, is incurred only when the switch fails on or before the replacement day, t_0. Thus,

$$E[C_{t_0}] = c_r + c_f Pr\{T \le t_0\} \,.$$

To obtain the denominator, let the random variable S denote the actual replacement time; that is $S = \min(T,t_0)$. The CDF of the random variable S is thus

$$G(s) = P\{S \le s\} = \begin{cases} F(s) & \text{if } s < t_0 \\ 1 & \text{if } s \ge t_0 \end{cases} ,$$

where $P\{T \le s\} = F(s)$, i.e., F is the CDF of T. (See Fig. 11.2 for a representation of G.) We now use the third statement of Property (1.11) to obtain the denominator; namely,

$$E[\min(T,t_0)] = E[S] = \int_0^\infty [1 - G(s)]ds = \int_0^{t_0} [1 - F(s)]ds \,.$$

Therefore, the expected long-run average cost associated with a particular replacement policy can be written as

$$z(t_0) = \frac{c_r + c_f F(t_0)}{\int_0^{t_0} [1 - F(s)]ds} \,. \tag{11.1}$$

Since our problem deals with a discrete random variable, we introduce some additional notation so that the computation of Eq. (11.1) will be more straight forward. Specifically, let $F_k = F(k) = Pr\{T \le k\}$; that is, F_k is the probability that the switch fails on or before day k. Also, let $\overline{F_k} = 1 - F(k)$; that is $\overline{F_k}$ is the probability that the switch fails *after* day k. The optimal replacement time is given as t_0^*, where t_0^* is the k that solves the following

$$\min_k z(k) = \frac{c_r + c_f F_k}{\sum_{i=0}^{k-1} \overline{F_i}} \,, \tag{11.2}$$

where $\overline{F_0} = 1$. It should also be observed that if replacement occurs on day t_0, then the sum in the denominator begins at zero and goes until $t_0 - 1$.

Unfortunately, Eq. (11.2) does not yield itself to a nice single equation from which the optimum can be determined; thus, we find the optimum by "brute force". The easiest way to do this is to make a table and continually evaluate the long-run average cost, $z(k)$ for successive values of k until the first optimum is obtained. The following table illustrates these computations:

The above computations are stopped when the first local minimum is identified, and we therefore set the optimum replacement time as $t_0^* = 22$. In doing this, some

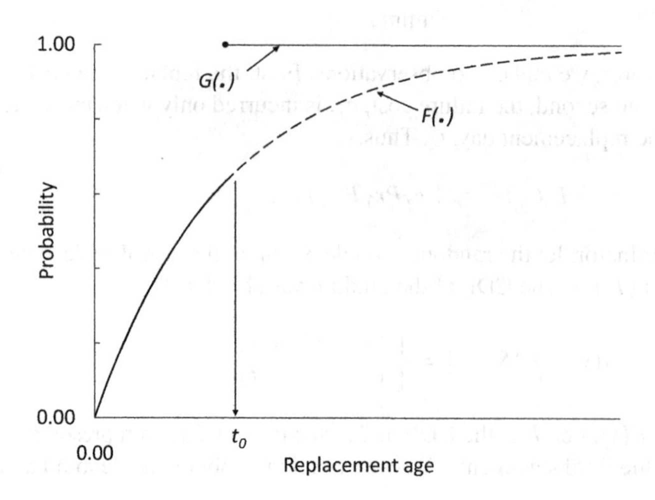

Fig. 11.2 The CDF $G(\cdot)$ of S derived from the CDF $F(\cdot)$ of T

Table 11.1 Computations for evaluating Eq. (11.2) applied to Example 11.1

k	f_k	F_k	$c_r + c_f F_k$	$\overline{F_k}$	$\sum_{i=0}^{k-1} \overline{F_i}$	$z(k)$
19	0.00	0.00	300	1.00	19.00	15.79
20	0.02	0.02	304	0.98	20.00	15.20
21	0.03	0.05	310	0.95	20.98	14.78
22	0.05	0.10	320	0.90	21.93	14.59
23	0.10	0.20	340	0.80	22.83	14.89

assumptions have been made about the shape of the long-run average cost function; namely, that the first local minimum is the global minimum. It is not appropriate at the level of this textbook to go into all the details about such an assumption; however, one concept that is useful to know when hoping to make such an assumption has to do with the hazard rate function. An important characteristic of random variables that relate to failure times is its hazard rate.

Definition 11.1. The function $h(\cdot)$ defined, for $t \geq 0$, by

$$h(t) = \frac{f(t)}{1 - F(t)}$$

is called the *hazard rate function*, where F is a CDF[1] and f is either a pdf or pmf depending on whether the random variable of interest is discrete or continuous. □

[1] It is also possible to obtain the distribution from the hazard rate through the relationship $F(t) = 1 - \exp\left\{-\int_0^t h(s)ds\right\}$.

Intuitively, the hazard rate evaluated at t gives the *rate* of failure at time t *given* that the item has not yet failed at t. A random variable is said to have an increasing failure rate (IFR) distribution if its hazard rate is increasing for all $t \geq 0$. Thus, a random variable that has an IFR distribution refers to an item that in some sense "gets worse" as it ages. As a general rule, replacement problems that deal with IFR distributions have costs that increase with age. Thus, cost functions associated with IFR distributions are usually well behaved, and Eq. (11.1) taken as a function of t_0 will have a unique minimum. Since the above example problem has an IFR distribution, we know that the local minimum is a global minimum, and the optimal policy is to replace the switch at the end of day 22 if failure does not occur during the first 22 days of operation.

- *Suggestion: Do Problems 11.1–11.4.*

11.1.2 Continuous Life Times

When the lifetime of the system being considered for a replacement policy can be described by a continuous random variable, it is possible to obtain a closed form expression for the optimum replacement time. The basic optimization equation (Eq. 11.1) is the same for both the continuous and discrete case. Therefore, we begin by taking the derivative of (11.1) with respect to t_0. It is not difficult to show that

$$\frac{dz(t_0)}{dt_0} = 0$$

yields the equation

$$c_f f(t_0) \int_0^{t_0} (1 - F(s))ds = (c_r + c_f F(t_0))(1 - F(t_0)) . \tag{11.3}$$

Before continuing with the derivative, we return to Definition 11.1. The hazard rate function for a distribution is an important descriptor and is often given instead of the distribution function to describe a random variable. In general, replacement policies are usually developed for systems that have IFR distributions, so we shall assume an increasing hazard rate function and convert the above equation into an expression using its hazard rate. We divide both sides of Eq. (11.3) by c_f and the factor $1 - F(t_0)$ to obtain

$$(1 - F(t_0)) + h(t_0) \int_0^{t_0} (1 - F(s))ds = 1 + \frac{c_r}{c_f} . \tag{11.4}$$

Before we can make definite statements regarding a solution to the above equation, we need to know how many solutions it has. We first observe that the right-hand-side is a constant; therefore, if the left-hand-side is increasing, then there can be at most one solution. The derivative of the left-hand-side with respect to t_0 yields

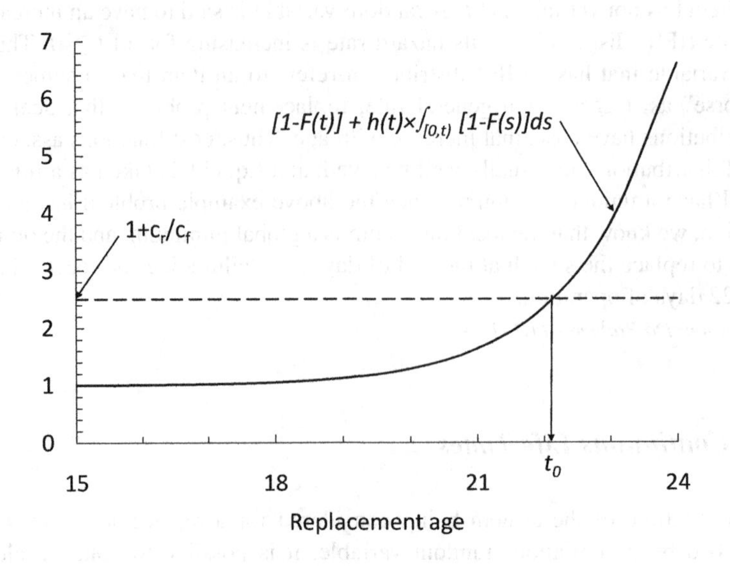

Fig. 11.3 A Graphical Solution of Eq. (11.4) for Example 11.2

$$-f(t_0) + (1 - F(t_0))h(t_0) + \frac{dh(t_0)}{dt_0} \int_0^{t_0} (1 - F(s))ds$$

which can be simplified to

$$\frac{dh(t_0)}{dt_0} \int_0^{t_0} (1 - F(s))ds \,.$$

Since $h(\cdot)$ was assumed to be increasing, the above expression is positive for all $t_0 \geq 0$, which implies that the left-hand-side of Eq. (11.4) is increasing. Therefore, we know that there is at most one solution to (11.4). (Figure 11.3 presents an example graph of the left-hand-side of Eq. (11.4) as a function of the replacement age.)

The optimal age-replacement policy can now be stated as follows: Let t_0 satisfy Eq. (11.4). If a failure occurs before t_0, replace at the failure time, if no failure occurs before t_0, replace at t_0.

Example 11.2. A smelting process involves steel pots that are lined with a ceramic lining. The ceramic lining is used because the electrolytic solution within the pots is corrosive; however, with age the lining will develop cracks which allows the electrolytic solution within the pot to find a path to the steel shell. If the lining fails (i.e., develops cracks) while the operation is in process, the solution will eat through the steel pot and spill on the floor which involves extra cleanup costs and extra repair time for the steel shells. Upon failure, or at a scheduled replacement, the steel shell is repaired and a new ceramic lining is placed in the shell. The lifetime of the lining is described by a Weibull distribution with scale parameter $\beta = 25.489$ months and shape parameter $\alpha = 16.5$ (see p. 19). Thus the expected lifetime of the lining is 24.685 months with a standard deviation of 1.84 months. Replacement cost is \$30,000 with an additional \$20,000 incurred if a failure occurs[2]; thus $c_r/c_f = 1.5$. The hazard rate function for the Weibull distribution is $h(t) = (\alpha/\beta)(t/\beta)^{\alpha-1}$ so that the Weibull has an IFR if $\alpha > 1$. Therefore, we need to find the value of t such that

$$\mathrm{e}^{-(t/\beta)^\alpha} + (\alpha/\beta)(t/\beta)^{\alpha-1} \int_0^t \mathrm{e}^{-(s/\beta)^\alpha} \mathrm{d}s = 2.5 \,.$$

The above equation cannot be solved analytically, but it can be evaluated numerically. Evaluating the above numerically yields $t_0 = 22.13$; therefore, any failure of a pot with an age less than 22.13 months will be immediately replaced, and any pot still working with an age of 22.13 months will be replaced even though no problems with the lining had yet become evident. The expected long-run cost for this policy is obtained by substituting the value of t_0 back into Eq. (11.1) to yield \$1,447.24 per month. □

Example 11.3. **Simulation.** To insure a good understanding of the replacement process, we simulate the failure of steel pots for Example 11.2. Since the Weibull distribution governs failure time, our first step is to obtain the inverse of the Weibull CDF. Using the inverse mapping method, the transformation from a random number to a Weibull random variate with scale and shape parameters β and α, respectively, is

$$T = \beta(-\ln(R))^{1/\alpha} \,.$$

For the simulation, we generate a random number and obtain a failure time. If the failure time is before the replacement time, t_0, then a failure cost is incurred; if the random failure time is larger than the replacement time, the failure cost is not incurred. After replacement (either at the failure time or replacement time), a new machine is instantaneously placed in service. We maintain the cumulative cost and the cumulative time. Table 11.2 shows the results of 10 replacements using an age-replacement policy of $t_0 = 22.13$. Based on this simulation, the estimated average cost for the replacement policy is \$1,453 per month.

For comparison, Table 11.3 shows the results of 10 replacements if a replacement policy of $t_0 = 24.7$ (i.e., the average life) had been used. The simulation would then yield an estimated average of \$1,488 per month. Of course, no conclusions can be drawn from these simulations, except that you now (hopefully) understand the dynamics of the process. Several replicates of longer runs need to be made if conclusions were to be drawn as to the preference of one policy over the other. This also emphasizes a weakness of simulation: a single calculation from the analytical approach is needed to obtain the optimal policy; whereas, many simulation runs must be made simply to obtain an estimate of the optimum. □

[2] These numbers are not very realistic, but they are given for an easy comparison with the discrete case of Example 11.1. Note that the means and standard deviations of the two examples are similar.

Table 11.2 Simulation of Example 11.2 with $t_0 = 22.13$

Time new machine starts	Random number	Failure time	Replacement time	Cost	Cumulative cost
0	0.7400	23.710	22.13	30,000	30,000
22.13	0.4135	25.318	22.13	30,000	60,000
44.26	0.1927	26.298	22.13	30,000	90,000
66.39	0.6486	24.241	22.13	30,000	120,000
88.52	0.0130	27.900	22.13	30,000	150,000
110.65	0.2967	25.815	22.13	30,000	180,000
132.78	0.0997	26.843	22.13	30,000	210,000
154.91	0.9583	21.044	21.04	50,000	260,000
175.91	0.0827	26.971	22.13	30,000	290,000
198.08	0.2605	25.976	22.13	30,000	320,000
220.21					

Table 11.3 Simulation of Example 11.2 with $t_0 = 24.7$

Time new machine starts	Random number	Failure time	Replacement time	Cost	Cumulative cost
0	0.7400	23.710	23.71	50,000	50,000
23.71	0.4135	25.318	24.70	30,000	80,000
48.41	0.1927	26.298	24.70	30,000	110,000
73.11	0.6486	24.241	24.24	50,000	160,000
97.35	0.0130	27.900	24.70	30,000	190,000
122.05	0.2967	25.815	24.70	30,000	220,000
146.75	0.0997	26.843	24.70	30,000	250,000
171.45	0.9583	21.044	21.04	50,000	300,000
192.49	0.0827	26.971	24.70	30,000	330,000
217.19	0.2605	25.976	24.70	30,000	360,000
241.89					

- *Suggestion: Do Problems 11.5, 11.6, and 11.8.*

11.2 Minimal Repair

Some types of failure do not always require replacement; it is sometimes possible to repair the failed equipment instead of completely replacing it. If a repair simply returns the equipment to service without any truly corrective work, it is called minimal repair; that is, a minimal repair is a repair that returns the equipment to a working condition with the remaining life of the equipment being the same as if it had never failed. For example, suppose you have a car and the tire has just blown out. You could buy a new car or you could replace the tire. If you choose the later approach and buy a new tire, the life of the car remains unaffected by the repair; such a repair is called a minimal repair. An example of a non-minimal repair would be a blown engine. If you overhaul the engine, that would most likely affect the life of the car and is not a minimal repair.

There are many variations possible when minimal repair is considered. In this section we consider two different scenarios. The two cases are presented to illustrate the modeling methodology and should not be viewed as *the* models to use whenever minimal repair is appropriate. In most situations, a new model would need to be developed that is designed uniquely for the given problem.

11.2.1 Minimal Repairs without Early Replacements

We again consider an age-replacement policy except that now whenever a failure occurs before the replacement age, a minimal repair is performed. Let t_0 designate the age limit at which point a working system will be replaced, let c_m designate the cost incurred for each minimal repair, and let c_r designate the replacement cost. Similar to the previous section, our first task is to determine the expected cost per system and the expected age per system. The objective function for the optimal policy is then the ratio of the two expected values.

The expected cost per system is the replacement cost, c_r, plus the expected cost for minimal repairs, c_m, times the expected number of failures before time t_0. It can be shown that the expected number of minimal repairs during the interval $(0,t]$ is equal to $\int_0^t h(s)\mathrm{d}s$, where $h(\cdot)$ is the hazard rate for the system. Since replacement occurs only at the replacement time, the age of the system is t_0. Thus, the value of the age-limit for replacement is given by

$$\min_{t_0} z(t_0) = \frac{c_r + c_m \int_0^{t_0} h(s)\mathrm{d}s}{t_0} . \tag{11.5}$$

To find the value of t_0 for which the above function is minimized, we take the derivative of the right-hand-side of Eq. (11.5). Setting the derivative equal to zero yields the equation

$$t_0 h(t_0) - \int_0^{t_0} h(s)\mathrm{d}s = \frac{c_r}{c_m} . \tag{11.6}$$

Again the question must be asked as to how many solutions might exist for Eq. (11.6). And, as before, as long as the model deals with a system having an IFR distribution, there will be at most one solution. The optimal minimal repair policy is, therefore, to determine the value of t_0 that satisfies Eq. (11.6), minimally repair any failures that occur before t_0 and replace at t_0.

Example 11.4. We return to Example 11.2 and assume that a procedure has been developed that can patch the cracks in the lining that caused the pot to fail. When a failure occurs, there is still the \$20,000 failure cost incurred, but the patch costs an additional \$4,000; thus, $c_m = \$24,000$. The integrals of Eqs. (11.5) and (11.6) involve the hazard rate function which is easily integrated for the Weibull distribution; in particular, $\int_0^t h(s)\mathrm{d}s = (t/\beta)^\alpha$. Therefore, the average long-run cost for the

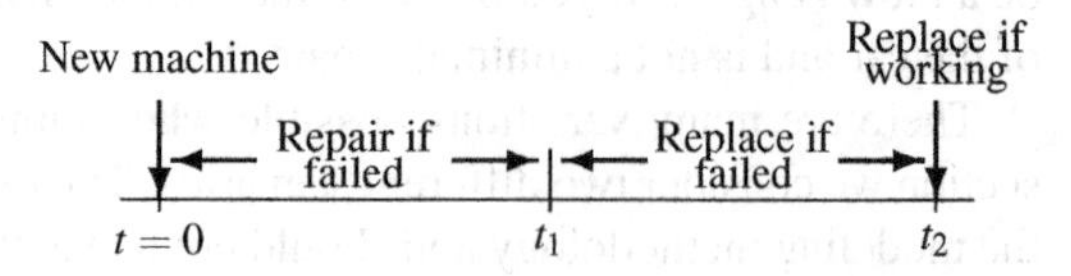

Fig. 11.4 Schematic representation for minimal-repair age-replacement policy

minimal repair model using a Weibull distribution is

$$z(t_0) = \frac{c_r + c_m (t_0/\beta)^\alpha}{t_0} \, ,$$

and Eq. (11.6) becomes

$$t_0^* = \beta \left(\frac{c_r}{(\alpha - 1) c_m} \right)^{1/\alpha} .$$

Since $\beta = 25.489$ and $\alpha = 16.5$, the optimum replacement time is $t_0^* = 21.88$ months, and the expected long-run average cost is \$1,459.45 per month. □

11.2.2 Minimal Repairs with Early Replacements

Another scenario when minimal repairs are a reasonable option is to establish two age-control points, t_1 and t_2. (See the schematic in Fig. 11.4.) Any failure before age t_1 is corrected by a minimal repair; a replacement is made for a system that fails between the ages of t_1 and t_2; and a preventative replacement is made at the age of t_2 if the system is still operating. There are three relevant costs: c_m is the cost for each minimal repair, c_r is the cost to replace the system, and c_f is an additional cost that must be added to c_r for a replacement due to a failure.

The expected cost per system is given by

$$c_r + c_m \int_0^{t_1} h(s) \mathrm{d}s + c_f \frac{F(t_2) - F(t_1)}{1 - F(t_1)} \, .$$

The first term is simply the cost of replacement which must always be incurred; the second term is the cost of a minimal repair times the expected number of minimal repairs during the interval $(0, t_1]$, and the third term is the cost of failure times the probability that the system fails in the interval $(t_1, t_2]$. Notice that the probability of failure is a conditional probability because we know that the system must be working at time t_1.

To obtain the expected life length of the system, let T be the random variable denoting the age at which the system is replaced, and as before, let $F(\cdot)$ denote the system's failure distribution. Then,

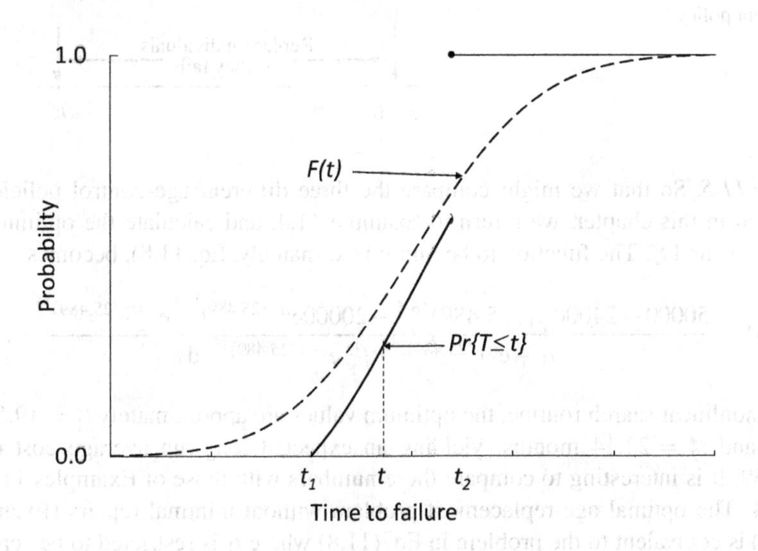

Fig. 11.5 Illustration of the replacement time distribution from Eq. (11.7)

$$\Pr\{T > t\} = \begin{cases} 1 & \text{if } t < t_1, \\ \frac{1-F(t)}{1-F(t_1)} & \text{if } t_1 \leq t < t_2, \\ 0 & \text{if } t \geq t_2 . \end{cases} \tag{11.7}$$

The expected life of the system is thus given by

$$t_1 + \frac{1}{1-F(t_1)} \int_{t_1}^{t_2} (1 - F(s))\mathrm{d}s .$$

(See the graph of the distribution functions in Fig. 11.5.)

The optimization problem for the (t_1, t_2) policy is again the expected cost per system divided by the expected life length per system. This yields a two-dimensional optimization problem as

$$\min_{t_1,t_2} z(t_1, t_2) = \tag{11.8}$$
$$\frac{c_r + c_m \int_0^{t_1} h(s)\mathrm{d}s + c_f (F(t_2) - F(t_1)) / (1 - F(t_1))}{t_1 + \int_{t_1}^{t_2} (1 - F(s)) / (1 - F(t_1))\mathrm{d}s} .$$

As a two-dimensional unconstrained optimization problem, we need to take the partial derivative of the right-hand-side of Eq. (11.8) with respect to t_1 and t_2. (The partials are not difficult, just tedious.) One of the standard nonlinear search techniques can then be used to find the optimum (t_1, t_2) policy.

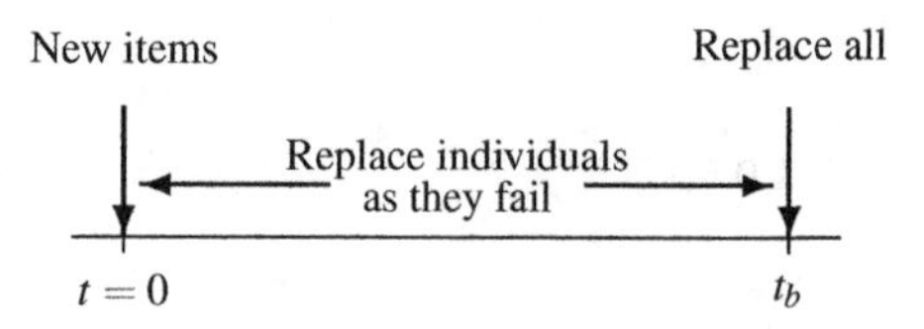

Fig. 11.6 Schematic representation for a block-replacement policy

Example 11.5. So that we might compare the three different age-control policies mentioned in this chapter, we return to Example 11.4, and calculate the optimum values for t_1 and t_2. The function to be minimized (namely, Eq. 11.8), becomes

$$z(t_1,t_2) = \frac{50000 + 24000(t_1/25.489)^{16.5} - 20000\mathrm{e}^{(t_1/25.489)^{16.5}}\mathrm{e}^{-(t_2/25.489)^{16.5}}}{t_1 + \mathrm{e}^{(t_1/25.489)^{16.5}}\int_{t_1}^{t_2}\mathrm{e}^{-(s/25.489)^{16.5}}ds}\,.$$

Using a nonlinear search routine, the optimum values are approximately $t_1^* = 19.50$ months and $t_2^* = 22.14$ months, yielding an expected long-run average cost of \$1,446.39. It is interesting to compare these numbers with those of Examples 11.2 and 11.4. The optimal age-replacement problem without minimal repairs (Example 11.2) is equivalent to the problem in Eq. (11.8) where t_1 is restricted to be zero, and the minimal repair problem without early replacement (Example 11.4) is equivalent to Eq. (11.8) where t_2 is restricted to be infinity. There is an almost insignificant difference between the optimal no minimal repair policy (Example 11.2) and the optimal (t_1,t_2) policy, but that does not imply minimal repair is never worthwhile. Since the optimal $t_1^* = 19.5$, the probability that a minimal repair is performed equals 0.012; thus, the *expected* cost savings is small. However, given the occurrence of an early failure, the actual cost savings may be significant. □

- *Suggestion: Do Problems 11.7 and 11.10.*

11.3 Block Replacement

One disadvantage of an age-replacement policy is that such a policy demands good record keeping. In a situation in which there are many items subject to failure, it may be advantageous to use a block replacement policy; that is, a replacement policy that replaces all items at fixed intervals of time independent of the age of the individual items (see Fig. 11.6 for a schematic representing a block replacement policy). We will further assume that the cost of the items is relatively low compared to a set-up cost needed for the replacement. Consider, for example, a city that must maintain its street lights. The major cost for replacement is the sending out of the crew. There is little difference in replacing one or two lights.

For modeling purposes, we shall return to the discrete period case and let f_m, for $m = 1,2,\cdots$, denote the failure probability mass function for an item (the items' failure times are assumed to be probabilistically identical), let c_r be the individual replacement costs, and let K be the set up cost. The policy under consideration is to establish a block replacement time, t_b; that is, at time 0 all items are assumed new, when an item fails it must be immediately replaced with a new part, then at time t_b all items are replaced, even those that had recently been replaced. Finally, we shall let $z(t_b)$ denote the expected long-run average cost for a replacement policy using a replacement time of t_b.

The difficult aspect of obtaining an expression for $z(t_b)$ is the determination of the number of individual replacements made. We will not be able to obtain a closed form expression for that number, but it can be easily obtained recursively. Let $\overline{n}_0$ denote the number of items in the system, and let $\overline{n}_k$ denote the expected number of items replaced during period k. These can be determined by the following recursive scheme:

$$
\begin{aligned}
\overline{n}_1 &= f_1\overline{n}_0 \\
\overline{n}_2 &= f_2\overline{n}_0 + f_1\overline{n}_1 \\
\overline{n}_3 &= f_3\overline{n}_0 + f_2\overline{n}_1 + f_1\overline{n}_2 \\
&\vdots \\
\overline{n}_m &= f_m\overline{n}_0 + f_{m-1}\overline{n}_1 + \cdots + f_1\overline{n}_{m-1} \, .
\end{aligned}
\tag{11.9}
$$

Notice that, for example, $\overline{n}_2$ involves the factor $f_1\overline{n}_1$ since those items that were new in time period one may fail in the following time period.

Utilizing the above definition, the expected long-run average cost associated with a block replacement time of t_b is given by

$$
z(t_b) = \frac{(K+\overline{n}_0 c_r) + (K+c_r)\sum_{m=1}^{t_b}\overline{n}_m}{t_b} \, .
$$

The first expression in the numerator on the right-hand-side of the above equation (namely, $K+\overline{n}_0 c_r$) reflects the cost incurred during each block replacement time, and the second term reflects the costs of all the individual replacements made. In the development of the second term, we assumed that as soon as a failure occurs, it is replaced immediately; thus, if three failures occur within a single period, the three failures would occur at separate times and would call for three individual replacements. Likewise, if failures occur in the last period (between times t_b-1 and t_b), the failed items are replaced at their failure time and then again when all items are replaced at time t_b.

Therefore, the optimal block replacement time is given as t_b^*, where t_b^* is the i that solves the following

$$
\min_i z(i) = \frac{1}{i}\left(K + \overline{n}_0 c_r + (K+c_r)\sum_{m=1}^{i}\overline{n}_m \right) . \tag{11.10}
$$

Sometimes it is possible for the costs to be such that it is never optimum to replace any items before failure. Therefore, it is important to always compare the minimal cost of the block replacement policy (i.e., the cost at the first local minimum) against the policy that never replaces ahead of time (i.e., the policy that replaces an item when it fails and never replaces working items). Such a policy has an average cost given by

$$z(\infty) = \frac{\overline{n}_0(K + c_r)}{E[T]}\,, \tag{11.11}$$

where $E[T] = \sum_{m=0}^{\infty} mf(m) = \sum_{m=0}^{\infty} \overline{F}_m$.

Example 11.6. A manufacturing firm produces aluminum powder. One of the uses of the powder is as an explosive, so care is taken during a major portion of the manufacturing process to insure that the powder is produced in an oxygen-free environment. Therefore, the firm has installed 50 sensors at various locations to record the presence of oxygen. As soon as a sensor stops working, it must be immediately replaced; however, the process must be shut down to replace a sensor costing approximately \$300 just for the shut-down and start-up period. The variable cost to replace a sensor is \$40, which includes the sensor itself as well as labor costs and a slight additional cost of lost production. The probability mass function describing the failure time of the sensors is as follows:

week	2	3	4	5	6	7	8	9	10	11	12
prob.	0.02	0.03	0.05	0.1	0.3	0.2	0.15	0.08	0.04	0.025	0.005

Thus, we have the following calculations:

For $i = 1$

$$\begin{aligned}\overline{n}_0 &= 50\\ \overline{n}_1 &= 50 \times 0 = 0\\ z(i) &= (300 + 50 \times 40) = 2,300\end{aligned}$$

For $i = 2$

$$\begin{aligned}\overline{n}_2 &= 50 \times 0.02 + 0 \times 0 = 1\\ z(i) &= \frac{1}{2}(300 + 50 \times 40 + 340 \times 1) = 1,320\end{aligned}$$

For $i = 3$

$$\begin{aligned}\overline{n}_3 &= 50 \times 0.03 + 0 \times 0.02 + 1 \times 0 = 1.5\\ z(i) &= \frac{1}{3}(300 + 50 \times 40 + 340 \times (1 + 1.5)) = 1,050\end{aligned}$$

For $i = 4$

$$\begin{aligned}\overline{n}_4 &= 50 \times 0.05 + 0 \times 0.03 + 1 \times 0.02 + 1.5 \times 0 = 2.52\\ z(i) &= \frac{1}{4}(300 + 50 \times 40 + 340 \times (1 + 1.5 + 2.52)) = 1,002\end{aligned}$$

For $i = 5$

$$\begin{aligned}\overline{n}_5 &= 50 \times 0.1 + 0 \times 0.05 + 1 \times 0.03 + 1.5 \times 0.02 + 2.52 \times 0 = 5.06\\ z(i) &= \frac{1}{5}(300 + 50 \times 40 + 340 \times (1 + 1.5 + 2.52 + 5.06)) = 1,145\,.\end{aligned}$$

Thus, the optimal block replacement time is at the end of the fourth week. We need to now compare this to the cost of only individual replacement (Eq. 11.11), which yields a weekly cost of $2,543. Therefore, the optimal block replacement cost of $1,002 is the best policy. □

- *Suggestion: Do Problems 11.9 and 11.11.*

Problems

11.1. This problem[3] is to consider the importance of keeping track of history when discussing the reliability of a machine. Let T be a random variable that indicates the time until failure for the machine. Assume that T has a uniform distribution from zero to two years and answer the question, "What is the probability that the machine will continue to work for at least three more months?"
(a) Assume the machine is new.
(b) Assume the machine is one year old and has not yet failed.
(c) Now assume that T has an exponential distribution with mean one year, and answer parts (a) and (b) again.
(d) Is it important to know how old the machine is in order to answer the question, "What is the probability that the machine will continue to work for at least three more months?"

11.2. An old factory uses a conveyor belt in its production process that seems to be always in need of repair. It costs $500 to have a repair crew "fix" the belt; however after it is fixed it still remains as unreliable as it was immediately after the last time it was fixed. After the belt is fixed, the probability that it will work for exactly k days is $0.7 \times 0.3^{k-1}$. When the belt fails, it costs an additional $600 due to the interruption to the production process. If the belt works during the day, the repair crew can be called in during the evening off-shift and make adjustments to the belt so that the probability law governing the next failure is the same as if the belt had failed during the day and the repair crew had fixed it that day. (When the repair-crew fixes the belt during the evenings, the $600 process interruption cost is saved.) Can you explain why the optimum replacement policy for this belt is to never replace before failure?

11.3. A specialized battery is a necessary component to a production machine. Purchase and installation of the battery costs $4,000. Although the battery is advertised as an "18-month" battery, the actual probability mass function governing its life length is

month	prob.	month	prob.	month	prob.	month	prob.	month	prob.
1	0.05	5	0.00	9	0.00	13	0.10	17	0.08
2	0.03	6	0.00	10	0.02	14	0.15	18	0.04
3	0.02	7	0.00	11	0.03	15	0.25	19	0.025
4	0.00	8	0.00	12	0.05	16	0.15	20	0.005

[3] This problem is from the first chapter and is included here in case it was skipped.

For purposes of this problem, assume that failure always occurs just before the end of the month. Replacement can be made at the end of a month without any disruption to the production schedule; however, if the battery fails before replacement, an additional cost of $1,000 is incurred. (Notice that the above distribution is not IFR because of the initial probability of failure. This is quite common as initial failures are usually due to manufacturing defects. After the "break-in" period, the distribution is IFR.)
(a) What is the optimal replacement time?
(b) Assume that the cost of record keeping is $10 per month. In other words, in order to follow the replacement policy, records need to be maintained regarding the life of the battery. A policy that ignores the life of the battery and simply replaces at failure has no monthly "record-keeping" costs. What is the optimal replacement policy?

11.4. The purpose of this problem is to consider warranties and to give you some practice in model development. You will need to adjust the numerator in Eq. (11.2) slightly in order to answer the questions. We continue with the above problem dealing with the specialized battery; that is, you should use the data from the previous problem assuming that there are no record-keeping costs.
(a) The battery comes with a six-month warranty; that is, if the battery fails in the first six months, a new one is installed free of charge. Does this change the optimal policy? (Assume that if battery failure occurs in the first six months, the warranty is renewed when the new battery is installed. Of course, the failure cost of $1000 is still incurred.)
(b) For an additional $450 in the purchase price of the battery, the six-month warranty can be extended to a 12-month warranty. Is it worthwhile? (Assume that if battery failure occurs in the first 12 months, a new 12-month warranty free of charge comes with the new battery.)

11.5. Simulate the situation described in Exercise 8.4 (a). Does your simulation agree with your analytical results?

11.6. Simulate the process described in Example 11.5. The theoretical optimum average value is approximately $1446. Is this within a 95% confidence interval derived from your simulation runs?

11.7. Show that a solution to Eq. (11.6) yields the unique minimum for Eq. (11.5) when the system has an increasing failure rate distribution.

11.8. The temperature control in an oven used within a manufacturing process must be periodically reset. Once the control has been reset, it will be accurate for a random length of time, which is described by the Weibull distribution with scale parameter $\beta = 100$ days and shape parameter $\alpha = 2$. To reset the control, the process must be stopped at a cost of $750. If the control is not reset, defective parts will be produced after the control loses its accuracy. As soon as the defective parts are identified, the process will be stopped and the control reset. The expected cost incurred due to the production of the defective parts is $550. What is the optimal age-replacement policy for the resetting of the control?

11.9. Part of the manufacturing process for powdered aluminum must be carried out in an oxygen free environment. There are 25 sensors that are used to insure that the environment stays free of oxygen. To replace one or more sensors, the process must be shut down at a cost of \$900 (including lost production). In addition, the sensors themselves cost \$75 to replace (including labor). The life of the sensors can be approximated by a Poisson distribution with a mean of 10 months. What is the optimum block-replacement policy?

11.10. Consider an alternative minimal-repair policy defined by the age t_1. Any failure before age t_1 is corrected by a minimal repair. A replacement is made at the first failure that occurs after t_1, i.e., no early replacement. Let c_m be the cost for each minimal repair, c_r the cost of replacement, and $F(\cdot)$ the system's failure distribution. Derive the long-run expected cost per unit time.

11.11. Equation (11.10) gives the optimum block replacement time, but to use it directly (as in Example 11.6) requires repetitive calculations. Show that a recursive approach can be taken where the long-run average costs are given by

$$z(1) = K(1+\overline{n}_1) + c_r(\overline{n}_0 + \overline{n}_1)$$

$$z(i) = z(i-1)\frac{i-1}{i} + \frac{(K+c_r)\overline{n}_i}{i} \text{ for } i = 2, 3, \cdots,$$

where $\overline{n}_i$ is given by Eq. (11.9).

Chapter 12
Markov Decision Processes[1]

Markov chains provide a useful modeling tool for determining expected profits or costs associated with certain types of systems. The key characteristic that allows for a Markov model is a probability law in which the future behavior of the system is independent of the past behavior given the present condition of the system. When this Markov property is present, the dynamics of the process can be described by a matrix containing the one-step transition probabilities and a vector of initial conditions. In some circumstances, the transition probabilities may depend on decisions made just before the transition time. Furthermore, not only the transition probabilities, but also associated costs or profits per transition may depend on decisions made at the transition times. For example, consider a slight variation of the first homework problem from Chap. 5. This problem had Joe and Pete playing a game of matching pennies (the pennies were biased), and the Markov model for the game used a state space representing the number of pennies in Joe's pocket. We shall generalize the previous homework problem by having two rules for the game instead of one: Rule 1 states that Joe wins when the coins match and Pete wins when coins do not match; Rule 2 is the opposite, namely Pete wins when the coins match and Joe wins when they do not match. Now, before each play of the game, the previous winner gets to decide which rule is to be used. The dynamics of this game can no longer be modeled as a simple Markov chain because we need to know how Joe and Pete will make their decisions before transition probabilities can be determined.

Processes that involve decisions which affect the transition probabilities often yield models in which optimization questions naturally arise. When the basic structure of Markov chains is combined with decision processes and optimization questions, a new model called Markov decision processes is formed. In Markov decision processes, in contrast to Markov chains, the future depends not only on the current state but also on the decisions made. The purpose of this chapter is to present some of the basic concepts and techniques of Markov decision theory and indicate the types of problems that are amenable to modeling as such processes.

[1] This chapter would normally be skipped in a one-semester undergraduate course.

R.M. Feldman, C. Valdez-Flores, *Applied Probability and Stochastic Processes*, 2nd ed.,
DOI 10.1007/978-3-642-05158-6_12,

12.1 Basic Definitions

The basic structure and dynamics of a Markov decision process will be introduced by way of an example. In this example, we introduce the basic elements of a Markov decision process; namely, a stochastic process with a state space denoted by E and a decision process with an action space denoted by A.

Example 12.1. Let $X = \{X_0, X_1, \cdots\}$ be a stochastic process with a four-state state space $E = \{a,b,c,d\}$. This process will represent a machine that can be in one of four operating conditions denoted by the states a through d indicating increasing levels of deterioration. As the machine deteriorates, not only is it more expensive to operate, but also production is lost. Standard maintenance activities are always carried out in states b through d so that the machine may improve due to maintenance; however, improvement is not guaranteed. In addition to the state space, there is an *action space* which gives the decisions possible at each step. (We sometimes use the words "decisions" and "actions" interchangeably referring to the elements of the action space.) In this example, we shall assume the action space is $A = \{1,2\}$; that is, at each step there are two possible actions: use an inexperienced operator (Action 1) or use an experienced operator (Action 2). To complete the description of a Markov decision problem, we need a cost vector and a transition matrix for each possible action in the action space. For our example, define the two cost vectors[2] and two Markov matrices as

$$\mathbf{f}_1 = (100, 125, 150, 500)^T ,$$

$$\mathbf{f}_2 = (300, 325, 350, 600)^T ,$$

$$\mathbf{P}_1 = \begin{bmatrix} 0.1 & 0.3 & 0.6 & 0.0 \\ 0.0 & 0.2 & 0.5 & 0.3 \\ 0.0 & 0.1 & 0.2 & 0.7 \\ 0.8 & 0.1 & 0.0 & 0.1 \end{bmatrix},$$

$$\mathbf{P}_2 = \begin{bmatrix} 0.6 & 0.3 & 0.1 & 0.0 \\ 0.75 & 0.1 & 0.1 & 0.05 \\ 0.8 & 0.2 & 0.0 & 0.0 \\ 0.9 & 0.1 & 0.0 & 0.0 \end{bmatrix}.$$

The dynamics of the process are illustrated in Fig. 12.1 and are as follows: if, at time n, the process is in state i and the decision k is made, then a cost of $f_k(i)$ is incurred and the probability that the next state will be j is given by $P_k(i,j)$. To illustrate, if $X_n = a$ and decision 1 is made, then a cost of \$100 is incurred (representing the operator cost, lost production cost, and machine operation cost) and $\Pr\{X_{n+1} = a\} = 0.1$; or, if $X_n = d$ and decision 2 is made, then a cost of \$600 is incurred (representing the operator cost, machine operation cost, major maintenance cost, and lost-production cost) and $\Pr\{X_{n+1} = a\} = 0.9$. □

---

[2] A superscript T denotes transpose.

Observe state $X_n = i$ ⟶ Take action $D_n = k$ ⟶ Incur cost $f_k(i)$ ⟶ Transition to next state $P_k(i,j)$

Fig. 12.1 Sequence of events in a Markov decision process

In general, the decision made at a given point in time is a random variable, which we shall denote by D_n; thus, there are two stochastic processes defined: the system description process, $X = \{X_0, X_1, \cdots\}$, and the decision process, $D = \{D_0, D_1, \cdots\}$. We can now give explicitly the form for the processes considered in this chapter.

Definition 12.1. Let X be a system description process with state space E and let D be a decision process with action space A. The process (X, D) is a *Markov Decision Process* if, for $j \in E$, and $n = 0, 1, \cdots$, the following holds

$$\Pr\{X_{n+1} = j | X_0, D_0, \cdots, X_n, D_n\} = \Pr\{X_{n+1} = j | X_n, D_n\} .$$

Furthermore, for each $k \in A$, let $\mathbf{f}_k$ be a cost vector and $\mathbf{P}_k$ be a Markov matrix. Then

$$\Pr\{X_{n+1} = j | X_n = i, D_n = k\} = P_k(i, j)$$

and the cost $f_k(i)$ is incurred whenever $X_n = i$ and $D_n = k$. □

An obvious question is, "How can decisions be made as to minimize costs?" A secondary question is, "What do you mean by minimize?" We first discuss the different ways decisions can be made.

Definition 12.2. A *policy* is any rule, using current information, past information, and/or randomization that specifies which action to take at each point in time. The set of all (decision) policies is denoted by $\mathscr{D}$. □

The following are some legitimate policies for the above problem:

Policy 1. Always chose action 1, independent of the state for X, i.e., let $D_n \equiv 1$ for all n.

Policy 2. If X_n is in state a or b, let $D_n = 1$; if X_n is in state c or d, let $D_n = 2$.

Policy 3. If X_n is in state a or b, let $D_n = 1$; if X_n is in state c, toss a (fair) coin and let $D_n = 1$ if the toss results in a head and let $D_n = 2$ if the toss results in a tail; if X_n is in state d, let $D_n = 2$.

Policy 4. Let $D_n \equiv 1$ for $n = 0$ and 1. For $n \geq 2$, if $X_n > X_{n-1}$ and $X_{n-2} = a$, let $D_n = 1$; if $X_n > X_{n-1}$, $X_{n-2} = b$, and $D_{n-1} = 2$ let $D_n = 1$; otherwise, let $D_n = 2$.

As you look over these different example policies, observe the wide range of possibilities: Policy 3 involves a randomization rule, and Policy 4 uses history. Once a policy is selected, the probability law governing the evolution of the process is determined. However, for an arbitrary policy, the Markov decision process is *not* necessarily a Markov chain, because we allow decisions to depend upon history.

For example, if Policy 4 is used, the decision maker needs X_n, X_{n-1}, and X_{n-2} in order to know which decision is to be made.

We are now ready to answer the question regarding the meaning of the term minimize. There are two common criteria used: (1) expected total discounted cost and (2) average long-run cost.

12.1.1 Expected Total Discounted Cost Criterion

The expected total discounted cost problem is equivalent to using a present worth calculation for the basis of decision making. Specifically, let α be a discount factor such that one dollar obtained at time $n = 1$ has a present value of α at time $n = 0$. (In traditional economic terms, if r is a rate of return (interest rate) specified by the management of a particular company, then $\alpha = 1/(1+r)$.) The expected total discounted cost for a particular Markov decision process is thus given by $E[\sum_{n=0}^{\infty} \alpha^n f_{D_n}(X_n)]$. For example, assume that in the above example, Policy 1 is chosen (i.e., the inexperienced operator is always used), and a discount factor of $\alpha = 0.95$ (equivalent to a rate of return of approximately 5.3% per period) is used. In that case, the example reduces to the computation of the total discounted cost of a standard Markov chain as was discussed in Chap. 5. The expected total discounted cost is calculated according to Property 5.14 and is given by $(\mathbf{I} - \alpha \mathbf{P}_1)^{-1} \mathbf{f}_1$ which yields the vector $\mathbf{v} = (4502, 4591, 4676, 4815)^T$. In other words, if the process starts in state a, the expected present value of all future costs is 4502. (It should be observed that when using a discount factor, the expected total discounted cost depends on the initial state.)

This example illustrates the fact that a specific policy needs to be selected before expectations can be taken. To designate this dependence on the policy, a subscript will be used with the expectation operator. Thus, $E_d[\cdot]$ denotes an expectation under the probability law specified by the policy $d \in \mathscr{D}$. The total discounted value of a Markov decision process under a discount factor of α using the policy $d \in \mathscr{D}$ will be denoted by v_d^α; that is,

$$v_d^\alpha(i) = E_d\left[\sum_{n=0}^{\infty} \alpha^n f_{D_n}(X_n) | X_0 = i\right]$$

for $i \in E$ and $0 < \alpha < 1$. Thus, the discounted cost optimization problem can be stated as: Find $d^\alpha \in \mathscr{D}$ such that $v_{d^\alpha}^\alpha(i) = v^\alpha(i)$ where the vector $\mathbf{v}^\alpha$ is defined, for $i \in E$, by

$$v^\alpha(i) = \min_{d \in \mathscr{D}} v_d^\alpha(i) . \tag{12.1}$$

It should be pointed out that the question of the existence of an optimal policy can be a difficult question when the state space is infinite; however, for the purposes of this text, we shall only consider problems in which its existence is assured by assuming that both the state space and action space are finite.

12.1.2 Average Long-Run Cost Criterion

Using an infinite horizon planning period, the total (undiscounted) cost may be infinite for all possible decisions so that total cost cannot be used to distinguish between alternative policies. However, if cost per transition is compared, then alternatives may be evaluated. Thus a commonly used criterion is $\lim_{m\to\infty} \frac{1}{m}\sum_{n=0}^{m-1} f_{D_n}(X_n)$. For example, we again assume that Policy 1 is used; thus, Action 1 is always chosen. Using the Markov chain results from the previous chapter, the long-run cost can be calculated according to Property 5.13. In other words, we first calculate the steady-state probabilities using the matrix $\mathbf{P}_1$, which yields $\boldsymbol{\pi} = (0.253, 0.167, 0.295, 0.285)$; then, the vector $\boldsymbol{\pi}$ is multiplied by the vector $\mathbf{f}_1$ yielding a long-run average cost of 232.925.

For a fixed policy $d \in \mathscr{D}$, the average long-run cost for the Markov decision process will be denoted by φ_d; in other words,

$$\varphi_d = \lim_{m\to\infty} \frac{f_{D_0}(X_0) + \cdots + f_{D_{m-1}}(X_{m-1})}{m} .$$

Thus, the optimization problem can be stated as: Find $d^* \in \mathscr{D}$ such that $\varphi_{d^*} = \varphi^*$, where φ^* is defined by

$$\varphi^* = \min_{d\in\mathscr{D}} \varphi_d . \tag{12.2}$$

As before, the existence question can be a difficult one for infinite state spaces. We leave such questions to the advanced textbooks.

12.2 Stationary Policies

The Markov decision problem as stated in Definitions 12.1 and 12.2 appears difficult because of the generality permitted by the policies. However, it turns out that under fairly general conditions, the optimum policy always has a very nice structure so that the search for an optimum can be limited to a much smaller set of policies. In particular, policies of the type exemplified by Policy 3 and Policy 4 above can be excluded in the search for an optimum. Consider the following two definitions.

Definition 12.3. An *action function* is a vector which maps the state space into the action space, i.e., an action function assigns an action to each state. □

In other words, if $\mathbf{a}$ is an action function, then $a(i) \in A$ for each $i \in E$. In the example policies given immediately after Definition 12.2, Policy 2 is equivalent to the action function $\mathbf{a} = (1, 1, 2, 2)$, where the action space is $A = \{1, 2\}$.

Definition 12.4. A *stationary policy* is a policy that can be defined by an action function. The stationary policy defined by the function $\mathbf{a}$ takes action $a(i)$ at time n if $X_n = i$, independent of previous states, previous actions, and time n. □

The key idea of a stationary policy is that it is independent of time and is a non-randomized policy that only depends on the current state of the process and, therefore, ignores history. Computationally, a stationary policy is nice in that the Markov decision process under a stationary policy is always a Markov chain. To see this, let the transition matrix $\mathbf{P}^a$ be defined, for $i, j \in E$, by

$$P^a(i,j) = P_{a(i)}(i,j) \tag{12.3}$$

and let the cost vector $\mathbf{f}^a$ be defined, for $i \in E$, by

$$f^a(i) = f_{a(i)}(i) . \tag{12.4}$$

Using the example from the previous section, we define the policy that uses the inexperienced operator whenever the machine is in state a or b and uses the experienced operator whenever the machine is in state c or d. The Markov decision process thus forms a chain with the Markov matrix

$$\mathbf{P}^a = \begin{bmatrix} 0.1 & 0.3 & 0.6 & 0.0 \\ 0.0 & 0.2 & 0.5 & 0.3 \\ 0.8 & 0.2 & 0.0 & 0.0 \\ 0.9 & 0.1 & 0.0 & 0.0 \end{bmatrix} ,$$

and cost vector given by

$$\mathbf{f}^a = (100, 125, 350, 600)^T .$$

Any stationary policy can be evaluated by forming the Markov matrix and cost vector associated with it according to Eqs. (12.3) and (12.4). The reason that this is important is that the search for the optimum can always be restricted to stationary policies as is given in the following property.

Property 12.1. *If the state space E is finite, there exists a stationary policy that solves the problem given in Eq. (12.1). Furthermore, if every stationary policy yields an irreducible Markov chain, there exists a stationary policy that solves the problem given in Eq. (12.2). (The optimum policy may depend on the discount factor and may be different for Eqs. (12.1) and (12.2).)*

For those familiar with linear programming, this property is analogous to the result that permits the Simplex Algorithm to be useful. Linear programming starts with an uncountably infinite set of feasible solutions, then by taking advantage of convexity, the property is established that allows the analyst to focus only on a finite set of solutions, namely the set of extreme points. The Simplex Algorithm is a procedure that starts at an easily defined extreme point and moves in a logical fashion to another extreme point in such a way that the solution is always improved. When no more improvement is possible, the optimum has been found. In Markov decision theory, we start with an extremely large class of possible policies, many of which produce processes that are not Markovian. Property 12.1 allows the analyst to focus on a much smaller set of policies, each one of which produces a Markov chain. In the remaining sections, algorithms will be developed that start with an easily defined stationary policy and moves to another stationary policy in such a way as to always improve until no more improvement is possible, in which case an optimum has been found.

12.3 Discounted Cost Algorithms

Three different procedures will be presented in this section for finding the optimal policy under an expected total discounted cost criterion. These procedures are based on a *fixed-point* property that holds for the optimal value function. In this section, we first discuss this key property.

In mathematics, a function is called *invariant* with respect to an operation if the operation does not vary the function. For example, the steady-state vector, $\boldsymbol{\pi}$, is also called the invariant vector for the Markov matrix $\mathbf{P}$ since the operation $\boldsymbol{\pi}\mathbf{P}$ does not vary the vector $\boldsymbol{\pi}$. For Markov decision processes the operation will be more complicated than simple matrix multiplication, but the basic idea of an invariant function will still hold. If the invariant function is unique, then it is called a fixed-point for the operation.

For those familiar with dynamic programming, the invariant property as applied to Markov decision processes will be recognized as the standard dynamic programming recursive relationship. The operation involves minimizing current costs plus all future costs, where the future costs must be discounted to the present in order to produce a total present value.

Property 12.2. ***Fixed-Point Theorem for Markov Decision Processes.*** *Let* $\mathbf{v}^\alpha$ *be the optimal value function as defined by Eq. (12.1) with* $0 < \alpha < 1$. *The function* $\mathbf{v}^\alpha$ *satisfies, for each* $i \in E$, *the following*

$$v^\alpha(i) = \min_{k \in A}\{f_k(i) + \alpha \sum_{j \in E} P_k(i,j) v^\alpha(j)\} . \tag{12.5}$$

Furthermore, it is the only function satisfying this property.

Property 12.2 provides a means to determine if a given function happens to be the optimal function. If we were given a function that turned out to be the optimal value function, it is easy to obtain the optimal policy through the next property[3].

[3] The term *argmin* used in the property refers to the *argument* that yields the minimum value.

Property 12.3. *Let* $\mathbf{v}^\alpha$ *be the optimal value function as defined by Eq. (12.1) with* $0 < \alpha < 1$*. Define an action function, for each* $i \in E$*, as follows:*

$$a(i) = \text{argmin}_{k \in A}\{f_k(i) + \alpha \sum_{j \in E} P_k(i,j) v^\alpha(j)\} .$$

The stationary policy defined by the action function **a** *is an optimal policy.*

At the moment, we do not know how to obtain $\mathbf{v}^\alpha$, but Property 12.3 tells how to obtain the optimal policy once $\mathbf{v}^\alpha$ is known. To illustrate, let us assert that the optimal value function for the machine problem given in Example 12.1 is $\mathbf{v}^\alpha = (4287, 4382, 4441, 4613)$ for $\alpha = 0.95$. Our first task is to verify the assertion through the Fixed-Point Theorem for Markov Decision Processes. The calculations needed to verify that the given vector is the optimal value function are as follows. (Note that the calculations are needed for each state in the state space.)

$$v^\alpha(a) = \min\{100 + 0.95\ (0.1, 0.3, 0.6, 0.0) \begin{pmatrix} 4287 \\ 4382 \\ 4441 \\ 4613 \end{pmatrix} ;$$

$$300 + 0.95\ (0.6, 0.3, 0.1, 0.0) \begin{pmatrix} 4287 \\ 4382 \\ 4441 \\ 4613 \end{pmatrix} \} = 4287 .$$

$$v^\alpha(b) = \min\{125 + 0.95\ (0.0, 0.2, 0.5, 0.3) \begin{pmatrix} 4287 \\ 4382 \\ 4441 \\ 4613 \end{pmatrix} ;$$

$$325 + 0.95\ (0.75, 0.1, 0.1, 0.05) \begin{pmatrix} 4287 \\ 4382 \\ 4441 \\ 4613 \end{pmatrix} \} = 4382 .$$

$$v^\alpha(c) = \min\{150 + 0.95\ (0.0, 0.1, 0.2, 0.7) \begin{pmatrix} 4287 \\ 4382 \\ 4441 \\ 4613 \end{pmatrix} ;$$

$$350 + 0.95\ (0.8, 0.2, 0.0, 0.0) \begin{pmatrix} 4287 \\ 4382 \\ 4441 \\ 4613 \end{pmatrix} \} = 4441 .$$

$$v^\alpha(d) = \min\{500 + 0.95\,(0.8, 0.1, 0.0, 0.1)\begin{pmatrix}4287\\4382\\4441\\4613\end{pmatrix};$$

$$600 + 0.95\,(0.9, 0.1, 0.0, 0.0)\begin{pmatrix}4287\\4382\\4441\\4613\end{pmatrix}\} = 4613\,.$$

Since, for each $i \in E$, the minimum of the two values yielded the asserted value of $v^\alpha(i)$, we know that it is optimum by Property 12.2. Looking back over the above calculations, we can also pick out the argument (i.e., action) that resulted in the minimum value. For State $i = a$, the first action yielded the minimum; for $i = b$, the first action yielded the minimum; for $i = c$, the second action yielded the minimum; and for $i = d$, the first action yielded the minimum. Therefore, from Property 12.3, the stationary optimal policy is defined by the action function $\mathbf{a} = (1, 1, 2, 1)$.

12.3.1 Value Improvement for Discounted Costs

The Fixed-Point Theorem for Markov Decision Processes (Property 12.2) allows for an easy iteration procedure that will limit to the optimal value function. The procedure starts with a guess for the optimal value function. That guess is then used as the value for $\mathbf{v}^\alpha$ in the right-hand-side of Eq. (12.5) and another value for $\mathbf{v}^\alpha$ is obtained. If the second value obtained is the same as the initial guess, we have an optimum; otherwise, the second value for $\mathbf{v}^\alpha$ is used in the right-hand-side of (12.5) to obtain a third, etc. The concept of a fixed-point is that by repeating such a successive substitution scheme, the fixed-point will be obtained.

Property 12.4. ***Value Improvement Algorithm.*** *The following iteration procedure will yield an approximation to the optimal value function as defined by Eq. (12.1).*

Step 1. *Make sure that $\alpha < 1$, choose a small positive value for ε, set $n = 0$, and let $v_0(i) = 0$ for each $i \in E$. (We set $\mathbf{v}_0 = \mathbf{0}$ for convenience; any initial solution is sufficient.)*

Step 2. *For each $i \in E$, define $v_{n+1}(i)$ by*

$$v_{n+1}(i) = \min_{k \in A}\{f_k(i) + \alpha \sum_{j \in E} P_k(i,j) v_n(j)\}\,.$$

Step 3. *Define δ by*

$$\delta = \max_{i \in E}\{|v_{n+1}(i) - v_n(i)|\}\,.$$

Step 4. *If* $\delta < \varepsilon$, *let* $\mathbf{v}^\alpha = \mathbf{v}_{n+1}$ *and stop; otherwise, increment n by one and return to Step 2.*

There are two major problems with the value improvement algorithm: (1) it can be slow to converge and (2) there is no simple rule for establishing a convergence criterion (i.e., setting a value for ε). In theory, it is true that as the number of iterations approaches infinity, the value function becomes the optimum; however, in practice we need to stop short of infinity and therefore the rule that says to stop when the change in the value function becomes negligible is commonly used. Exactly what is negligible is a judgment call that is sometimes difficult to make.

Another aspect of the value improvement algorithm is that the intermediate values produced by the algorithm do not give any indication of what the optimal policy is; in other words, the values for $\mathbf{v}_n$ are not helpful for determining the optimal (long-run) policy unless the algorithm has converged. (However, $\mathbf{v}_n$ does represent the optimal value associated with a finite horizon problem when $\mathbf{v}_0 = 0$; that is, $\mathbf{v}_n$ gives the minimal discounted cost assuming there are n transitions remaining in the life of the process.) The optimal long-run policy is obtained by taking the final value function from the algorithm and using Property 12.3. The major advantage of the Value Improvement Algorithm is its computational simplicity.

When the Value Improvement Algorithm is applied to the example machine problem defined in the previous section (using $\alpha = 0.95$), the following sequence of values are obtained:

$$\begin{aligned}
\mathbf{v}_0 &= (0,0,0,0) \\
\mathbf{v}_1 &= (100,125,150,500) \\
\mathbf{v}_2 &= (230.62,362.50,449.75,635.38) \\
\mathbf{v}_3 &= (481.58,588.59,594.15,770.07) \\
&\;\;\vdots
\end{aligned}$$

12.3.2 Policy Improvement for Discounted Costs

The algorithm of the previous section focused on the value function. In this section, we present an algorithm that focuses on the policy and then calculates the value associated with that particular policy. The result is that convergence is significantly faster, but there are more calculations for each iteration. Specifically, the Policy Improvement Algorithm involves an inverse routine (see Step 3 below) which can be time consuming for large problems and subject to round-off errors. However, if the problem is such that an accurate inverse is possible, then Policy Improvement is preferred over Value Improvement.

Property 12.5. ***Policy Improvement Algorithm.*** *The following iteration procedure will yield the optimal value function as defined by Eq. (12.1) and its associated optimal stationary policy.*

Step 1. *Make sure that $\alpha < 1$, set $n = 0$, and define the action function $\mathbf{a}_0$ by*

$$a_0(i) = \text{argmin}_{k \in A} f_k(i)$$

for each $i \in E$.

Step 2. *Define the matrix $\mathbf{P}$ and the vector $\mathbf{f}$ by*

$$f(i) = f_{a_n(i)}(i)$$
$$P(i,j) = P_{a_n(i)}(i,j)$$

for each $i, j \in E$.

Step 3. *Define the value function $\mathbf{v}$ by*

$$\mathbf{v} = (\mathbf{I} - \alpha \mathbf{P})^{-1} \mathbf{f}\,.$$

Step 4. *Define the action function $\mathbf{a}_{n+1}$ by*

$$a_{n+1}(i) = \text{argmin}_{k \in A} \{ f_k(i) + \alpha \sum_{j \in E} P_k(i,j) v(j) \}$$

for each $i \in E$.

Step 5. *If $\mathbf{a}_{n+1} = \mathbf{a}_n$, let $\mathbf{v}^\alpha = \mathbf{v}$, $\mathbf{a}^\alpha = \mathbf{a}_n$, and stop; otherwise, increment n by one and return to Step 2.*

The basic concept for the Policy Improvement Algorithm is to take a stationary policy, calculate the cost vector and transition matrix associated with that policy (Step 2, see Eqs. 12.3 and 12.4), then determine the expected total discounted cost associated with that cost vector and transition matrix (Step 3, see Property 5.14). If that policy was an optimal policy, then we would get that policy again through the use of Property 12.3 and the value function just calculated (Step 4). The reason this algorithm works is that if the policy was not the optimum, we are guaranteed that the policy formed in Step 4 will have a value associated with it that is better (not worse) than the previous one. To illustrate this algorithm we shall outline the results obtained from applying the algorithm to our example problem:

Iteration I. Step 1.

$$\mathbf{a}_0 = (1,1,1,1)$$

Step 2.

$$\mathbf{f} = (100, 125, 150, 500)^T$$

$$\mathbf{P} = \begin{bmatrix} 0.1 & 0.3 & 0.6 & 0.0 \\ 0.0 & 0.2 & 0.5 & 0.3 \\ 0.0 & 0.1 & 0.2 & 0.7 \\ 0.8 & 0.1 & 0.0 & 0.1 \end{bmatrix}$$

Step 3.

$$\mathbf{v} = (4502, 4591, 4676, 4815)^T$$

Step 4.

$$a_1(a) = \text{argmin}\{100 + 0.95\ (0.1, 0.3, 0.6, 0.0) \begin{pmatrix} 4502 \\ 4591 \\ 4676 \\ 4815 \end{pmatrix};$$

$$300 + 0.95\ (0.6, 0.3, 0.1, 0.0) \begin{pmatrix} 4502 \\ 4591 \\ 4676 \\ 4815 \end{pmatrix}\} = 1\,.$$

$$a_1(b) = \text{argmin}\{125 + 0.95\ (0.0, 0.2, 0.5, 0.3) \begin{pmatrix} 4502 \\ 4591 \\ 4676 \\ 4815 \end{pmatrix};$$

$$325 + 0.95\ (0.75, 0.1, 0.1, 0.05) \begin{pmatrix} 4502 \\ 4591 \\ 4676 \\ 4815 \end{pmatrix}\} = 1\,.$$

$$a_1(c) = \text{argmin}\{150 + 0.95\ (0.0, 0.1, 0.2, 0.7) \begin{pmatrix} 4502 \\ 4591 \\ 4676 \\ 4815 \end{pmatrix};$$

$$350 + 0.95\ (0.8, 0.2, 0.0, 0.0) \begin{pmatrix} 4502 \\ 4591 \\ 4676 \\ 4815 \end{pmatrix}\} = 2\,.$$

$$a_1(d) = \text{argmin}\{500 + 0.95\,(0.8, 0.1, 0.0, 0.1)\begin{pmatrix}4502\\4591\\4676\\4815\end{pmatrix};$$

$$600 + 0.95\,(0.9, 0.1, 0.0, 0.0)\begin{pmatrix}4502\\4591\\4676\\4815\end{pmatrix}\} = 1\,.$$

Thus, $\mathbf{a}_1 = (1, 1, 2, 1)$.

Step 5. Since $\mathbf{a}_1 \neq \mathbf{a}_0$, repeat the above Steps 2 through 5 using the stationary policy defined by $\mathbf{a}_1$.

Iteration II.

Step 2.

$$\mathbf{f} = (100, 125, 350, 500)^T$$

$$\mathbf{P} = \begin{bmatrix} 0.1 & 0.3 & 0.6 & 0.0 \\ 0.0 & 0.2 & 0.5 & 0.3 \\ 0.8 & 0.2 & 0.0 & 0.0 \\ 0.8 & 0.1 & 0.0 & 0.1 \end{bmatrix}$$

Step 3.

$$\mathbf{v} = (4287, 4382, 4441, 4613)^T$$

Step 4. The calculations for Step 4 will be a repeat of those calculations immediately following Property 12.3 since the value function is the same. Therefore, the result will be

$$\mathbf{a}_2 = (1, 1, 2, 1)\,.$$

Step 5. The algorithm is finished since $\mathbf{a}_2 = \mathbf{a}_1$; therefore, the results from the most recent Steps 3 and 4 are optimum.

As a final computational note, it should be pointed out that there are more efficient numerical routines for obtaining the value of $\mathbf{v}$ than through the inverse as it is written in Step 3. Numerical procedures for solving a system of linear equations are usually written without finding the inverse; thus, if the Policy Improvement Algorithm is written into a computer program, a numerical routine to solve

$$(\mathbf{I} - \alpha\mathbf{P})\mathbf{v} = \mathbf{f}$$

as a linear system of equations should be used.

12.3.3 Linear Programming for Discounted Costs

It is possible to solve Markov decision processes using linear programming, although it is more limited than the other approaches. For ease of presentation, we restricted ourselves to finite states and action spaces; however, the other two algorithms work for infinite state spaces, whereas linear programming can only be used for finite dimensional problems. The key to the linear programming formulation is the following property.

Property 12.6. ***Lemma for Linear Programming.*** *Let* $\mathbf{v}^\alpha$ *be the optimal value function as defined by Eq. (12.1) with* $0 < \alpha < 1$, *and let* $\mathbf{u}$ *be another real-valued function on the (finite) state space E. If* $\mathbf{u}$ *is such that*

$$u(i) \le \min_{k\in A}\{f_k(i) + \alpha \sum_{j\in E} P_k(i,j)u(j)\}$$

for all $i \in E$, *then* $\mathbf{u} \le \mathbf{v}^\alpha$.

Consider a set made up of all functions that satisfy the inequality of Property 12.6, then from Eq. (12.5) we know that the optimal value function, $\mathbf{v}^\alpha$, would be included in that set. The force of this lemma for linear programming is that the largest of all functions in that set is $\mathbf{v}^\alpha$ (see Property 12.2). In other words, $\mathbf{v}^\alpha$ is the solution to the following mathematical programming problem (stated in matrix notation):

max $\mathbf{u}$

subject to:

$$\mathbf{u} \le \min_{k\in A}\{\mathbf{f}_k + \alpha \mathbf{P_k u}\} \quad .$$

The maximization in the above problem is a component by component operation; therefore, *max* $\mathbf{u}$ is equivalent to *max* $\sum_{i\in E} u(i)$. Furthermore, the inequality being true for the minimum over all $k \in A$ is equivalent to it holding for each $k \in A$. Therefore, we have the following:

Property 12.7. ***Linear Programming for Discounted Costs.*** *The optimal solution to the following linear program gives the minimum value function,* $\mathbf{v}^\alpha$, *with* $0 < \alpha < 1$, *for the problem defined by Eq. (12.1).*

max $\sum_{i\in E} u(i)$

subject to:

$$u(i) \le f_k(i) + \alpha \textstyle\sum_{j\in E} P_k(i,j)u(j) \quad \text{for each } i \in E \text{ and } k \in A.$$

The optimal policy is to choose an action k for state i such that $s_{i,k} = 0$, *where* $s_{i,k}$ *is the slack variable associated with the equation corresponding to state i and action k.*

Notice that the variables in the linear programming formulation are unrestricted as to sign. Many of the software packages available for linear programming assume that all variables are restricted to be nonnegative; however, this is easy to remedy without doubling the size of the problem by the standard technique of letting unrestricted variables be the difference of two nonnegative variables. If $f_k(i) \geq 0$ for all $k \in A$ and $i \in E$, the optimal solution will be nonnegative so that nothing will be lost by a nonnegative restriction. If some of the $\mathbf{f}_k$ components are negative, then let δ be the absolute value of the most negative and add δ to all values of $\mathbf{f}_k$. Then the linear program using the new values for $\mathbf{f}_k$ with the variables restricted will yield the proper optimum, and the value of the objective function will be too high by the amount $\delta/(1-\alpha)$.

To illustrate the linear programming formulation, we again use the machine problem from the previous section yielding

$$\max z = u_a + u_b + u_c + u_d$$

subject to:

$$\begin{array}{lllll}
u_a \leq 100 & +0.095u_a & +0.285u_b & +0.57u_c & \\
u_a \leq 300 & +0.57u_a & +0.285u_b & +0.095u_c & \\
u_b \leq 125 & & +0.19u_b & +0.475u_c & +0.285u_d \\
u_b \leq 325 & +0.7125u_a & +0.095u_b & +0.095u_c & +0.0475u_d \\
u_c \leq 150 & & +0.095u_b & +0.19u_c & +0.665u_d \\
u_c \leq 350 & +0.76u_a & +0.19u_b & & \\
u_d \leq 500 & +0.76u_a & +0.095u_b & & +0.095u_d \\
u_d \leq 600 & +0.855u_a & +0.095u_b & &
\end{array}$$

Before solving this system, we moved the variables on the right-hand-side to the left and added slack variables. Then, the solution to this problem yields $u_a = 4287, u_b = 4382, u_c = 4441, u_d = 4613$ and the slacks associated with the first, third, sixth, and seventh equations are zero.

The linear programming algorithm used a maximizing objective function for the minimum cost problem; it would use a minimizing objective function for the maximum profit problem, and in addition, the inequalities in the constraints must be reversed. In other words, the programming formulation for a discounted maximum *profit* problem would be to *minimize* $\mathbf{u}$ subject to $\mathbf{u} \geq \mathbf{f}_k + \alpha \mathbf{P}_k \mathbf{u}$ for each $k \in A$. Again, the decision variables are unrestricted as to sign.

12.4 Average Cost Algorithms

The average cost criterion problem is slightly more difficult than the discounted cost criterion problem because it was the discount factor that produced a fixed-point. However, there is a recursive equation that is analogous to Property 12.2 as follows:

Property 12.8. *Assume that every stationary policy yields a Markov chain with only one irreducible set. There exists a scalar φ^* and a vector* **h** *such that, for all $i \in E$,*

$$\varphi^* + h(i) = \min_{k \in A}\{f_k(i) + \sum_{j \in E} P_k(i,j)h(j)\} . \tag{12.6}$$

The scalar φ^* is the optimal cost as defined by Eq. (12.2), and the optimal action function is defined by

$$a(i) = \text{argmin}_{k \in A}\{f_k(i) + \sum_{j \in E} P_k(i,j)h(j)\} .$$

The vector **h** *is unique up to an additive constant.*

Let us use this property to determine if the optimal policy previously determined for a discount factor of 0.95 is also the optimal policy under the average cost criterion. In other words, we would like to determine if the stationary policy defined by $\mathbf{a} = (1,1,2,1)$ is optimal using the long-run average cost criterion. The first step is to solve the following system of equations:

$$\begin{aligned}
\varphi^* + h_a &= 100 + 0.1h_a + 0.3h_b + 0.6h_c \\
\varphi^* + h_b &= 125 \qquad\qquad + 0.2h_b + 0.5h_c + 0.3h_d \\
\varphi^* + h_c &= 350 + 0.8h_a + 0.2h_b \\
\varphi^* + h_d &= 500 + 0.8h_a + 0.1h_b \qquad\qquad + 0.1h_d \ .
\end{aligned}$$

An immediate difficulty in the above system might be seen in that there are four equations and five unknowns; however, this was expected since the **h** vector is not unique. This will be discussed in more detail later; for now simply set $h_a = 0$ and solve for the remaining four unknowns. The solution should yield $\varphi^* = 219.24$ and $\mathbf{h} = (0.0, 97.10, 150.18, 322.75)^T$. In other words, these values have been determined so that for each $k = a(i)$, the following equality holds:

$$\varphi^* + h(i) = f_k(i) + \mathbf{P_k h}(i) .$$

Now, to determine optimality, we must verify that for each $k \neq a(i)$ the following inequality holds:

$$\varphi^* + h(i) \leq f_k(i) + \mathbf{P_k h}(i),$$

that is, the following must hold for optimality to be true:

$$\begin{aligned}
\varphi^* + h_a &\leq 300 + 0.6h_a + 0.3h_b + 0.1h_c \\
\varphi^* + h_b &\leq 325 + 0.75h_a + 0.1h_b + 0.1h_c + 0.05h_d \\
\varphi^* + h_c &\leq 150 \qquad\qquad + 0.1h_b + 0.2h_c + 0.7h_d \\
\varphi^* + h_d &\leq 600 + 0.9h_a + 0.1h_b \ .
\end{aligned}$$

Since the above holds, we have optimality.

It should also be observed that once the optimal policy is known, the value for φ^* can be obtained using Property 5.13. First, obtain the long-run probabilities associated with the Markov matrix given by

$$\mathbf{P}^a = \begin{bmatrix} 0.1 & 0.3 & 0.6 & 0.0 \\ 0.0 & 0.2 & 0.5 & 0.3 \\ 0.8 & 0.2 & 0.0 & 0.0 \\ 0.8 & 0.1 & 0.0 & 0.1 \end{bmatrix}.$$

These probabilities are $\boldsymbol{\pi} = (0.3630, 0.2287, 0.3321, 0.0762)$. Secondly, multiply those probabilities by the cost $\mathbf{g}^a = (100, 125, 350, 500)$, which yields $\varphi^* = 219.24$.

There is a close connection between the discounted cost problem and the long-run average cost problem. To see some of these relationships, we present the following table containing some information derived from the value function associated with the optimal policy for $\alpha = 0.95, \alpha = 0.99$, and $\alpha = 0.999$.

Table 12.1 Comparison of the discounted cost and long-run average cost problems

discount	vector	$i=a$	$i=b$	$i=c$	$i=d$
	$\mathbf{v}^\alpha$	4287	4382	4441	4613
$\alpha = 0.95$	$(1-\alpha)\mathbf{v}^\alpha$	214.35	219.10	222.05	230.65
	$\mathbf{v}^\alpha - v^\alpha(a)$	0	94	154	326
	$\mathbf{v}^\alpha$	21827	21923	21978	22150
$\alpha = 0.99$	$(1-\alpha)\mathbf{v}^\alpha$	218.27	219.23	219.78	221.50
	$\mathbf{v}^\alpha - v^\alpha(a)$	0	97	151	323
	$\mathbf{v}^\alpha$	219136	219233	219286	219459
$\alpha = 0.999$	$(1-\alpha)\mathbf{v}^\alpha$	219.13	219.23	219.28	219.45
	$\mathbf{v}^\alpha - v^\alpha(a)$	0	97	150	323

The value for each component in the vector $(1-\alpha)\mathbf{v}^\alpha$ approaches φ^* as the discount factor approaches one. To understand this, remember that the geometric series is $\sum_{n=0}^{\infty} \alpha^n = 1/(1-\alpha)$; thus, if a cost of c is incurred every period with a discount factor of α, its total value would be $v = c/(1-\alpha)$. Or conversely, a total cost of v is equivalent to an average per period cost of $c = (1-\alpha)v$. This gives the intuitive justification behind the following property.

Property 12.9. *Let $\mathbf{v}^\alpha$ be the optimal value function defined by Eq. (12.1), let φ^* be the optimal cost defined by Eq. (12.2), and assume that every stationary policy yields a Markov chain with only one irreducible set. Then*

$$\lim_{\alpha \to 1} (1-\alpha) v^\alpha(i) = \varphi^*$$

for any $i \in E$.

Referring back to the above table, the third row under each discount factor gives the relative difference in total cost for starting in the different states. In other words, there is a \$97 advantage in starting in state a instead of b under a discount factor of $\alpha = 0.99$; thus, it follows that $h(b) = 97$ because the $\mathbf{h}$ vector gives these relative differences. Namely, if $h(a) = 0$, then $h(i)$ for $i \in E$ gives the additional cost of starting in state i instead of state a. When we initially solved the above system of equations to determine the values for φ^* and $\mathbf{h}$, we set $h(a) = 0$. If we had started by setting $h(b) = 0$, we would have obtained $h(a) = -97.10, h(c) = 53.08$, and $h(d) = 225.65$. This is stated explicitly in the following:

Property 12.10. *Let $\mathbf{v}^\alpha$ be the optimal value function defined by Eq. (12.1), let φ^* be the optimal cost defined by Eq. (12.2), and let $\mathbf{h}$ be the vector defined by Property 12.8. Then*

$$\lim_{\alpha \to 1} v^\alpha(i) - v^\alpha(j) = h(i) - h(j)$$

for $i, j \in E$.

This property is the justification for arbitrarily setting $h(a) = 0$ when we solve for φ^* and $\mathbf{h}$. In fact, it is legitimate to pick any single state and set its $\mathbf{h}$ value equal to any given number.

12.4.1 Policy Improvement for Average Costs

Property 12.8 enables the design of an algorithm similar to the algorithm of Property 12.5. The procedure begins with an arbitrary policy, determines the φ^* and $\mathbf{h}$ values associated with it, and then either establishes that the policy is optimal or produces a better policy. The specifics are as follows:

Property 12.11. ***Policy Improvement Algorithm.*** *The following iteration procedure will yield the optimal value function as defined by Eq. (12.2) and its associated optimal stationary policy.*

Step 1. *Set $n = 0$, let the first state in the state space be denoted by the number 1, and define the action function $\mathbf{a}_0$ by*

$$a_0(i) = \operatorname{argmin}_{k \in A} f_k(i)$$

for each $i \in E$.

Step 2. *Define the matrix $\mathbf{P}$ and the vector $\mathbf{f}$ by*

$$f(i) = f_{a_n(i)}(i)$$
$$P(i,j) = P_{a_n(i)}(i,j)$$

for each $i, j \in E$.

Step 3. Determine values for φ and $\mathbf{h}$ by solving the system of equations given by

$$\varphi + \mathbf{h} = \mathbf{f} + \mathbf{P}\mathbf{h} ,$$

where $h(1) = 0$.

Step 4. Define the action function $\mathbf{a}_{n+1}$ by

$$a_{n+1}(i) = \operatorname{argmin}_{k \in A}\{f_k(i) + \sum_{j \in E} P_k(i,j)h(j)\}$$

for each $i \in E$.

Step 5. If $\mathbf{a}_{n+1} = \mathbf{a}_n$, let $\varphi^* = \varphi, \mathbf{a}^* = \mathbf{a}_n$, and stop; otherwise, increment n by one and return to Step 2.

To illustrate this algorithm we shall outline the results obtained from applying the algorithm to our example problem:

Iteration I. Step 1.

$$\mathbf{a}_0 = (1,1,1,1)$$

Step 2.

$$\mathbf{f} = (100, 125, 150, 500)^T$$
$$\mathbf{P} = \begin{bmatrix} 0.1 & 0.3 & 0.6 & 0.0 \\ 0.0 & 0.2 & 0.5 & 0.3 \\ 0.0 & 0.1 & 0.2 & 0.7 \\ 0.8 & 0.1 & 0.0 & 0.1 \end{bmatrix}$$

Step 3. Solve the following (where h_a has been set to zero):

$$\begin{aligned} \varphi &= 100 + 0.3h_b + 0.6h_c \\ \varphi + h_b &= 125 + 0.2h_b + 0.5h_c + 0.3h_d \\ \varphi + h_c &= 150 + 0.1h_b + 0.2h_c + 0.7h_d \\ \varphi + h_d &= 500 + 0.1h_b \qquad\qquad + 0.1h_d \end{aligned}$$

to obtain $\varphi = 232.86$ and $\mathbf{h} = (0, 90.40, 176.23, 306.87)^T$.

Step 4.

$$a_1(a) = \operatorname{argmin}\{100 + (0.1, 0.3, 0.6, 0.0)\begin{pmatrix} 0 \\ 90.40 \\ 176.23 \\ 306.87 \end{pmatrix};$$

$$300 + (0.6, 0.3, 0.1, 0.0) \begin{pmatrix} 0 \\ 90.40 \\ 176.23 \\ 306.87 \end{pmatrix} \} = 1 .$$

$$a_1(b) = \text{argmin}\{125 + (0.0, 0.2, 0.5, 0.3) \begin{pmatrix} 0 \\ 90.40 \\ 176.23 \\ 306.87 \end{pmatrix} ;$$

$$325 + (0.75, 0.1, 0.1, 0.05) \begin{pmatrix} 0 \\ 90.40 \\ 176.23 \\ 306.87 \end{pmatrix} \} = 1 .$$

$$a_1(c) = \text{argmin}\{150 + (0.0, 0.1, 0.2, 0.7) \begin{pmatrix} 0 \\ 90.40 \\ 176.23 \\ 306.87 \end{pmatrix} ;$$

$$350 + (0.8, 0.2, 0.0, 0.0) \begin{pmatrix} 0 \\ 90.40 \\ 176.23 \\ 306.87 \end{pmatrix} \} = 2 .$$

$$a_1(d) = \text{argmin}\{500 + (0.8, 0.1, 0.0, 0.1) \begin{pmatrix} 0 \\ 90.40 \\ 176.23 \\ 306.87 \end{pmatrix} ;$$

$$600 + (0.9, 0.1, 0.0, 0.0) \begin{pmatrix} 0 \\ 90.40 \\ 176.23 \\ 306.87 \end{pmatrix} \} = 1 .$$

Thus, $\mathbf{a}_1 = (1, 1, 2, 1)$.

Step 5. Since $\mathbf{a}_1 \neq \mathbf{a}_0$, repeat the above using the policy defined by $\mathbf{a}_1$.

Iteration II. Step 2.

$$\mathbf{f} = (100, 125, 350, 500)^T$$

$$\mathbf{P} = \begin{bmatrix} 0.1 & 0.3 & 0.6 & 0.0 \\ 0.0 & 0.2 & 0.5 & 0.3 \\ 0.8 & 0.2 & 0.0 & 0.0 \\ 0.8 & 0.1 & 0.0 & 0.1 \end{bmatrix}$$

Step 3. Solve the system of equations on page 338 by first setting $h(a) = 0$ to obtain $\varphi = 219.24$ and $\mathbf{h} = (0.0, 97.10, 150.18, 322.75)^T$.
Step 4.

$$a_2(a) = \operatorname{argmin}\{100 + (0.1, 0.3, 0.6, 0.0)\begin{pmatrix} 0 \\ 97.10 \\ 150.18 \\ 322.75 \end{pmatrix};$$

$$300 + (0.6, 0.3, 0.1, 0.0)\begin{pmatrix} 0 \\ 97.10 \\ 150.18 \\ 322.75 \end{pmatrix}\} = 1\,.$$

$$a_2(b) = \operatorname{argmin}\{125 + (0.0, 0.2, 0.5, 0.3)\begin{pmatrix} 0 \\ 97.10 \\ 150.18 \\ 322.75 \end{pmatrix};$$

$$325 + (0.75, 0.1, 0.1, 0.05)\begin{pmatrix} 0 \\ 97.10 \\ 150.18 \\ 322.75 \end{pmatrix}\} = 1\,.$$

$$a_2(c) = \operatorname{argmin}\{150 + (0.0, 0.1, 0.2, 0.7)\begin{pmatrix} 0 \\ 97.10 \\ 150.18 \\ 322.75 \end{pmatrix};$$

$$350 + (0.8, 0.2, 0.0, 0.0)\begin{pmatrix} 0 \\ 97.10 \\ 150.18 \\ 322.75 \end{pmatrix}\} = 2\,.$$

$$a_2(d) = \operatorname{argmin}\{500 + (0.8, 0.1, 0.0, 0.1)\begin{pmatrix} 0 \\ 97.10 \\ 150.18 \\ 322.75 \end{pmatrix};$$

$$600 + (0.9, 0.1, 0.0, 0.0) \begin{pmatrix} 0 \\ 97.10 \\ 150.18 \\ 322.75 \end{pmatrix} \} = 1 .$$

Thus, $\mathbf{a}_2 = (1, 1, 2, 1)$.

Step 5. The algorithm is finished since $\mathbf{a}_2 = \mathbf{a}_1$; therefore, the results from the most recent Steps 3 and 4 are optimum.

As a final computational note, a matrix form for the equations to be solved in Step 3 can be given. Let $\mathbf{S}$ be a matrix equal to $\mathbf{I} - \mathbf{P}$ except that the first column is all ones; that is

$$S(i,j) = \begin{cases} 1 & \text{if } j = 1, \\ 1 - P(i,j) & \text{if } i = j \text{ and } j \neq 1, \\ -P(i,j) & \text{if } i \neq j \text{ and } j \neq 1, \end{cases}$$

where we have identified the first state in the state space as State 1. As mentioned at the end of Sect. 12.3.2, numerical procedures that specifically solve linear systems of equations are more efficient than first determining the inverse and then multiplying the inverse times the right-hand-side vector. The system that must be solved for the average cost criterion problem is

$$\mathbf{Sx} = \mathbf{f} .$$

The values for the quantities needed in the algorithm are then obtained as $h(1) = 0, \varphi^* = x(1)$, and $h(i) = x(i)$ for $i > 1$.

12.4.2 Linear Programming for Average Costs

The linear programming formulation for the long-run average cost problem takes a completely different approach than the formulation for the discounted cost problem, because there is no analogous property to Property 12.6 for average costs. The decision variables for the discounted cost linear program are components of the value function with a maximizing objective function. The decision variables for the long-run average cost linear program are components of the policy function with a minimizing objective function.

In order for the feasible set of possible policies to be convex, the linear program for the long-run average cost problem will consider not only stationary policies but also randomized policies. A randomized policy (see the example Policy 3 on page 325) is a policy that assigns a probability mass function over the action space for each state in the state space. Specifically, let $\nu_i(k)$ for $k \in A$ be a probability mass function for each $i \in E$; that is,

$$\nu_i(k) = \Pr\{D_n = k | X_n = i\} .$$

Of course, a stationary policy defined by the action function $\mathbf{a}$ is a subset of these randomized policies, where $v_i(k) = 1$ if $k = a(i)$ and zero otherwise.

Each randomized policy will yield a steady-state probability vector, $\boldsymbol{\pi}$. Multiplying the steady-state probabilities with the randomized policy produces a joint probability mass function of the state and action taken in the long-run, which we shall denote as $x(i,k)$ for $i \in E$ and $k \in A$; that is,

$$x(i,k) = v_i(k)\pi(i) = \lim_{n\to\infty} \Pr\{X_n = i, D_n = k\} . \tag{12.7}$$

For a fixed $i \in E$, the function $\mathbf{v}_i$ is a probability mass function so that it sums to one; therefore

$$\sum_{k\in A} x(i,k) = \sum_{k\in A} v_i(k)\pi(i) = \pi(i) . \tag{12.8}$$

The decision variables for the linear program will be the joint probabilities designated by $x(i,k)$ for $i \in E$ and $k \in A$. For a fixed sequence of the joint probabilities, the expected value of the average cost for the associated randomized policy is

$$\varphi = \sum_{i\in E}\sum_{k\in A} x(i,k) f_k(i) ,$$

and the three conditions that must be fulfilled are:

1. As a mass function, the probabilities must sum to one.
2. As a mass function, the probabilities must be nonnegative.
3. As steady-state probabilities, the equation $\boldsymbol{\pi}\mathbf{P} = \boldsymbol{\pi}$ must hold.

Conditions 1 and 2 are straightforward, but the third condition needs some discussion. Specifically, we need to obtain the vector $\boldsymbol{\pi}$ and the matrix $\mathbf{P}$ from the given joint probabilities. It should not be difficult to show (see the homework problems) that the appropriate matrix for a fixed randomized policy is given by

$$P(i,j) = \sum_{k\in A} v_i(k) P_k(i,j)$$

for $i \in E$ and $k \in A$. Therefore, the equation $\boldsymbol{\pi}\mathbf{P} = \boldsymbol{\pi}$ becomes

$$\pi(j) = \sum_{i\in E} \pi(i) \sum_{k\in A} v_i(k) P_k(i,j)$$

for $j \in E$. We can now combine Eqs. (12.7) and (12.8) with the above equation to write $\boldsymbol{\pi}\mathbf{P} = \boldsymbol{\pi}$ in terms of the joint probabilities:

$$\sum_{k\in A} x(j,k) = \sum_{i\in E}\sum_{k\in A} x(i,k) P_k(i,j) . \tag{12.9}$$

With the above background, these relationships can be structured as the following linear program, the solution of which yields the optimal policy for the long-run average cost criterion problem.

Property 12.12. ***Linear Programming for Average Costs.*** *The optimal solution to the following linear program gives the minimum value for the problem defined by Eq. (12.2).*

$$\min \varphi = \sum_{i \in E} \sum_{k \in A} x(i,k) f_k(i)$$

subject to:

$$\sum_{k \in A} x(j,k) = \sum_{i \in E} \sum_{k \in A} x(i,k) P_k(i,j) \quad \text{for each } j \in E$$

$$\sum_{i \in E} \sum_{k \in A} x(i,k) = 1$$

$$x(i,k) \geq 0 \text{ for each } i \in E \text{ and } k \in A,$$

where the first equation in the constraint set is redundant and may be deleted. The optimal policy is to choose an action k for state i such that $x(i,k) > 0$.

Because we know that a (nonrandomized) stationary policy is optimal, there will only be one positive value of $x(i,k)$ for each $i \in E$; therefore, the optimal action function is to let $a(i)$ be the value of k for which $x(i,k) > 0$. It should also be observed that the steady-state probabilities under the optimal policy are given as $\pi(i) = x(i,a(i))$ for $i \in E$.

To illustrate the linear programming formulation, we again return to our machine problem example (here we use x_{ik} for $x(i,k)$ to aid in the readability of the equations):

$$\begin{aligned} \min \varphi = &\ 100x_{a1} + 125x_{b1} + 150x_{c1} + 500x_{d1} \\ &+300x_{a2} + 325x_{b2} + 350x_{c2} + 600x_{d2} \end{aligned}$$

subject to:

$$\begin{aligned} x_{a1} + x_{a2} &= 0.1x_{a1} + 0.6x_{a2} + 0.75x_{b2} + 0.8x_{c2} + 0.8x_{d1} + 0.9x_{d2} \\ x_{b1} + x_{b2} &= 0.3x_{a1} + 0.3x_{a2} + 0.2x_{b1} + 0.1x_{b2} + 0.1x_{c1} + 0.2x_{c2} \\ &\quad +0.1x_{d1} + 0.1x_{d2} \\ x_{c1} + x_{c2} &= 0.6x_{a1} + 0.1x_{a2} + 0.5x_{b1} + 0.1x_{b2} + 0.2x_{c1} \\ x_{d1} + x_{d2} &= 0.3x_{b1} + 0.05x_{b2} + 0.7x_{c1} + 0.1x_{d1} \\ x_{a1} + x_{b1} + x_{c1} &+ x_{d1} + x_{a2} + x_{b2} + x_{c2} + x_{d2} = 1 \\ x_{ik} &\geq 0 \text{ for all } i \text{ and } k. \end{aligned}$$

The solution to the linear program yields an objective function value of 219.24 with the decision variables being $x_{a1} = 0.363, x_{b1} = 0.229, x_{c2} = 0.332, x_{d1} = 0.076$ and all other variables zero. Thus, the optimal action function is $\mathbf{a} = (1,1,2,1)$; its average per period cost is $\varphi^* = 219.24$; and the steady-state probabilities for the resulting Markov chain are $\boldsymbol{\pi} = (0.363, 0.229, 0.332, 0.076)$.

12.5 The Optimal Stopping Problem

The final section of this chapter is a special case of the previous material; however, because of its special structure it is more efficient to solve the problem using its own formulation. The context for the optimal stopping problem is that we have a Markov chain $X_0, X_1, \cdots$, with state space E, Markov matrix $\mathbf{P}$, and *profit* function $\mathbf{f}$. Instead of a profit occurring at each visit, the profit will only be realized once, when the process is stopped. The process works as follows: (1) The Markov chain starts in an initial state. (2) The decision-maker considers the state of the process and decides whether to continue the process or to stop. (3) If the decision to continue is made, no cost or profit is made and a transition of the chain occurs according to the Markov matrix $\mathbf{P}$ and again the decision-maker decides whether to continue or to stop after the new state is observed. (4) If the decision is made to stop, a profit is made according to the profit function $\mathbf{f}$ and the process terminates.

In order to state the problem mathematically, we first define a type of stopping rule. Let S be a subset of the state space E, and define the random variable N_S by

$$N_S = \min\{n \geq 0 : X_n \in S\} . \tag{12.10}$$

The random variable N_S is thus the first time the Markov chain enters the set of states designated by S, and we shall call the set S a *stopping set*. Once a stopping set has been identified, its associated stopping rule random variable is defined (through Eq. 12.10), and the stopping problem using that particular stopping rule has a value given by

$$w_S(i) = E[f(X_{N_S})|X_0 = i] .$$

In other words, $w_S(i)$ is the expected profit made if the process starts in state i and it is continued until the first time that the process reaches a state in the set S. Thus the optimal stopping problem is to find the set $S^* \subseteq E$ such that $w_{S^*}(i) = w(i)$ where the vector $\mathbf{w}$ is defined, for $i \in E$, by

$$w(i) = \max_{S \subseteq E} w_S(i) . \tag{12.11}$$

Example 12.2. As an example, consider the Markov chain (Fig. 12.2) with state space $E = \{a,b,c,d,e\}$, Markov matrix

$$\mathbf{P} = \begin{bmatrix} 1.0 & 0.0 & 0.0 & 0.0 & 0.0 \\ 0.5 & 0.0 & 0.3 & 0.2 & 0.0 \\ 0.0 & 0.2 & 0.6 & 0.2 & 0.0 \\ 0.0 & 0.1 & 0.5 & 0.2 & 0.2 \\ 0.0 & 0.0 & 0.0 & 0.0 & 1.0 \end{bmatrix} ,$$

and *profit* function

$$\mathbf{f} = (1,2,5,8,10)^T .$$

First, observe the structure for this matrix and corresponding state diagram given in Fig. 12.2: States a and e are absorbing, and States b, c, and d are transient. If a stopping set does not contain states a and e, there is a positive probability the process would never stop, thus yielding a profit of zero; therefore, we would only consider stopping sets containing the absorbing states. However, the inclusion of states b, c, and d in the stopping set forms the crux of the optimal stopping problem.

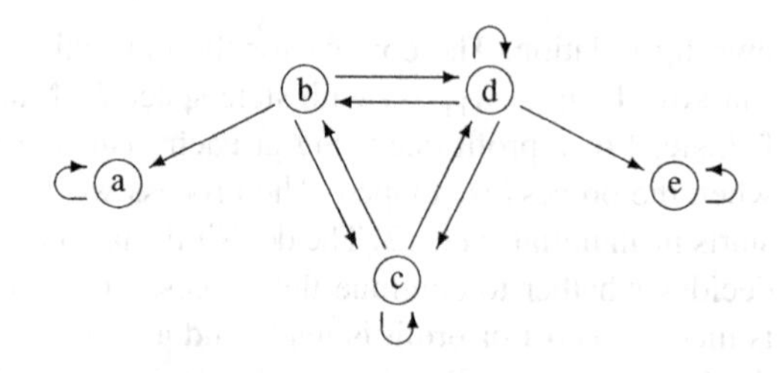

Fig. 12.2 State diagram for the Markov chain of Example 12.2

Consider the stopping set $S = \{a,c,d,e\}$. If the process starts in any of the states within S, it stops immediately; therefore $w(a) = 1, w(c) = 5, w(d) = 8$, and $w(e) = 10$. Before calculating $w(b)$, observe that at the next transition for the Markov chain, the stopping rule designated by S will force the chain to stop; therefore we simply need to take the expected value over the possible paths of this chain to determine $w(b)$; thus, $w(b) = 0.5w(a) + 0.3w(c) + 0.2w(d) = 3.6$. (Of course, in general it is harder to determine the values for $\mathbf{w}$, but this example should serve to illustrate the dynamics of the optimal stopping problem.) Note that although $w(b) = 3.6$, sometimes the process will actually yield a profit of 1, sometimes it will yield a profit of 5, and sometimes it will yield a profit of 8 under the assumption that $X_0 = b$. In general, $w_S(i)$ represents the expected profit if the Markov chain starts in state i and is stopped on the first visit to the set S. If the Markov chain starts in an irreducible set, the maximum profit that could be made is equal to the maximum profit in the irreducible set; thus, absorbing states always belong to optimal stopping sets. □

It is also easy to incorporate a discount factor into the optimal stopping problem by discounting the final profit back to a present value. Thus we can state the discounted problem similarly as finding the set $S^* \subseteq E$ such that $w^{\alpha}_{S^*}(i) = w^{\alpha}(i)$ where the vector $\mathbf{w}^{\alpha}$ is defined, for $i \in E$, by

$$w^{\alpha}(i) = \max_{S \subseteq E} E[\alpha^{N_S} f(X_{N_S}) | X_0 = i] \,, \tag{12.12}$$

for $0 < \alpha \leq 1$.

The basis for the linear programming formulation for the discounted cost Markov decision process was the lemma given in Property 12.6. A similar lemma allows the formulation of the optimal stopping problem as a linear program.

Property 12.13. ***Lemma for Linear Programming.*** *Let $\mathbf{w}^{\alpha}$ be the optimal value function as defined by Eq. (12.12) with $0 < \alpha \leq 1$, and let $\mathbf{u}$ be another real-valued function on the (finite) state space E. If $\mathbf{u}$ is such that $\mathbf{u} \geq \mathbf{f}$ and*

$$u(i) \geq \alpha \sum_{j \in E} P(i,j) u(j)$$

for all $i \in E$, then $\mathbf{u} \geq \mathbf{w}^{\alpha}$.

It can also be shown that the optimum function $\mathbf{w}^\alpha$ has the properties listed in the lemma; therefore, $\mathbf{w}^\alpha$ is the minimum such function. In order for $w^\alpha(i)$ to be optimal, the expected profit for state i (namely, $w^\alpha(i)$) must also be no smaller than the payoff, $f(i)$, if the process stops on state i. This results in the algorithm given in the following property.

Property 12.14. ***Linear Programming for Optimal Stopping.*** *The optimal solution to the following linear program gives the maximal value function, $\mathbf{w}^\alpha$, with $0 < \alpha \le 1$, for the problem defined by Eq. (12.12).*

$$\min \sum_{i \in E} u(i)$$

subject to:

$$u(i) \ge \alpha \sum_{j \in E} P(i,j) u(j) \quad \text{for each } i \in E$$
$$u(i) \ge f(i) \quad \text{for each } i \in E$$

The optimal stopping set is given by $S^* = \{i \in E : \mathbf{w}^\alpha(i) = f(i)\}$.

Taking advantage of this algorithm to solve the (undiscounted or $\alpha = 1$) example problem of this section yields the following (here we use u_i for $u(i)$ to aid in the readability of the equations):

$$\min z = u_a + u_b + u_c + u_d + u_e$$

subject to:

$$\begin{aligned}
u_b &\ge 0.5u_a + 0.3u_c + 0.2u_d \\
u_c &\ge +0.2u_b + 0.6u_c + 0.2u_d \\
u_d &\ge +0.1u_b + 0.5u_c + 0.2u_d + 0.2u_e \\
u_a &\ge 1,\ u_b \ge 2,\ u_c \ge 5,\ u_d \ge 8,\ u_e \ge 10\,,
\end{aligned}$$

where the trivial restrictions $u_a \ge u_a$ and $u_e \ge u_e$ have been excluded.

The solution to the above linear program yields $w(a) = 1.0, w(b) = 3.882, w(c) = 5.941, w(d) = 8.0, w(e) = 10.0$, which gives the optimal stopping set as $S^* = \{a, d, e\}$.

Problems

12.1. Using the maintenance problem given in Example 12.1 (p. 324), we wish to maximize profits. Profits are determined by revenue minus costs, where revenue is a function of the state but not of the decision and is given by the function $\mathbf{r} = (900, 400, 450, 750)^T$. The costs and transition probabilities are as presented in the example.

(a) Compute the profit function $\mathbf{g}_1$ associated with the stationary policy that uses

action 1 in every state.
(b) Using a discount factor $\alpha = 0.95$, verify that the vector

$$\mathbf{v}^\alpha = (8651.88, 8199.73, 8233.37, 8402.65)^T$$

is the optimal value that maximizes the total discounted profit. (You need to use Property 12.2 with a minor modification since this is a maximizing problem.)
(c) Using a discount factor of $\alpha = 0.7$, use the value iteration algorithm to find the optimal value function that maximizes the total discounted profit.
(d) Using the policy improvement algorithm, find the policy that maximizes the total discounted profit if the discount factor is such that \$1 today is worth \$1.12 after one time period.
(e) Setup the linear programming formulation that solves the problem in part (d).

12.2. Use the data from Problem 12.1 to answer the following questions:
(a) Find the average (undiscounted) profit per transition for the policy $\mathbf{a} = (2,1,2,1)$.
(b) Show that the policy $\mathbf{a} = (2,1,2,1)$ is not the policy that maximizes the profit per transition. (Use Property 12.8 after replacing the "min" operator with the "max" operator.)
(c) Using the policy improvement algorithm, find the policy that maximizes the average profit per transition.
(d) Find the policy that maximizes the average profit per transition using the linear programming formulation.

12.3. Joe recently graduated with a degree in operations research emphasizing stochastic processes. He wants to use his knowledge to advise people about presidential candidates. Joe has collected data on the past presidents according to their party (the two major parties[4] are the Labor Party and the Worker's Choice Party) and has determined that if the economy is good, fair, or bad, the selection of a president from a specific party will have varying effects on the economy after his/her term. The data collected shows that if the state of the economy is good and a candidate is elected as president from the Labor Party, 3.2 million new jobs are created and a total of 4.5 trillion dollars are spent during the presidential term, whereas a Worker's Choice candidate only creates 3.0 million jobs and spends 4 trillion dollars. However, a candidate from the Labor Party that receives the country in good economic conditions will leave the country after his/her term in good, fair, or bad economic condition with probability of 0.75, 0.20 and 0.05, respectively, whereas the Worker's Choice counterpart will leave the country in those conditions with probabilities 0.80, 0.15 and 0.05, respectively.

If the initial state of the economy is fair a Labor Party president creates 2 million jobs, spends \$3.5 trillion and the probabilities of leaving a good, fair, or bad economy are 0.3, 0.5, and 0.2, respectively. A Worker's Choice president creates

[4] We use fictitious party names to emphasize that this problem is for illustrative purposes only. We do not mean to imply that such decisions should be made on purely short-term economic issues removed from moral issues.

2.3 million jobs, spends \$4.5 trillion and the probabilities of leaving a good, fair, or bad economy are 0.2, 0.6, and 0.2, respectively.

If the initial state of the economy is bad a Labor president creates 1 million jobs, spends \$2.5 trillion and the probabilities of leaving a good, fair, or bad economy are 0.2, 0.2, and 0.6, respectively. A Worker's Choice president creates 1.2 million jobs, spends \$3 trillion and the probabilities of leaving a good, fair, or bad economy are 0.1, 0.3, and 0.6, respectively.

When the state of the economy is bad, independent candidates try to capitalize on the bad economy and run for president. If an independent candidate is elected president 1.5 million jobs are created, \$3.3 trillion are spent and the economy is left in good, fair and poor condition with probabilities 0.05, 0.40, and 0.55, respectively.
(a) Use Markov decision processes to determine the optimal voting strategy for presidential elections if the average number of new jobs per presidential term are to be maximized.
(b) Use Markov decision processes to determine how to vote if the objective is to minimize the total discounted spending if the discount factor is $\alpha = 0.9$ per year.
(c) Find the policy that minimizes the total discounted spending per job created if the discount factor is $\alpha = 0.9$ per year.

12.4. Consider a gambling machine that has five light bulbs labeled "a" through "e". When a person starts the game one of the five bulbs will light up. At this point, the player has two options: (1) the game can be continued or (2) the game can be stopped. If the game is continued, then another bulb will be lit according to the Markov matrix below. When the game is stopped, a payoff is received corresponding to the current bulb that is lit.

Let reward function (in pennies) and one-step transition probabilities of the Markov chain with state space $E = \{a,b,c,d,e\}$ be given by,

$$\mathbf{f} = (35, 5, 10, 20, 25)^T ,$$

$$\mathbf{P} = \begin{bmatrix} 0.3 & 0 & 0 & 0.7 & 0 \\ 0 & 1 & 0 & 0 & 0 \\ 0.2 & 0.5 & 0.3 & 0 & 0 \\ 0.2 & 0 & 0 & 0.8 & 0 \\ 0.1 & 0.3 & 0.2 & 0.1 & 0.3 \end{bmatrix} .$$

(a) Determine the optimal strategy.
(b) Formulate the problem as a Markov decision process specifying the state space E, action space A, the $P_k(i,j)$ for each $i,j \in E$ and $k \in A$, and the reward functions, f_k.
(c) Now assume that the game is started by inserting some nickels, and, as long as the game is continued, no additional coins are needed. After the coins are inserted, the first bulb is lit according to a uniform distribution. What is the minimum amount that should be charged per game if the machine is to make a profit in the long run. (For simplicity assume that operating and maintenance costs are negligible.)

12.5. Formulate the optimal stopping time problem as a Markov decision process using linear programming and show that the formulation reduces to the linear programming problem formulation of the optimal stopping time problem.

12.6. Consider the following optimal stopping time problem with payoff function **f**, transition probabilities **P**, state space $E = \{a,b,c,d,e\}$, and discount factor $\alpha = 0.5$, where

$$\mathbf{f} = (50, 40, 10, 20, 30)^T ,$$

$$\mathbf{P} = \begin{bmatrix} 0.3 & 0 & 0 & 0.7 & 0 \\ 0 & 1 & 0 & 0 & 0 \\ 0.5 & 0.2 & 0.3 & 0 & 0 \\ 0.6 & 0 & 0 & 0.4 & 0 \\ 0.2 & 0.1 & 0.1 & 0.3 & 0.3 \end{bmatrix} .$$

Besides the payoff, there is a continuation fee given by the vector

$$\mathbf{h} = (1, 1, 2, 2, 3)^T .$$

In other words, if the Markov chain is in state i at time n and the decision to continue is made, than a fee of $h(i)$ is paid, regardless of the time n.
(a) Use a Markov decision process to find the optimal strategy that a gambler should follow to maximize his profit. That is, find the set of states where his decision should be to stop.
(b) Show that the linear programming formulation of this problem can be reduced to an optimal stopping time problem.

12.7. The custodian of an inventory system for a store counts the number of items on hand and the end of each week and places an order for additional items. Orders can only be made Friday night and arrive early the following Monday morning. (The store is closed during the weekends.)

The ordering cost for j additional units is given by

$$c(j) = \begin{cases} 0 & \text{if } j = 0 \\ 100 + 500j & \text{if } j > 0 . \end{cases}$$

The demand is $0, 1, 2, 3, 4$ or 5 items per week, with probabilities 0.05, 0.15, 0.22, 0.22, 0.18 and 0.18, respectively. Items are sold for \$750 each. If a customer asks for an item after all the inventory has been sold, then the manager of the store will buy the item from a competitor for a cost of \$800 and sell it to the customer thus taking a loss but maintaining good will. The company's cost of inventory is affected by a discount factor of 0.90.
(a) Set up the problem as a Markov Decision Process. (Let the action space be $A = \{0, 1, 2, 3, 4, 5\}$, where $k \in A$ refers to an order up-to quantity of size k, so that if the inventory level is i and $i < k$, then the quantity ordered is $k - i$.)
(b) Find the optimal number of items that should be ordered at the end of the week to minimize the total discounted cost.
(c) Assume that there is an additional holding cost of \$10 per item remaining at the

end of the week. Find the optimal number of items that should be ordered at the end of the week to minimize the total discounted cost.
(d) Assume that instead of buying the item from a competitor, the store manager gives the customer a 10% reduction in the purchase price and places the item on backorder (i.e., promises to deliver it to the customer the following Monday). What is the optimum ordering policy now? (Assume the $10 holding cost from part (c) is still relevant.)
(e) How does the above answer change using an average cost criterion (i.e., no discount factor)?

12.8. Consider the following discrete time queueing system. One, two, or three customers arrive to the system at the beginning of each day with probabilities 0.25, 0.60, or 0.15, respectively. The system can hold at most five customers, so any customers in excess of five leave the system, never to return. (In other words, if at the start of the day, the system contains four customers and two customers arrive, then only one will join the system and the other one will disappear.) At the start of the day, after the arrival of the day's customers, a decision must be made as to whether one or two workers will be used for the day. If one worker is used, there is a 50% chance that only one customer will be server during the day and a 50% chance that two customers will be served during the day. If two workers are used, there is a 60% chance that two customers will be served, a 30% chance that three customers will be served, and a 10% chance that four customers will be served. (If there are only two customers present when two workers are used, then obviously there is a 100% chance that two customers will be served. A similar adjustment in the percentages needs to be made when three customers are present.) Each worker used costs $75 per day. At the end of the day, any customers that must wait until the next day will cost the system $30. Using Markov decision theory with the average cost criterion, determine how many workers should be used at the start of each day.

12.9. A machine produces items that are always inspected. The inspection is perfect in that the inspector always makes a correct determination as to whether the item is good or bad. The machine can be in one of two states: good or bad. If the machine is in the good state, it will produce a good item with a probability of 95%; if the machine is in the bad state, it will always produce bad items. If the machine is in the good state, there is a 2% probability that it will transition to the bad state for the next produced item. If the machine is in the bad state, it will stay in the bad state. The raw material and production costs are $500 per item and they are sold for $1,000 per item if good, and scrapped with a salvage value of $25 per item if bad. It costs $2,500 to overhaul the machine (i.e., instantaneously change it from the bad state to the good state). The state of the machine is not known directly, only through the inspection of the produced items. Each time a part is produced, that part is inspected and a decision is made to continue with the machine as is or to overhaul the machine.
(a) Consider the Markov chain $X = \{X_0, X_1, \cdots\}$ which represents the state of the machine. Since the state of X is not observable, we must work with the Markov chain $Z = \{Z_0, Z_1, \cdots\}$ where Z is the probability that the machine is bad after the

item is inspected; that is, $Z_n = P\{X_n = \text{"bad"}\}$, which has the state space given by the continuous interval [0,1]. The process Z depends on the inspection process $I = \{I_0, I_1, \cdots\}$ where $I_n = 0$ if the nth part was inspected and found to be bad and $I_n = 1$ if the nth part was inspected and found to be good. (Thus the dynamics is as follows: a decision is made based on Z_n and I_n, the X process makes a transition, an observation is made, and the cycle repeats.) If $Z_n = p$, what is the probability that $I_{n+1} = 0$?
(b) If $Z_n = p$, what is the probability that $I_{n+1} = 1$?
(c) If $Z_n = p$, what are the possible values of Z_{n+1}?
(d) The goal of this problem is to determine when to fix the machine. Formulate this as a Markov Decision Process with state space [0,1] and with a discount factor of 0.9. In other words, write out the relationship (Eq. 12.5) that the optimal function must satisfy.
(e) The optimal policy can be expressed as a control limit policy. That is, there exists a value p^* such that if $Z_n = p < p^*$, the decision is to leave the machine alone; if $p \geq p^*$, the decision is to fix the machine. Using this fact and your expression from part (d), write a computer code that finds the optimal policy.

Chapter 13
Advanced Queues[1]

The previous chapter introduced the concept of modeling processes using queueing theory and indicated the wide variety of applications possible through the use of queues. However, a major drawback to the application of the queueing theory discussed so far is its dependence on the Poisson and exponential assumptions regarding the probability laws for the arrival and service processes. One of the properties of the exponential distribution is that its mean equals its standard deviation, which indicates a considerable amount of variation; therefore, for many service processes, the exponential distribution is not a good approximation. However, the memorylessness property of the exponential distribution is the key to building mathematically tractable models. Without the memorylessness property, the Markovian property is lost and the ability to build tractable models is significantly reduced.

There are two separate (although related) problems for the analyst who is using queueing processes for modeling: (1) to develop a model that adequately approximates the physical system being studied and (2) to analyze the model and obtain the desired measures of effectiveness.

The first part of this chapter introduces methods for analyzing queueing systems more complicated than the simple systems discussed so far. Over the past twenty or twenty-five years, there has been an increasing emphasis on practical queueing models and the development of computationally tractable methods for these models; much of this emphasis has been due to extensive work by M. F. Neuts (we recommend [1] as a starting point). The main feature of these models is that matrix methods, which are easily written into computer routines, are used in the queueing analysis. This should be contrasted with the old way of queueing analysis which was through transform techniques which are not easily written into computer code.

After introducing these matrix geometric techniques, we shall present a class of distributions called phase-type distributions which will allow the modeling of nonexponential processes while at the same time maintaining a Markovian structure. Methods to compute measures of effectiveness for queueing models with phase-type distributions are also presented.

---

[1] This chapter would normally be skipped in a one-semester undergraduate course.

R.M. Feldman, C. Valdez-Flores, *Applied Probability and Stochastic Processes*, 2nd ed.,
DOI 10.1007/978-3-642-05158-6_13,

13.1 Difference Equations

Before presenting the matrix geometric technique for solving queueing models, we return once again to the simple M/M/1 queue. We shall present an alternative method for obtaining the steady-state probabilities, and then expand on that method to obtain a matrix geometric procedure. As you recall, the M/M/1 model yields the system of equations given by Eq. (7.1) that must be solved to obtain the steady-state probabilities. The solution method used in Chap. 7 to solve that system of equations was a successive substitution procedure. In this section, we study the characteristic equation method for linear difference equations. For further reading, almost any introductory textbook on numerical analysis will contain these procedures.

We first present the general theory for solving difference equations of order two and then apply this knowledge to our queueing problem. Suppose we wish to find expressions for $x_0, x_1, \cdots$ that satisfy the following system of difference equations:

$$\begin{aligned} a_0 x_0 + a_1 x_1 + a_2 x_2 &= 0 \\ a_0 x_1 + a_1 x_2 + a_2 x_3 &= 0 \\ &\vdots \\ a_0 x_{n-1} + a_1 x_n + a_2 x_{n+1} &= 0 \\ &\vdots \end{aligned}$$

where a_0, a_1, a_2 are constants. The first step in obtaining an expression for x_n is to form the *characteristic function* defined by

$$f(z) = a_0 + a_1 z + a_2 z^2 .$$

As long as $a_0 \neq 0$, $a_2 \neq 0$, and the roots of $f(\cdot)$ are real, the solutions to the system of difference equations are easy to represent. Suppose the characteristic function has two distinct roots[2], called z_1 and z_2. Then

$$x_n = c_1 z_1^n + c_2 z_2^n ,$$

is a solution to the system of difference equations, where c_1 and c_2 are constants to be determined from some other conditions (boundary conditions) that fully specify the system of equations. Suppose the characteristic function has a single root (of multiplicity 2), called z. Then

$$x_n = c_1 z^n + c_2 n z^{n-1} ,$$

is a solution to the system of difference equations.

One of the issues that is sometimes confusing is being able to determine the limits on the index when the general expression is written. If the recursive equation is

[2] The value z is a root if $f(z) = 0$.

$x_{n+1} = x_n + x_{n-1}$ for $n = n_0, n_0 + 1, \cdots$, then the general expression must be true for all x_n with n starting at $n_0 - 1$ because that is the first term that appears in the recursive equations.

Example 13.1. **Fibonacci Sequence.** We wish to find values of x_n that satisfy the following:

$$x_2 = x_0 + x_1$$
$$x_3 = x_1 + x_2$$
$$x_4 = x_2 + x_3$$
$$\vdots$$

The characteristic function is $f(z) = z^2 - z - 1$, and the quadratic equation gives the roots as $z = (1 \pm \sqrt{5})/2$; therefore a general solution for these difference equations is $x_n = c_1((1+\sqrt{5})/2)^n + c_2((1-\sqrt{5})/2)^n$. The Fibonacci numbers are those numbers that satisfy the recursion $x_{n+1} = x_n + x_{n-1}$ for $n = 1, 2, \cdots$ with $x_0 = 0$ and $x_1 = 1$. These boundary conditions are the information needed to obtain specific values for the constants; namely, c_1 and c_2 are found by rewriting the boundary conditions using the general form of x_n:

$$c_1 \left(\frac{1+\sqrt{5}}{2} \right)^0 + c_2 \left(\frac{1-\sqrt{5}}{2} \right)^0 = 0$$

$$c_1 \left(\frac{1+\sqrt{5}}{2} \right)^1 + c_2 \left(\frac{1-\sqrt{5}}{2} \right)^1 = 1 .$$

This yields

$$c_1 + c_2 = 0$$
$$c_1(1+\sqrt{5}) + c_2(1-\sqrt{5}) = 2 ;$$

thus, $c_1 = -c_2 = 1/\sqrt{5}$ and the general expression for the Fibonacci numbers is

$$x_n = \frac{1}{\sqrt{5}} \left[\left(\frac{1+\sqrt{5}}{2} \right)^n - \left(\frac{1-\sqrt{5}}{2} \right)^n \right] ,$$

for $n = 0, 1, \cdots$. The fact that these are integers for all values of n is surprising, but if you check out some of these values, the sequence $0, 1, 1, 2, 3, 5, 8, 13 \cdots$ should be obtained. □

Example 13.2. **The M/M/1 Queue.** The M/M/1 queue produces a difference equation (from Eq. 7.1) of the form

$$\lambda p_{n-1} - (\lambda + \mu) p_n + \mu p_{n+1} = 0$$

for $n = 1, 2, \cdots$; thus the characteristic function for the M/M/1 queue is $f(z) = \lambda - (\lambda + \mu)z + \mu z^2$. From the quadratic equation we easily get that the roots[3] of the characteristic function are $z_1 = \rho$ and $z_2 = 1$, where $\rho = \lambda/\mu$; therefore the general form of the solution is

$$p_n = c_1 \rho^n + c_2$$

for $n = 0, 1, \cdots$. To obtain specific values for c_1 and c_2 we next look at the norming equation; namely, $\sum_{n=0}^{\infty} p_n = 1$. The first observation from the norming equation is that c_2 must be set equal to zero; otherwise, the infinite sum would always diverge. Once $c_2 = 0$ is fixed, the steps for finding c_1 are identical to Eq. (7.5) where c_1 takes the place of p_0 yielding $c_1 = 1 - \rho$, as long as $\lambda < \mu$. □

13.2 Batch Arrivals

We begin the discussion of the matrix geometric approach by investigating a queueing system in which *batches* of customers arrive together; we designate such a system as an $M^{[X]}$/M/1 queueing system. For example, customers might arrive to the queueing system in cars, and the number of people in each car is a random variable denoted by X. Or, in a manufacturing setting, items might arrive to a processing center on a pallet, but then the items must be processed individually; this would also lead to a batch arrival queueing system. As usual, we let the mean arrival rate for the Poisson process (i.e., arrival rate of batches) be λ, then the rate at which the process makes a transition from state n to $n + i$ is

$$\lambda_i = \lambda \Pr\{X = i\} ,$$

where X denotes the size of a batch. Finally, we assume that the server treats customers individually with a mean (exponential) service rate of μ.

To illustrate the structure of this process, we use the $M^{[X]}$/M/1 queueing system assuming a maximum batch size of two as an example. Thus, $\lambda = \lambda_1 + \lambda_2$ and the generator matrix, $\mathbf{Q}$, is given by

$$\begin{array}{c|cc|cc|cc|c}
0 & 1 & 2 & 3 & 4 & 5 & 6 & \cdots \\
-\lambda & \lambda_1 & \lambda_2 & & & & & \\
\hline
\mu & -(\lambda+\mu) & \lambda_1 & \lambda_2 & 0 & & & \\
0 & \mu & -(\lambda+\mu) & \lambda_1 & \lambda_2 & & & \\
\hline
 & 0 & \mu & -(\lambda+\mu) & \lambda_1 & \lambda_2 & 0 & \\
 & 0 & 0 & \mu & -(\lambda+\mu) & \lambda_1 & \lambda_2 & \\
\hline
 & & & 0 & \mu & -(\lambda+\mu) & \lambda_1 & \cdots \\
 & & & 0 & 0 & \mu & -(\lambda+\mu) & \cdots \\
\hline
 & & & & & \ddots & \ddots & \ddots
\end{array}$$

[3] Note that $(\lambda + \mu)^2 - 4\lambda\mu = (\lambda - \mu)^2$.

where blanks within the matrix are interpreted as zeros and the row above the matrix indicates the states as they are associated with the individual columns of the matrix.

The $M^{[X]}/M/1$ generator matrix can also be represented using submatrices for the elements within the various partitions. The advantage of doing so is that the generator $\mathbf{Q}$ can then be easily written for a general system with a maximum batch size of m customers. This gives

$$\mathbf{Q} = \begin{bmatrix} -\lambda & \overline{\boldsymbol{\lambda}} & & & & \\ \overline{\boldsymbol{\mu}} & \mathbf{A} & \boldsymbol{\Lambda}_0 & & & \\ & \mathbf{M} & \mathbf{A} & \boldsymbol{\Lambda}_0 & & \\ & & \mathbf{M} & \mathbf{A} & \boldsymbol{\Lambda}_0 & \\ & & & \mathbf{M} & \mathbf{A} & \cdots \\ & & & & \vdots & \ddots \end{bmatrix}, \tag{13.1}$$

where $\overline{\boldsymbol{\lambda}}$ is the row vector $(\lambda_1, \lambda_2, \cdots)$; $\overline{\boldsymbol{\mu}}$ is the column vector $(\mu, 0, 0, \cdots)^T$; $\mathbf{M}$ is a matrix of zeros except for the top right element which is μ; $\mathbf{A}$ is a matrix with $-(\lambda+\mu)$ on the diagonal, μ on the subdiagonal, and λ_k on the kth superdiagonal; and $\boldsymbol{\Lambda}_0$ is a matrix made up of zeroes and the λ_k terms. (If the largest possible batch has m customers, then the kth column of $\boldsymbol{\Lambda}_0$ has $k-1$ zeros and then $m-k+1$ values of λ_i starting with λ_m and ending with λ_k.)

Our goal is to find the values for the steady-state probabilities denoted by $\mathbf{p} = (p_0|p_1, p_2|p_3, p_4|\cdots) = (p_0, \mathbf{p}_1, \mathbf{p}_2, \cdots)$, where $\mathbf{p}_n = (p_{2n-1}, p_{2n})$ for $n = 1, 2, \cdots$. In other words, the vector $\mathbf{p}$ is partitioned in the same fashion as $\mathbf{Q}$. Combining Eq. (13.1) with the equation $\mathbf{pQ} = \mathbf{0}$ yields the following system of equations written using the submatrices from our partitioning:

$$\begin{aligned} -p_0\lambda + \mathbf{p}_1\overline{\boldsymbol{\mu}} &= 0 \\ p_0\overline{\boldsymbol{\lambda}} + \mathbf{p}_1\mathbf{A} + \mathbf{p}_2\mathbf{M} &= \mathbf{0} \\ \mathbf{p}_1\boldsymbol{\Lambda}_0 + \mathbf{p}_2\mathbf{A} + \mathbf{p}_3\mathbf{M} &= \mathbf{0} \\ \mathbf{p}_2\boldsymbol{\Lambda}_0 + \mathbf{p}_3\mathbf{A} + \mathbf{p}_4\mathbf{M} &= \mathbf{0} \\ &\vdots \end{aligned} \tag{13.2}$$

The above system of equations can be solved using the matrix geometric technique, which is simply the matrix analogue to the solution procedures of Sect. 13.1. The recursive section from the equations listed in (13.2) reduces to the following problem: find the vectors $\mathbf{p}_1, \mathbf{p}_2, \cdots$ such that

$$\mathbf{p}_{n-1}\boldsymbol{\Lambda}_0 + \mathbf{p}_n\mathbf{A} + \mathbf{p}_{n+1}\mathbf{M} = \mathbf{0}$$

for $n = 2, 3, \cdots$. The first step in the solution procedure is to form the *matrix* characteristic function and find its root. That is, we must find the matrix $\mathbf{R}$ such that

$$\boldsymbol{\Lambda}_0 + \mathbf{RA} + \mathbf{R}^2\mathbf{M} = \mathbf{0}\,. \tag{13.3}$$

Once the characteristic root matrix $\mathbf{R}$ is found[4], it follows that

$$\mathbf{p}_n = \mathbf{c}\mathbf{R}^n , \tag{13.4}$$

for $n = 1, 2, \cdots$, where $\mathbf{c}$ is a vector of constants which must be determined from the boundary conditions, i.e., the first two equations from Eq. (13.2).

13.2.1 Quasi-Birth-Death Processes

The characteristic function of Eq. (13.3) represents the main work of the matrix geometric method so we pause momentarily from our discussion of the batch arrival queue and discuss the general solution method for this type of matrix geometric problem. You should notice that the generator matrix in Eq. (13.1) is a tridiagonal block matrix. That is, the generator $\mathbf{Q}$ has submatrices along the diagonal, the subdiagonal, the superdiagonal, and zeros everywhere else. Thus the general form for the matrix is

$$\mathbf{Q} = \begin{bmatrix} \mathbf{B}_{00} & \mathbf{B}_{01} & & & & \\ \mathbf{B}_{10} & \mathbf{B}_{11} & \mathbf{A}_0 & & & \\ & \mathbf{A}_2 & \mathbf{A}_1 & \mathbf{A}_0 & & \\ & & \mathbf{A}_2 & \mathbf{A}_1 & \mathbf{A}_0 & \\ & & & \mathbf{A}_2 & \mathbf{A}_1 & \cdots \\ & & & & \vdots & \ddots \end{bmatrix} . \tag{13.5}$$

Any queueing system that gives rise to a generator of this form is called a *quasi-birth-death process*. In finding the steady-state probabilities for a quasi-birth-death process, the characteristic function given by

$$\mathbf{A}_0 + \mathbf{R}\mathbf{A}_1 + \mathbf{R}^2\mathbf{A}_2 = \mathbf{0} \tag{13.6}$$

will arise, where $\mathbf{R}$ is the unknown matrix and must be obtained. In some systems, the matrices $\mathbf{A}_0$, $\mathbf{A}_1$, and $\mathbf{A}_2$ have simple structures so that $\mathbf{R}$ can be obtained analytically; however, in many systems, an expression for $\mathbf{R}$ is impossible. However, what makes the matrix geometric approach attractive is that numerical methods to compute $\mathbf{R}$ are readily available. To present one approach, let us rearrange the terms in Eq. (13.6) to obtain

$$\mathbf{R} = -\left(\mathbf{A}_0 + \mathbf{R}^2\mathbf{A}_2\right)\mathbf{A}_1^{-1} . \tag{13.7}$$

Equation (13.7) leads to a successive substitution scheme where a guess is made for $\mathbf{R}$, that guess is used in the right-hand-side of the equation to produce a new estimate for $\mathbf{R}$; this new estimate is used again in the right-hand-side to obtain another estimate of $\mathbf{R}$; and so on. Specifically, we have the following algorithm.

---

[4] There are actually two matrices satisfying Eq. (13.3), but Neuts [1] showed that exactly one matrix exists that will satisfy the boundary (norming) conditions.

Property 13.1. Successive Substitution Algorithm. *Let $\mathbf{A}_0$, $\mathbf{A}_1$, and $\mathbf{A}_2$ be submatrices from the generator $\mathbf{Q}$ as defined by Eq. (13.5), and let $\mathbf{A}_s = \mathbf{A}_0 + \mathbf{A}_1 + \mathbf{A}_2$. Define the vector $\mathbf{v}$ such that $\mathbf{v}\mathbf{A}_s = \mathbf{0}$ and $\mathbf{v}\mathbf{1} = 1$, and define the traffic intensity as $\rho = \mathbf{v}\mathbf{A}_0\mathbf{1}/\mathbf{v}\mathbf{A}_2\mathbf{1}$. If $\rho < 1$, the following iteration procedure will yield an approximation to the matrix $\mathbf{R}$ that satisfies Eq. (13.6).*

Step 1. *Check that the steady-state conditions hold, i.e., check that $\rho < 1$.*
Step 2. *Let $\mathbf{R}_0 = \mathbf{0}$ (i.e., a matrix of all zeros), let $n = 0$, and fix ε to be a small positive real value.*
Step 3. *Define the matrix $\mathbf{R}_{n+1}$ by*

$$\mathbf{R}_{n+1} = -\left(\mathbf{A}_0 + \mathbf{R}_n^2\mathbf{A}_2\right)\mathbf{A}_1^{-1}.$$

Step 4. *Define δ by*

$$\delta = \max_{i,j}\{|R_{n+1}(i,j) - R_n(i,j)|\}.$$

Step 5. *If $\delta < \varepsilon$, let $\mathbf{R} = \mathbf{R}_{n+1}$ and stop; otherwise, increment n by one and return to Step 3.*

The Successive Substitution Algorithm is often slow to converge, especially if the traffic intensity, ρ, is close to one; however, the algorithm is very easy to build into a computer code so that it becomes a convenient method for obtaining the matrix $\mathbf{R}$. Once the matrix $\mathbf{R}$ is determined, Eq. (13.4) gives an expression for the probability vectors $\mathbf{p}_n$ for $n \geq 1$. To obtain the values for the constant $\mathbf{c}$ and the probability vector $\mathbf{p}_0$, the norming equation and the boundary conditions must be used.

The boundary conditions are taken from the first columns of submatrices from the generator in Eq. (13.5). These equations are

$$\begin{aligned}\mathbf{p}_0\mathbf{B}_{00} + \mathbf{p}_1\mathbf{B}_{10} &= \mathbf{0} \\ \mathbf{p}_0\mathbf{B}_{01} + \mathbf{p}_1\mathbf{B}_{11} + \mathbf{p}_2\mathbf{A}_2 &= \mathbf{0}.\end{aligned} \tag{13.8}$$

The norming equation is

$$\sum_{n=0}^{\infty}\sum_{k} p_{nk} = \sum_{n=0}^{\infty}\mathbf{p}_n\mathbf{1} = 1.$$

A closed form expression for this equation is obtained by taking advantage of the geometric progression result for matrices[5] together with the matrix geometric form for the probabilities (Eq. 13.4). Thus, we have

[5] $\sum_{n=0}^{\infty}\mathbf{R}^n = (\mathbf{I}-\mathbf{R})^{-1}$ if the absolute value of all eigenvalues for $\mathbf{R}$ is less than one.

$$\begin{aligned}1 &= \mathbf{p}_0\mathbf{1} + \sum_{n=1}^{\infty} \mathbf{p}_n\mathbf{1} \\ &= \mathbf{p}_0\mathbf{1} + \sum_{n=1}^{\infty} \mathbf{c}\mathbf{R}^n\mathbf{1} \\ &= \mathbf{p}_0\mathbf{1} + \mathbf{c}\mathbf{R}\left(\sum_{n=1}^{\infty} \mathbf{R}^{n-1}\right)\mathbf{1} \\ &= \mathbf{p}_0\mathbf{1} + \mathbf{p}_1(\mathbf{I}-\mathbf{R})^{-1}\mathbf{1}\,.\end{aligned} \tag{13.9}$$

In conclusion, the long-run probabilities for quasi-birth-death processes are found by first finding the matrix $\mathbf{R}$, possibly using the Successive Substitution Algorithm (Property 13.1), and then finding the values for $\mathbf{p}_0$ and $\mathbf{p}_1$ by solving the system of equations

$$\begin{aligned}\mathbf{p}_0\mathbf{B}_{00} + \mathbf{p}_1\mathbf{B}_{10} &= \mathbf{0} \\ \mathbf{p}_0\mathbf{B}_{01} + \mathbf{p}_1\mathbf{B}_{11} + \mathbf{p}_1\mathbf{R}\mathbf{A}_2 &= \mathbf{0} \\ \mathbf{p}_0\mathbf{1} + \mathbf{p}_1(\mathbf{I}-\mathbf{R})^{-1}\mathbf{1} &= 1\,.\end{aligned} \tag{13.10}$$

Notice that the system of equations (13.10) is simply a combination of the Eqs. (13.8) and (13.9) together with the substitution[6] $\mathbf{p}_2 = \mathbf{p}_1\mathbf{R}$. As is common with all irreducible recurrent systems, one of the equations in the system (13.10) will be redundant.

13.2.2 Batch Arrivals (continued)

Before proceeding with the batch arrival queueing system, we should check the steady-state conditions; namely, check that the traffic intensity is less than one (Step 1 from Property 13.1). For our example with a maximum batch size of two, the matrix $\mathbf{A}_s$ is

$$\mathbf{A}_s = \begin{bmatrix} -(\lambda_1+\mu) & \lambda_1+\mu \\ \lambda_1+\mu & -(\lambda_1+\mu) \end{bmatrix}$$

and thus the distribution from $\mathbf{A}_s$ is $\nu_1 = 0.5$ and $\nu_2 = 0.5$. Therefore, the traffic intensity is given by

$$\rho = \frac{\nu\Lambda_0\mathbf{1}}{\nu\mathbf{M}\mathbf{1}} = \frac{0.5\lambda_2 + 0.5\lambda}{0.5\mu} = \frac{\lambda_1 + 2\lambda_2}{\mu} = \frac{\lambda E[X]}{\mu}\,,$$

where $E[X]$ is the mean batch size.

For purposes of a numerical example, let $\mu = 5$ per hour, $\lambda_1 = 2$ per hour and $\lambda_2 = 1$ per hour. This yields $\rho = 0.8$ so that it makes sense to find the steady-state probabilities. The matrix $\mathbf{R}$ is the solution to the equation

---

[6] If the substitution is not obvious, observe that Eq. (13.4) implies that $\mathbf{p}_{n+1} = \mathbf{p}_n\mathbf{R}$.

$$\mathbf{R}^2 \begin{bmatrix} 0 & 5 \\ 0 & 0 \end{bmatrix} + \mathbf{R} \begin{bmatrix} -8 & 2 \\ 5 & -8 \end{bmatrix} + \begin{bmatrix} 1 & 0 \\ 2 & 1 \end{bmatrix} = \mathbf{0},$$

and the Successive Substitution Algorithm yields the following sequence

$$\mathbf{R}_0 = \begin{bmatrix} 0 & 0 \\ 0 & 0 \end{bmatrix}$$

$$\mathbf{R}_1 = \begin{bmatrix} 0.148 & 0.037 \\ 0.389 & 0.222 \end{bmatrix}$$

$$\mathbf{R}_2 = \begin{bmatrix} 0.165 & 0.064 \\ 0.456 & 0.329 \end{bmatrix}$$

$$\vdots$$

$$\mathbf{R} = \begin{bmatrix} 0.2 & 0.12 \\ 0.6 & 0.56 \end{bmatrix}.$$

With the determination of $\mathbf{R}$, there are three unknowns that need to be obtained in order to fully define the steady-state probabilities; specifically, p_0 and the constant vector $\mathbf{c}$ from Eq. (13.4). Alternatively, we could say there are three unknowns, p_0, p_1, and p_2, and these must be determined through the boundary conditions given by the system of equations (13.2) and the norming equation (13.9). (Remember that $\mathbf{p}_1 = (p_1, p_2)$ is a vector. Also, the vector $\mathbf{p}_0$ given in the general quasi-birth-death system equates to the scalar p_0 in the batch arrival system.)

The norming equation can be combined with the second equation from (13.2) to obtain a system defining the constant vector $\mathbf{c}$; thus,

$$1 - \mathbf{p}_1(\mathbf{I} - \mathbf{R})^{-1}\mathbf{1} = p_0$$

$$p_0\boldsymbol{\lambda} + \mathbf{p}_1\mathbf{A} + \mathbf{p}_1\mathbf{R}\mathbf{M} = \mathbf{0}.$$

The above system is sufficient to determine the unknown probabilities; however, it can be simplified somewhat using a result based on Little's formula. All single-server systems that process customers one-at-a-time have the same form for the probability that the system is empty; p_0 is always $1 - \lambda_e/\mu$ where λ_e is the mean arrival rate of customers *into* the system and μ is the mean service rate. Therefore,

$$p_0 = 1 - \frac{\lambda E[X]}{\mu}$$

$$\mathbf{p}_1 = -p_0\overline{\boldsymbol{\lambda}}(\mathbf{A} + \mathbf{R}\mathbf{M})^{-1},$$

where X is the random variable indicating the size of each batch.

Returning again to the numerical example for the batch arrival process, we have

$$\mathbf{A} + \mathbf{R}\mathbf{M} = \begin{bmatrix} -8.0 & 3.0 \\ 5.0 & -5.0 \end{bmatrix}$$

yielding $\mathbf{p}_1 = (0.12, 0.112)$.

Before going to another example of the use of the matrix geometric approach, we present some background in phase-type distributions. These distributions will allow for a significant amount of generality over the exponential distribution while taking advantage of the computational tractability afforded by the matrix geometric approach.

13.3 Phase-Type Distributions

The Erlang distribution defined in Chap. 1 (page 18) is an example of a phase-type distribution, because it is the sum of independent exponential distributions. That is, an Erlang random variable could be thought of as the length of time required to go through a sequence of phases or steps where the time to go through each phase requires an exponentially distributed length of time. For example, suppose we wish to model a single server queueing system with a Poisson arrival process, but all we know about its service distribution is that it has a mean service time of 10 minutes and a standard deviation of 5 minutes. The temptation might be to use an exponential distribution with parameter $\mu = 0.1$ for service times. Such an assumption would clearly be wrong because the standard deviation would be twice too big even though the mean would be correct. A much better modeling distribution would be the Erlang distribution (see Eq. 1.16) with parameters $k = 4$ and $\beta = 10$; such a distribution yields a mean of 10 and a standard deviation of 5. This Erlang distribution is the sum of four exponentials, each exponential having a mean of 2.5; thus, a schematic diagram for the Erlang is contained in the figure below.

In other words, instead of modeling the Erlang server as a single system, we model it as a four-phase process (see Fig. 13.1). To describe the dynamics of this phase-type model, assume there is a queue outside the server, and the server has just completed a service. Immediately, the customer at the head of the queue enters the phase 1 box. After an exponentially distributed length of time, the customer leaves phase 1 and enters phase 2 and remains in that box for another exponentially distributed length of time. This is repeated until the customer passes through all four boxes, and as long as the customer is in any of the four boxes, no additional customer may enter the server. As soon as the customer leaves phase 4, the next customer in the queue enters phase 1. Thus, we are able to model the server using the exponential distribution even though the server itself does not have an exponential but has an Erlang distribution with a standard deviation half of its mean. In modeling such a system, we emphasize that the phases are *fictitious*. Whether or not customers actually enter individual physical phases is not relevant; the important question is whether or not the resultant distribution of the server adequately mimics the real process.

The exponential nature imbedded within the Erlang distribution allows for a Markov process to be used for its distribution. To demonstrate, let $\{Y_t; t \geq 0\}$ be a Markov process with state space $E = \{1, 2, 3, 4, \Delta\}$, generator matrix

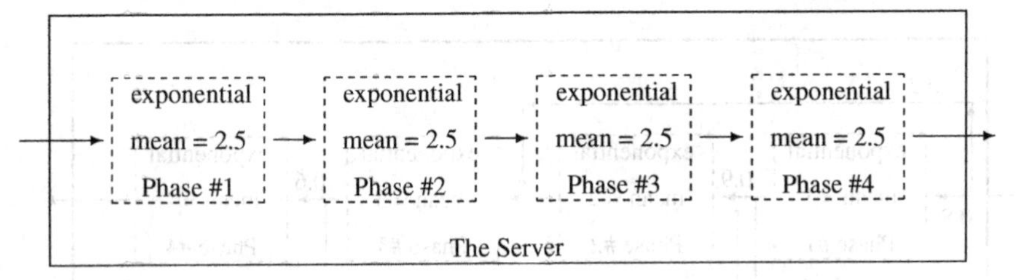

Fig. 13.1 Schematic for the Erlang distribution

$$\mathbf{G} = \begin{bmatrix} -0.4 & 0.4 & 0.0 & 0.0 & 0.0 \\ 0.0 & -0.4 & 0.4 & 0.0 & 0.0 \\ 0.0 & 0.0 & -0.4 & 0.4 & 0.0 \\ 0.0 & 0.0 & 0.0 & -0.4 & 0.4 \\ 0.0 & 0.0 & 0.0 & 0.0 & 0.0 \end{bmatrix}$$

and initial probability vector $\boldsymbol{\alpha} = (1,0,0,0,0)$. In other words, Y always starts in state 1 and sequentially passes through all the states until being absorbed in the final state, Δ. Mathematically, this can be expressed by defining the random variable T as the first passage time to state Δ, that is,

$$T = \min\{t \geq 0 : Y_t = \Delta\}. \tag{13.11}$$

The distribution of T is the sum of four independent exponential distributions, each with mean 2.5; therefore, T has a four-phase Erlang distribution with mean 10. This example is an illustration of the property that any Erlang distribution can be represented as the first passage time for a Markov process.

A k-phase Erlang has the property that the mean divided by the standard deviation equals the square root of k. If we have a non-exponential server, one approach for finding an appropriate distribution for it is to take the "best" Erlang distribution based on the mean and standard deviation; in other words, if we square the ratio of the mean over the standard deviation and obtain something close to an integer, that would give the appropriate number of phases for the Erlang. For example, if the mean were 10 and standard deviation were 5.77, we would use a 3-phase Erlang model (because $(10/5.77)^2 = 3.004$). Or, as another example, consider a server that has a mean service time of 10 and standard deviation of 5.48 (observe that $(10/5.48)^2 = 3.33$). We now have a dilemma since the number of phases for the Erlang must be an integer. However, if you look again at Fig. 13.1, you should realize that the assumptions for the Erlang are unnecessarily (from a modeling point of view) restrictive. If the purpose of using the Erlang is to reduce everything to exponentials, we must ask the question, "Why do the means of each phase need to be the same?" The answer is, "They do not!" If in Fig. 13.1, the first phase was exponential with mean 1, the second phase was exponential with mean 2, the third phase was exponential with mean 3, and the fourth phase was exponential with mean 4, then the overall mean would be 10 and the standard deviation would be 5.48 (review Property 1.6 for verifying the standard deviation).

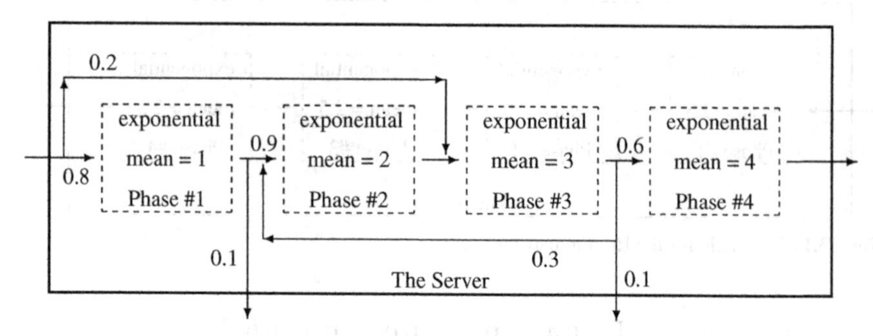

Fig. 13.2 Schematic for a general phase-type distribution

The idea of phase-type distributions is to generalize the Erlang concept to allow non-equal means for the phases and to allow more general paths than simply a "straight-through" path. For example, consider the phase-type distribution illustrated in Fig. 13.2. The distribution of the service time illustrated in that figure can again be represented by a Markov process. Let $\{Y_t; t \geq 0\}$ be a Markov process with state space $E = \{1, 2, 3, 4, \Delta\}$, generator matrix

$$\mathbf{G} = \begin{bmatrix} -1.0 & 0.9 & 0.0 & 0.0 & 0.1 \\ 0.0 & -0.5 & 0.5 & 0.0 & 0.0 \\ 0.0 & 0.1 & -0.3333 & 0.2 & 0.0333 \\ 0.0 & 0.0 & 0.0 & -0.25 & 0.25 \\ 0.0 & 0.0 & 0.0 & 0.0 & 0.0 \end{bmatrix}, \tag{13.12}$$

and initial probability vector $\boldsymbol{\alpha} = (0.8, 0, 0.2, 0, 0)$. Consider a customer, who upon entering the server, begins in phase 1 with probability 0.8 and begins in phase 3 with probability 0.2. After entering a phase, the customer spends an exponential length of time in that phase and then moves according to a Markov chain as illustrated in Fig. 13.2. In other words, the length of time that the customer spends in the server as illustrated in the figure is equivalent to the first passage time to state Δ for the Markov process defined by the generator of Eq. (13.12). This furnishes the motivation for phase-type distributions: namely, any distribution that can be represented by a first passage time of a Markov process will be called a phase-type distribution.

The generator matrix of Eq. (13.12) can be partitioned into four key submatrices as

$$\mathbf{G} = \left[\begin{array}{cccc|c} -1.0 & 0.9 & 0.0 & 0.0 & 0.1 \\ 0.0 & -0.5 & 0.5 & 0.0 & 0.0 \\ 0.0 & 0.1 & -0.3333 & 0.2 & 0.0333 \\ 0.0 & 0.0 & 0.0 & -0.25 & 0.25 \\ \hline 0.0 & 0.0 & 0.0 & 0.0 & 0.0 \end{array}\right] = \left[\begin{array}{c|c} \mathbf{G}_* & \mathbf{G}_\Delta \\ \hline \mathbf{0} & 0 \end{array}\right],$$

and the initial probability vector can be partitioned as

$$\boldsymbol{\alpha} = (0.8, 0.0, 0.2, 0.0 \mid 0.0) = (\boldsymbol{\alpha}_* \mid 0.0) .$$

Such partitioning is always possible for Markov processes that are to be used in defining phase-type distributions. The matrix $\mathbf{G}_*$ and the vector $\boldsymbol{\alpha}_*$ are sufficient to define a phase-type distribution as is specified in the following:

Definition 13.1. A random variable has a *phase-type distribution* with parameters m, $\mathbf{G}_*$, and $\boldsymbol{\alpha}_*$ if it can be expressed as a first-passage time random variable (Eq. 13.11) for a Markov process with state space $E = \{1, \cdots, m, \Delta\}$, generator

$$\mathbf{G} = \left[\begin{array}{c|c} \mathbf{G}_* & \mathbf{G}_\Delta \\ \hline \mathbf{0} & 0 \end{array} \right] ,$$

and initial probability vector $\boldsymbol{\alpha} = (\boldsymbol{\alpha}_* | 0)$. □

First-passage time random variables have well known properties that can be easily described mathematically, but not necessarily easy to compute. These properties are:

Property 13.2. *Let T be a random variable with a phase-type distribution from Definition 13.1. Its cumulative distribution function and first two moments are given by*

$$F(t) = 1 - \boldsymbol{\alpha}_* \exp\{t\mathbf{G}_*\}\mathbf{1} \text{ for } t \ge 0,$$
$$E[T] = -\boldsymbol{\alpha}_* \mathbf{G}_*^{-1} \mathbf{1} ,$$
$$E[T^2] = 2\boldsymbol{\alpha}_* \mathbf{G}_*^{-2} \mathbf{1} ,$$

where ${\mathbf{G}_*}^{-2}$ indicates the square of the inverse and **1** *is a vector of ones.*

Property 13.2 gives an easy-to-calculate formula for determining the mean and variance; however, the distribution is somewhat harder. You should be familiar with a scalar times a matrix; namely, the matrix $t\mathbf{G}_*$ is the matrix $\mathbf{G}_*$ with all components multiplied by t. However, a matrix in the exponent may be unfamiliar. Specifically, the matrix $e^{\mathbf{A}}$ is defined by the power series[7]

$$e^{\mathbf{A}} = \sum_{n=0}^{\infty} \frac{\mathbf{A}^n}{n!} .$$

Most linear algebra textbooks will discuss matrices of the form $e^{\mathbf{A}}$; therefore, we will not discuss its computational difficulties here. For the purposes of this text, we will only use the moment formulas from Property 13.2.

[7] Remember, for scalars, the exponential power series is $e^a = \sum_{n=0}^{\infty} a^n/n!$.

One caveat worth mentioning is that our example has focused only on the mean and standard deviation as if those two properties adequately described the randomness inherent in the server. This, of course, is an oversimplification, although at times it may be all the analyst has to work with. It is obviously best to try to fit the entire distribution to empirical data; however, goodness-of-fit tests require considerable data which are not always available. When data are not available to give a good estimate for the full distribution function, then moments are often used. The first moment gives the central tendency, and the second moment is used to measure variability. The third moment is useful as a measure of skewness (see p. 22). For phase-type distributions, the nth moment is given as

$$E[T^n] = (-1)^n n! \boldsymbol{\alpha}_* \mathbf{G}_*^{-n} \mathbf{1}\,.$$

As a final comment regarding the choice of distributions for modeling purposes, we note that it is often the tails of distributions that are the key to behavioral characteristics of queueing systems. Therefore, fitting the tail percentiles is an additional task that is worthwhile for the analyst in obtaining a proper distribution function for modeling purposes.

13.4 Systems with Phase-Type Service

The service distribution often describes a process designed and controlled by people who view variability as being bad; therefore, the exponential assumption for service times is too limiting because of its large variability. In this section we discuss generalizing the service time distribution. With single-server systems, using phase-type distributions presents little difficulty. With multiple-server systems, tractability becomes an important issue and places limitations on the usability of the matrix geometric methods.

13.4.1 The M/Ph/1 Queueing System

We begin our discussion by looking at a specific example of an M/Ph/1 system. Our example will use Poisson arrivals and a single server with a phase-type service distribution having two phases. The state space will have to be two-dimensional because not only do we need to know the number of customers in the system, but also the phase number of the customer being served; therefore, let

$$E = \{0, (1,1), (1,2), (2,1), (2,2), (3,1), (3,2), (4,1), (4,2), \cdots\}\,.$$

If the queueing system is in state (n,k), there are n customers in the system with $n-1$ customers being in the queue, and the customer that is being served is in the kth phase. The mean arrival rate is denoted by λ, and the phase-type distribution has

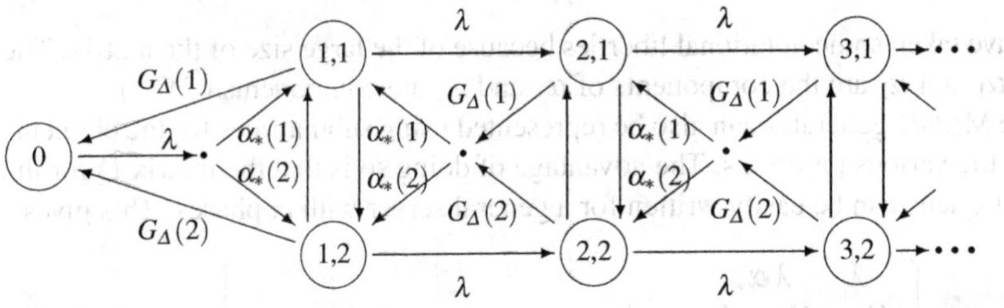

Fig. 13.3 State diagram for an M/Ph/1 system; rates associated with vertical arrows are $G_*(1,2)$ going down and $G_*(2,1)$ going up

parameters $m = 2$, $\boldsymbol{\alpha}_*$ and $\mathbf{G}_*$. Thus, for $k, j = 1,2$, we have the following interpretations: $\alpha_*(k)$ is the probability a customer who enters the server will begin in phase k, $G_*(k,j)$ is the rate at which the customer in service moves from phase k to phase j, and $G_\Delta(k)$ is the rate at which a customer who is in phase k leaves the server. The state diagram for this system is given in Fig. 13.3.

Thus, for an M/Ph/1 queueing system, an empty system will make a transition to a system containing one customer with a mean rate of λ; however, there are two states representing a system with one customer. When a customer enters the server, there is a probability of $\alpha_*(1)$ that the customer will begin in phase 1 and a probability of $\alpha_*(2)$ that the customer will begin in phase 2. Therefore, the rate of a transition from state 0 to state (1,1) is $\lambda\alpha_*(1)$, and the rate of a transition from state 0 to state (1,2) is $\lambda\alpha_*(2)$. If the process is in state (n,k), an arriving customer joins the queue so that the rate of transition from (n,k) to $(n+1,k)$ is λ. If there are two or more customers in the system, then a service completion occurs with rate $G_\Delta(k)$ if the customer who is in service is occupying phase k. Immediately after departure, the customer at the head of the queue will enter service and join phase k with probability $\alpha_*(k)$; therefore, the rate of transition from state (n,k) to $(n-1,j)$ is $G_\Delta(k)\alpha_*(j)$. With this description (and with the help of Fig. 13.3), the generator matrix for the M/Ph/1 queueing model can be formed; we shall call it $\mathbf{Q}$ to keep it distinct from the generator $\mathbf{G}$ which is associated with the description of the Markov process representing the server. Thus, $\mathbf{Q}$ is defined by

$$
\begin{array}{c} \\ \mathbf{Q} = \end{array}
\begin{array}{c|cc|cc|cc|c}
0 & (1,1) & (1,2) & (2,1) & (2,2) & (3,1) & (3,2) & \cdots \\
\hline
-\lambda & \lambda\alpha_1 & \lambda\alpha_2 & & & & & \\
\hline
G_\Delta(1) & g_{11}-\lambda & g_{12} & \lambda & 0 & & & \\
G_\Delta(2) & g_{21} & g_{22}-\lambda & 0 & \lambda & & & \\
\hline
 & G_\Delta(1)\alpha_1 & G_\Delta(1)\alpha_2 & g_{11}-\lambda & g_{12} & \lambda & 0 & \\
 & G_\Delta(2)\alpha_1 & G_\Delta(2)\alpha_2 & g_{21} & g_{22}-\lambda & 0 & \lambda & \\
\hline
 & & & G_\Delta(1)\alpha_1 & G_\Delta(1)\alpha_2 & g_{11}-\lambda & g_{12} & \cdots \\
 & & & G_\Delta(2)\alpha_1 & G_\Delta(2)\alpha_2 & g_{21} & g_{22}-\lambda & \cdots \\
\hline
 & & & & & \ddots & \ddots & \ddots
\end{array}
$$

where blanks within the matrix are interpreted as zeros and the row above the matrix indicates the states as they are associated with the individual columns of the matrix.

(We have taken some notational liberties because of the large size of the matrix. The terms α_1 and α_2 are the components of $\boldsymbol{\alpha}_*$ and g_{ij} are components of $\mathbf{G}_*$.)

The M/Ph/1 generator can also be represented using submatrices for the elements within the various partitions. The advantage of doing so is that the matrix $\mathbf{Q}$ for the M/Ph/1 queue can be easily written for a general server with m phases. This gives

$$\mathbf{Q} = \begin{bmatrix} -\lambda & \lambda\boldsymbol{\alpha}_* & & & & \\ \mathbf{G}_\Delta & \mathbf{G}_* - \boldsymbol{\Lambda} & \boldsymbol{\Lambda} & & & \\ & \mathbf{G}_\Delta\boldsymbol{\alpha}_* & \mathbf{G}_* - \boldsymbol{\Lambda} & \boldsymbol{\Lambda} & & \\ & & \mathbf{G}_\Delta\boldsymbol{\alpha}_* & \mathbf{G}_* - \boldsymbol{\Lambda} & \boldsymbol{\Lambda} & \\ & & & \mathbf{G}_\Delta\boldsymbol{\alpha}_* & \mathbf{G}_* - \boldsymbol{\Lambda} & \cdots \\ & & & & \vdots & \ddots \end{bmatrix}, \tag{13.13}$$

where $\boldsymbol{\Lambda} = \lambda\mathbf{I}$ (i.e., a matrix with λ's on the diagonal and zeros elsewhere). It should also be noted that $\mathbf{G}_\Delta$ is a *column* vector and $\boldsymbol{\alpha}_*$ is a *row* vector; therefore, the matrix product of $\mathbf{G}_\Delta\boldsymbol{\alpha}_*$ is an m by m matrix, the i-j element of which is the product $G_\Delta(i)\alpha_*(j)$. Our goal is to find the values for the steady-state probabilities denoted by $\mathbf{p} = (p_0|p_{11}, p_{12}, \cdots|p_{21}, p_{22}, \cdots|\cdots) = (p_0, \mathbf{p}_1, \mathbf{p}_2, \cdots)$, where $\mathbf{p}_n = (p_{n1}, p_{n2}, \cdots)$ for $n = 1, 2, \cdots$. In other words, the vector $\mathbf{p}$ is partitioned in the same fashion as the matrix $\mathbf{Q}$. Combining Eq. (13.13) with the equation $\mathbf{pQ} = \mathbf{0}$ yields the following system of equations written using the submatrices from the partitioning of Eq. (13.13):

$$\begin{aligned} -\lambda p_0 + \mathbf{p}_1\mathbf{G}_\Delta &= 0 \\ \lambda\boldsymbol{\alpha}_* p_0 + \mathbf{p}_1(\mathbf{G}_* - \boldsymbol{\Lambda}) + \mathbf{p}_2\mathbf{G}_\Delta\boldsymbol{\alpha}_* &= \mathbf{0} \\ \mathbf{p}_1\boldsymbol{\Lambda} + \mathbf{p}_2(\mathbf{G}_* - \boldsymbol{\Lambda}) + \mathbf{p}_3\mathbf{G}_\Delta\boldsymbol{\alpha}_* &= \mathbf{0} \\ \mathbf{p}_2\boldsymbol{\Lambda} + \mathbf{p}_3(\mathbf{G}_* - \boldsymbol{\Lambda}) + \mathbf{p}_4\mathbf{G}_\Delta\boldsymbol{\alpha}_* &= \mathbf{0} \\ &\vdots \end{aligned} \tag{13.14}$$

The matrix in Eq. (13.13) is from a quasi-birth-death process, so we know the solution to the system given in (13.14) involves the characteristic matrix equation given by

$$\boldsymbol{\Lambda} + \mathbf{R}(\mathbf{G}_* - \boldsymbol{\Lambda}) + \mathbf{R}^2\mathbf{G}_\Delta\boldsymbol{\alpha}_* = \mathbf{0}\,. \tag{13.15}$$

Once the characteristic root matrix $\mathbf{R}$ is found, we can again take advantage of the matrix geometric form for the probabilities as we repeat Eq. (13.4); namely, we have

$$\mathbf{p}_n = \mathbf{c}\mathbf{R}^n\,, \tag{13.16}$$

for $n = 1, 2, \cdots$, where $\mathbf{c}$ is a vector of constants.

Since this is a single-server system, traffic intensity is always given by $\rho = \lambda/\mu$, where $1/\mu$ is the mean service time. Thus, from Property 13.2, the traffic intensity is $\rho = -\lambda\boldsymbol{\alpha}_*\mathbf{G}_*{}^{-1}\mathbf{1}$. Assuming that ρ is less than one, it makes sense to find the matrix $\mathbf{R}$ that satisfies Eq. (13.15). It turns out that for this M/Ph/1 characteristic function, it is possible [1, p. 84] to find a closed form expression to the equation and obtain

$$\mathbf{R} = \lambda(\boldsymbol{\Lambda} - \lambda \mathbf{1}\boldsymbol{\alpha}_* - \mathbf{G}_*)^{-1}, \tag{13.17}$$

where $\mathbf{1}$ is a column vector of ones so that the product $\mathbf{1}\boldsymbol{\alpha}_*$ is a matrix, each row of which is the vector $\boldsymbol{\alpha}_*$.

It can also be shown that the constant vector $\mathbf{c}$ from Eq. (13.16) is given by the vector $p_0\boldsymbol{\alpha}_*$ with $p_0 = 1 - \rho$, where $\rho = \lambda/\mu$. (To see this, we rewrite the second equation from the system of equations (13.14) using Eq. (13.17) as a substitute for $\mathbf{p}_1$ and $\mathbf{p}_2$ and observe that $\lambda\boldsymbol{\alpha}_* = \boldsymbol{\alpha}_*\boldsymbol{\Lambda}$. This yields

$$\begin{aligned}
&\lambda\boldsymbol{\alpha}_* p_0 + \mathbf{p}_1(\mathbf{G}_* - \boldsymbol{\Lambda}) + \mathbf{p}_2\mathbf{G}_\Delta\boldsymbol{\alpha}_* \\
&= \boldsymbol{\alpha}_*\boldsymbol{\Lambda} p_0 + p_0\boldsymbol{\alpha}_*\mathbf{R}(\mathbf{G}_* - \boldsymbol{\Lambda}) + p_0\boldsymbol{\alpha}_*\mathbf{R}^2\mathbf{G}_\Delta\boldsymbol{\alpha}_* \\
&= \boldsymbol{\alpha}_* p_0 \left[\boldsymbol{\Lambda} + \mathbf{R}(\mathbf{G}_* - \boldsymbol{\Lambda}) + \mathbf{R}^2\mathbf{G}_\Delta\boldsymbol{\alpha}_*\right] = 0
\end{aligned}$$

where the final equality is obtained by recognizing that the quantity in the square brackets is the same as Eq. (13.15).) Thus, the M/Ph/1 queueing system with a mean arrival rate of λ and a service distribution defined by parameters m, $\boldsymbol{\alpha}_*$, and $\mathbf{G}_*$ has steady-state probabilities given, for $\rho < 1$, by

$$p_{n,k} = \begin{cases} 1-\rho & \text{for } n = 0, \\ (1-\rho)(\boldsymbol{\alpha}_*\mathbf{R}^n)(k) & \text{for } n = 1, 2, \cdots, \text{and } k = 1, \cdots, m, \end{cases}$$

where $\rho = -\lambda\boldsymbol{\alpha}_*\mathbf{G}_*^{-1}\mathbf{1}$, and $\mathbf{R}$ is defined by Eq. (13.17). It also follows that the expected number of customers in the system and in the queue are given by

$$L = (1-\rho)\boldsymbol{\alpha}_*\mathbf{R}(\mathbf{I} - \mathbf{R})^{-2}\mathbf{1} \tag{13.18}$$

$$L_q = (1-\rho)\boldsymbol{\alpha}_*\mathbf{R}^2(\mathbf{I} - \mathbf{R})^{-2}\mathbf{1} \tag{13.19}$$

Example 13.3. Let us consider an M/Ph/1 queueing system such that the arrival rate is $\lambda = 0.079$ and the distribution function governing the service time random variable is the one illustrated in Fig. 13.2 with generator matrix given by Eq. (13.12). Thus, the mean and standard deviation of the service time is 10.126 and 7.365, respectively. Thus the queueing system has a traffic intensity of 0.8 which implies that a steady-state analysis is possible. The matrix that must be inverted (Eq. 13.17) to obtain $\mathbf{R}$ is

$$\boldsymbol{\Lambda} - \lambda\mathbf{1}\boldsymbol{\alpha}_* - \mathbf{G}_* = \begin{bmatrix} 1.0158 & -0.9000 & -0.0158 & 0.0000 \\ -0.0632 & 0.5790 & -0.5158 & 0.0000 \\ -0.0632 & 0.1000 & 0.3965 & -0.2000 \\ -0.0632 & 0.0000 & -0.0158 & 0.3290 \end{bmatrix},$$

thus, the matrix $\mathbf{R}$ is

$$\mathbf{R} = \begin{bmatrix} 0.1298 & 0.2633 & 0.3563 & 0.2166 \\ 0.0579 & 0.2946 & 0.3951 & 0.2402 \\ 0.0490 & 0.1453 & 0.3999 & 0.2431 \\ 0.0273 & 0.0576 & 0.0876 & 0.2934 \end{bmatrix}.$$

The probability that the system is empty is 0.2; the probability that there is exactly one customer in the system is 0.188. The expected number in the system and the expected number in the queue are 3.24 and 2.44, respectively. As a practice in the use of Little's formula, we can also derive L_q knowing only that $L = 3.24$. With Little's formula we have that $W = 3.24/0.079 = 41.013$. Since the system only has one server, it follows that $W = W_q + 10.126$ or $W_q = 30.887$. Using Little's formula one more time yields $L_q = \lambda W_q = 0.079 \times 30.887 = 2.44$. □

Example 13.4. **Simulation.** In order to provide as much help as possible in understanding the dynamics of a phase-type queueing system, we step through a simulation of Example 13.3. There will be three types of events denoted as follows: "a" signifies an arrival, "p" signifies the completion of a sojourn time within a phase, "q" signifies the movement of a queued entity into the server. (The example simulations of Chap. 9 did not have a separate event for the transfer from the queue to the server, but because of the extra complication of the movement within the server, it is easier to distinguish between the "p" event and the "q" event.) The transformation from the random number, R, in Column 3 of Table 13.1 to the arrival time in Column 4 is

$$\text{Clock Time} - \ln(R)/0.079\,.$$

An arriving entity goes into Phase 1 if the random number in Column 5 is less than or equal to 0.8; otherwise, the entity goes into Phase 3. An entity that is leaving Phase 1 will enter Phase 2 if the random number in Column 5 is less than or equal to 0.9 (as at clock time 13.04 and 28.45 in the table); otherwise, the entity will leave the system (as at clock time 0.03, 6.52, etc.). An entity that is leaving Phase 2 will always proceed into Phase 3. An entity that is leaving Phase 3 will enter Phase 4 if the random number is less than or equal to 0.6; it will enter Phase 2 if the random number is greater than 0.6 and less than or equal to 0.9; otherwise it will leave the system. An entity leaving Phase 4 will always leave the system. The random number in Column 7 is used to determine the random variate representing the sojourn time within a phase; therefore, its transformation depends on the phase in which the entity resides. The simulation results contained in Table 13.1 should contain enough of the dynamics of the system to illustrate these concepts. □

13.4.2 The M/Ph/c Queueing System

There are two approaches for modeling the multiple-server phase-type system: (1) let the state space contain information regarding the current phase number for each customer in each server or (2) let the state space contain information regarding the

Table 13.1 Simulation of an M/Ph/1 queueing system

(1) Clock time	(2) Type	(3) RN	(4) Time next arrival	(5) RN	(6) Into which phase	(7) RN	(8) Within phase time	(9) Time leave phase	(10) No. in sys.
0	a	0.7156	4.24	0.5140	1	0.9734	0.03	0.03	1
0.03	p	—	—	0.9375	out	—	—	—	0
4.24	a	0.7269	8.28	0.2228	1	0.1020	2.28	6.52	1
6.52	p	—	—	0.9991	out	—	—	—	0
8.28	a	0.9537	8.88	0.6292	1	0.4340	0.83	9.11	1
8.88	a	0.2416	26.86	queued	—	—	—	—	2
9.11	p	—	—	0.9946	out	—	—	—	1
9.11	q	—	—	0.6799	1	0.0197	3.93	13.04	1
13.04	p	—	—	0.3988	2	0.4703	1.51	14.55	1
14.55	p	—	—	—	3	0.8872	0.36	14.91	1
14.91	p	—	—	0.5035	4	0.2335	5.82	20.73	1
20.73	p	—	—	—	out	—	—	—	0
26.86	a	0.6576	32.17	0.5989	1	0.2047	1.59	28.45	1
28.45	p	—	—	0.4328	2	0.2419	2.84	31.29	1
31.29	p	—	—	—	3	0.1215	6.32	37.61	1
32.17	a	0.6602	37.43	queued	—	—	—	—	2
37.43	a	0.7957	40.32	queued	—	—	—	—	3
37.61	p	—	—	0.9518	out	—	—	—	2
37.61	q	—	—	0.1589	1	0.5695	0.56	38.17	1

number of customers in each phase. Consider an M/Ph/3 system with a phase-type distribution containing two phases. If we model this system using the first approach, the vector $\mathbf{p}_n$ would be defined by

$$\mathbf{p}_n = (p_{n111}, p_{n112}, p_{n121}, p_{n122}, p_{n211}, p_{n212}, p_{n221}, p_{n222}),$$

for $n \geq 3$. In this case, the state $nijk$ refers to n customers in the system, the customer in the first server being in phase i, the customer in the second server being in phase j, and the customer in the third server being in phase k. If we model the three-server system with a two-phase service distribution using the second approach, the vector $\mathbf{p}_n$ would be defined by

$$\mathbf{p}_n = (p_{n30}, p_{n21}, p_{n12}, p_{n03}),$$

for $n \geq 3$. In this second case, the state nij refers to there being n customers in the system, i customers being in their first phase, and j customers being in their second phase. For a system of c servers and a phase-type distribution of m phases, the first approach yields submatrices containing m^c rows and columns, and the second approach yields submatrices containing $(c+m-1)!/(c!(m-1)!)$ rows and columns. Thus, the advantage of the second approach is that it can yield significantly smaller submatrices, especially when the number of phases is small. However, the first approach can be used to model a system that contains servers with different service time distributions.

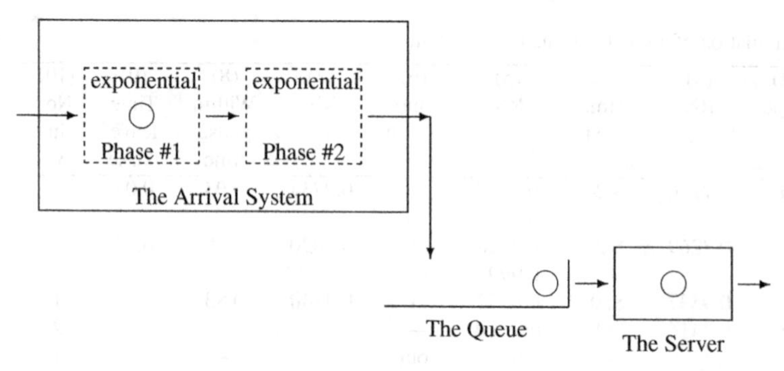

Fig. 13.4 Schematic for the Erlang arrival process showing two customers in the queueing system

13.5 Systems with Phase-Type Arrivals

Another example of a queueing system for which the matrix geometric procedure permits computational results for a problem otherwise difficult is a system with exponential service and phase-type arrivals.

For the phase-type arrival process, there is always a customer in an "arrival" phase. Consider first an Erlang-type 2 system (see Fig. 13.4). A customer (not yet in the queueing system) starts in phase one of the arrival system. After an exponential length of time, the customer moves to phase two. After an additional exponential length of time, the customer leaves phase two of the arrival system and enters the queueing system. As soon as the first customer enters the queueing system, the next customer instantaneously begins phase one in the arrival system.

We now generalize from Erlang distributions to phase-type distributions. Assume for illustrative purposes that the distribution governing interarrival times is a phase-type distribution with parameters $m = 2$, $\boldsymbol{\alpha}_*$, and $\mathbf{G}_*$. Also let the mean service time be $1/\mu$. As soon as a customer leaves the arrival system and enters the queueing system, the next customer begins in phase k of the arrival system with probability $\alpha_*(k)$. The customer moves around the phases according to the rate matrix $\mathbf{G}_*$ and then enters the queueing system upon leaving the arrival system. Such a system gives rise to the following generator matrix:

$$
\begin{array}{c} \\ \mathbf{Q} = \end{array}
\begin{array}{c}
\begin{array}{cccccccc}
(1,0) & (2,0) & (1,1) & (2,1) & (1,2) & (2,2) & \cdots
\end{array} \\
\left[\begin{array}{cc|cc|cc|c}
g_{11} & g_{12} & G_\Delta(1)\alpha_1 & G_\Delta(1)\alpha_2 & & & \\
g_{21} & g_{22} & G_\Delta(2)\alpha_1 & G_\Delta(2)\alpha_2 & & & \\
\hline
\mu & 0 & g_{11}-\mu & g_{12} & G_\Delta(1)\alpha_1 & G_\Delta(1)\alpha_2 & \\
0 & \mu & g_{21} & g_{22}-\mu & G_\Delta(2)\alpha_1 & G_\Delta(2)\alpha_2 & \\
\hline
 & & \mu & 0 & g_{11}-\mu & g_{12} & \cdots \\
 & & 0 & \mu & g_{21} & g_{22}-\mu & \cdots \\
\hline
 & & & & \ddots & \ddots & \ddots
\end{array}\right]
\end{array}
$$

The above representation of the Ph/M/1 system for a two-phase distribution provides the pattern to generalize the generator for the general phase-type distribution. In this case, the state (k,n) indicates that there are n customers in the system, and the customer in the arrival system is in phase k. In matrix notation, we have

$$\mathbf{Q} = \begin{bmatrix} \mathbf{G}_* & \mathbf{G}_\Delta \boldsymbol{\alpha}_* & & & \\ \mu\mathbf{I} & \mathbf{G}_* - \mu\mathbf{I} & \mathbf{G}_\Delta \boldsymbol{\alpha}_* & & \\ & \mu\mathbf{I} & \mathbf{G}_* - \mu\mathbf{I} & \mathbf{G}_\Delta \boldsymbol{\alpha}_* & \\ & & \mu\mathbf{I} & \mathbf{G}_* - \mu\mathbf{I} & \mathbf{G}_\Delta \boldsymbol{\alpha}_* \\ & & & \mu\mathbf{I} & \mathbf{G}_* - \mu\mathbf{I} & \cdots \\ & & & & \vdots & \ddots \end{bmatrix}. \tag{13.20}$$

(We remind you again that the vector product $\mathbf{G}_\Delta \boldsymbol{\alpha}_*$ yields a square matrix whose i-j element is $G_\Delta(i)\alpha_*(j)$.)

The phase-type arrival system does not yield itself to a closed form solution to the characteristic function; therefore, the matrix $\mathbf{R}$ must be obtained numerically. That is, the algorithm of Property 13.1 can be used to find the $\mathbf{R}$ that satisfies the equation

$$\mu\mathbf{R}^2 + \mathbf{R}(\mathbf{G}_* - \mu\mathbf{I}) + \mathbf{G}_\Delta \boldsymbol{\alpha}_* = \mathbf{0}\,. \tag{13.21}$$

After $\mathbf{R}$ is found, the vector $\mathbf{p}_0$ can be found by solving the system of equations given by

$$\begin{aligned} \mathbf{p}_0(\mathbf{G}_* + \mu\mathbf{R}) &= \mathbf{0} \\ \mathbf{p}_0(\mathbf{I} - \mathbf{R})^{-1}\mathbf{1} &= 1\,. \end{aligned} \tag{13.22}$$

This chapter has briefly touched on a rich topic for numerically analyzing queueing systems. There are many further examples of queueing systems that can be profitably analyzed through matrix geometric procedures. There are also many aspects of the matrix geometric method that we have not discussed and interested students should find pursuing this topic rewarding. To help the student gain experience in modeling with these techniques, several further queueing systems are given as homework for which matrix geometric models must be specified.

Problems

To obtain the maximum benefit from these homework problems, it would be best to have a computer available. As a good starting point, you should write a program implementing the Successive Substitution Algorithm. The only difficult part is the matrix inversion procedure. If you do not have a subroutine available that inverts a matrix, the appendix at the end of the book contains a description of an algorithm for matrix inversion.

13.1. Consider the sequence of numbers: 0, 1, 2, 5, 12, 29, $\cdots$.
(a) What is the next number in the series?
(b) What is the characteristic function that defines the general expression for the numbers in the series?
(c) Without writing out the next several numbers, give the tenth number in the series.

13.2. Consider a batch arrival queueing system. The batches arrive according to a Poisson process with mean rate of one per hour, and every batch contains exactly two customers. Service is one-at-a-time according to an exponential distribution with a mean of 16 minutes. The purpose of this problem is to practice the use of (scalar) difference equations, so you should analyze this queueing system without using the matrix geometric concepts.
(a) Form the generator matrix and write out the characteristic function that must be solved to obtain a general expression for p_n. (Note that the characteristic function is a cubic with one root equal to 1.)
(b) Show that

$$p_n = c_1 \left(\frac{2}{3}\right)^n + c_2 \left(-\frac{2}{5}\right)^n$$

for $n = 0, 1, \cdots$.
(c) Show that the probability that there are more than three in the system is 17.6%.
(d) What is the expected number in the system?

13.3. Consider a batch service system where the arrival process is Poisson (i.e., one-at-a-time) with a mean arrival rate of one per hour . The service mechanism is an oven which can hold two items at once and service never starts unless two items are ready for service; in other words, if only one item is in the system, that item will wait in the queue until the second item arrives. The service time is exponentially distributed with a mean batch service rate 0.8284 per hour. Again, this problem is to emphasize the use of the difference equation methodology, so you should analyze this system without using the matrix geometric technique.
(a) Form the generator matrix and write out the characteristic function that must be solved to obtain a general expression for p_n. (Note that the characteristic function is a cubic with one root equal to 1.)
(b) Show that

$$p_n = p_1 (0.7071)^{n-1}$$

for $n = 1, 2, \cdots$.
(c) Show that the probability that there are more than three in the system is approximately 30.2%.
(d) What is the expected number in the system?

13.4. Consider an $M^{[X]}/M/1$ queueing system with a mean arrival rate of 3 batches per hour and a mean service time of 7.5 minutes. The probability that the batch size is one, two, or three is 0.3, 0.5, or 0.2, respectively.
(a) What is the matrix characteristic equation for this system?
(b) Using the algorithm of Property 13.1, determine the matrix **R**.

(c) What is the probability that the queue is empty?
(d) What is the value of L?

13.5. Consider an $M^{[X]}/M/2$ queueing system with a mean arrival rate for batches of λ and a mean service rate of μ. Assume the maximum batch size is two and show how this system can be analyzed using the matrix geometric methodology. In other words, give the relevant matrices for the matrix characteristic equation and the necessary boundary equations.

13.6. Consider the following batch service system. Arrivals occur one-at-a-time with mean rate $\lambda = 1$. The service mechanism is an oven, and after service is started, it cannot be interrupted. If only one item is in the system, that item is placed in the oven, and service is immediately started. If two or more are in the system and the oven is empty, two items are placed in the oven and service is started. The length of time that the oven takes to complete its operation is an exponential random variable with mean rate $\mu = 0.8284$ whether one or two items are in the oven.
(a) Write the generator matrix using a two-dimensional state space where the state (n,k) denotes n items in the *queue* and k items in the oven. Also, give the (cubic) matrix characteristic equation.
(b) Solve for $\mathbf{R}$ analytically, i.e., without using the Successive Substitution Algorithm. (Hint: observe that since all the matrices in the characteristic equation are upper triangular, the matrix $\mathbf{R}$ will be upper triangular; thus

$$\mathbf{R}^3 = \begin{bmatrix} r_{11}^3 & r_{21}^{(3)} \\ 0 & r_{22}^3 \end{bmatrix},$$

where $r_{21}^{(3)}$ can be written in terms of the elements of $\mathbf{R}$, and r_{ii}^3 is the cube of the r_{ii} term.)
(c) What is the probability that the system is empty?

13.7. A two-channel communication node can take two types of calls: data calls and voice calls. Data calls arrive according to a Poisson process with mean rate λ_d and the length of time a data call uses a channel is exponentially distributed with a mean time of $1/\mu_d$. If both channels are busy, there is a buffer that can hold the data calls until a channel is free. The buffer has an unlimited capacity. Voice calls arrive according to a Poisson process with mean rate λ_v and the length of time a voice call uses a channel is exponentially distributed with a mean time of $1/\mu_v$. However, if both channels are busy when a voice call arrives, the call is lost.
(a) This system can be modeled as a quasi-birth-death process. Give the matrix characteristic equation that must be solved to find $\mathbf{R}$, and write out the boundary equations.
(b) Data calls come in at a rate of one per minute and the average service time is one minute. Voice calls come in at a rate of 4 per hour and last an average of 12 minutes. Assume that voice calls are *non-preemptive*. Find the numerical values for the matrix $\mathbf{R}$ analytically instead of with the successive substitution algorithm. (Hint: since the matrices $\mathbf{A}_0, \mathbf{A}_1$, and $\mathbf{A}_2$ are triangular, the matrix $\mathbf{R}$ will also be triangular; thus,

the diagonal elements of $\mathbf{R}^2$ are the square of the diagonal elements of $\mathbf{R}$.)
(c) What is the expected number of data calls in the system?
(d) What is the probability that both channels are occupied by voice calls?
(e) Given that both channels are occupied by voice calls, what is the expected number of data calls in the system?

13.8. Answer the question of the previous problem assuming that the voice call have *preemptive priority*. In other words, if a voice call arrived and both channels were busy with at least one channel being occupied by a data call, then that data call would be placed back in the buffer and the voice call would take over the channel; if both channels were occupied by voice calls, the newly arrived voice call would be lost. (For part b, use the algorithm of Property 13.1 to find the matrix $\mathbf{R}$.)

13.9. Consider an M/M/1 system with a mean arrival rate of 10 per hour and a mean service time of 5 minutes. A cost is incurred for each item that is in the system at a rate of $50 per hour, and an operating cost of $100 per hour is incurred whenever the server is busy. Through partial automation, the variability of the service time can be reduced, although the mean will be unaffected by the automation. The engineering department claims that they will be able to cut the standard deviation of the service time in half. What are the potential savings due to this reduction?

13.10. Develop a model for an M/E_2/2 queueing system.

13.11. Consider the E_2/M/1 model with a mean arrival rate of 10 per hour and a mean service time of 5 minutes.
(a) Solve for the matrix $\mathbf{R}$ analytically using the fact that the first row of $\mathbf{R}$ is zero.
(b) What is the probability that a queue is present?
(c) What is the expected number in the system?

13.12. Derive the formula $L = \mathbf{p}_0\mathbf{R}(\mathbf{I}-\mathbf{R})^{-2}\mathbf{1}$ for the Ph/M/1 queueing system. What is the formula for L_q?

13.13. Consider a Ph/M/2 queueing system with a mean arrival rate of λ and a mean service rate of μ. Show how this system can be analyzed using the matrix geometric methodology. In other words, give the relevant matrices for the matrix characteristic equation and the necessary boundary equations.

13.14. Let $X = \{X_t; t \geq 0\}$ denote a Markov process with state space $E = \{\ell, m, h\}$ and generator matrix

$$G = \begin{bmatrix} -0.5 & 0.5 & 0 \\ 1 & -2 & 1 \\ 0 & 1 & -1 \end{bmatrix}.$$

The process X is called the *environmental* process and the three states will refer to low traffic conditions, medium traffic conditions, and high traffic conditions.

Let $N = \{N_t; t \geq 0\}$ denote a single server queueing system. The arrivals to the system form a Poisson process whose mean arrival rate is dependent on the state of the environmental process. The mean arrival rate at time t is 2 per hour during

periods of low traffic, 5 per hour during periods of medium traffic, and 10 per hour during heavy traffic conditions. (In other words, at time t the mean arrival rate is 2, 5, or 10 per hour depending on X_t being ℓ, m, or h, respectively.) The mean service time is 10 minutes, independent of the state of the environment.
(a) Give the generator matrix for this system, where a state (n,k) denotes that there are n customers in the system and the environmental process is in state k.
(b) Give the matrix characteristic equation and the boundary equations for this system.
(c) Show that this system satisfies the steady-state conditions.
(d) Derive an expression for L, the mean number in the system.
(e) Derive an expression for L_k, the *conditional* mean number in the system given that the environment is in state k.
(f) Use the algorithm of Property 13.1 to obtain numerical values for L and L_k.

13.15. Write a spreadsheet simulation that generates 1,000 random variates of the phase-type distribution used for service times in Example 13.3. Treating the 1,000 random variates as a random sample, obtain a 95% confidence interval for the mean and standard deviation. Compare these intervals with the true values.

References

1. Neuts, M.F. (1981). *Matrix-Geometric Solutions in Stochastic Models*, The Johns Hopkins University Press, Baltimore.

Appendix A
Matrix Review

Within the body of the textbook, we assumed that students were familiar with matrices and their basic operations. This appendix has been written for those students for whom a review would be helpful.

Definition A.1. An $m \times n$ *matrix* is an array of numbers in m rows and n columns as follows:

$$\mathbf{A} = \begin{bmatrix} a_{11} & a_{12} & \cdots & a_{1n} \\ a_{21} & a_{22} & \cdots & a_{2n} \\ \vdots & \vdots & & \vdots \\ a_{m1} & a_{m2} & \cdots & a_{mn} \end{bmatrix}.$$

□

Notice that the row index is always mentioned first, so that the a_{ij} element refers to the element in row i[1] and column j. A matrix that consists of only one row is called a row vector. Similarly, a matrix that consists of only one column is called a column vector.

Definition A.2. If $m = n$, the matrix is a *square matrix* of order n. □

Definition A.3. The *identity matrix* is a square matrix with ones on the diagonal and zeros elsewhere. That is,

$$\mathbf{I} = \begin{bmatrix} 1 & 0 & \cdots & 0 \\ 0 & 1 & \cdots & 0 \\ \vdots & \vdots & & \vdots \\ 0 & 0 & \cdots & 1 \end{bmatrix}.$$

□

The following sections present several useful properties of matrices. The final section of the appendix deals with derivatives of matrices. We include it because

[1] If you have trouble remembering that the row is always indexed first and then the column, repeat the phrase "Rabbits eat Carrots" indicating that R (row) is before C (column).

R.M. Feldman, C. Valdez-Flores, *Applied Probability and Stochastic Processes*, 2nd ed., DOI 10.1007/978-3-642-05158-6,

the topic is not often included in introductory matrix courses; however, we have found it useful for some optimization problems. There are obviously many topics that could be covered that we have chosen to ignore. In particular, some of the numerical analysis topics related to inverses, eigenvectors, and computing matrices of the form $\exp\{\mathbf{A}\}$ would be helpful but are beyond the scope of this text.

A.1 Matrix Addition and Subtraction

Two matrices can be added or subtracted if the number of rows and columns are the same for both matrices. The resulting matrix has the same number of rows and columns as the matrices being added or subtracted.

Definition A.4. The *sum* (*difference*) of two $m \times n$ matrices $\mathbf{A}$ and $\mathbf{B}$ is an $m \times n$ matrix $\mathbf{C}$ with elements $c_{ij} = a_{ij} + b_{ij}$ ($c_{ij} = a_{ij} - b_{ij}$). Thus to obtain $\mathbf{A}+\mathbf{B}$ ($\mathbf{A}-\mathbf{B}$) we add (subtract) corresponding elements. Thus,

$$\mathbf{A}+\mathbf{B}=\mathbf{C}=\begin{bmatrix} a_{11}+b_{11} & a_{12}+b_{12} & \cdots & a_{1n}+b_{1n} \\ a_{21}+b_{21} & a_{22}+b_{22} & \cdots & a_{2n}+b_{2n} \\ \vdots & \vdots & & \vdots \\ a_{m1}+b_{m1} & a_{m2}+b_{m2} & \cdots & a_{mn}+b_{mn} \end{bmatrix}$$

and

$$\mathbf{A}-\mathbf{B}=\mathbf{C}=\begin{bmatrix} a_{11}-b_{11} & a_{12}-b_{12} & \cdots & a_{1n}-b_{1n} \\ a_{21}-b_{21} & a_{22}-b_{22} & \cdots & a_{2n}-b_{2n} \\ \vdots & \vdots & & \vdots \\ a_{m1}-b_{m1} & a_{m2}-b_{m2} & \cdots & a_{mn}-b_{mn} \end{bmatrix}.$$

□

A.2 Matrix Multiplication

Two matrices can be multiplied if the number of columns of the first matrix is equal to the number of rows of the second matrix. The resulting matrix then has the same number of rows as the first matrix and the same number of columns as the second matrix.

Definition A.5. The *product* $\mathbf{AB}$ of an $m \times n$ matrix $\mathbf{A}$ and an $n \times \ell$ matrix $\mathbf{B}$ results in an $m \times \ell$ matrix $\mathbf{C}$ with elements defined by

$$c_{ij} = \sum_{k=1}^{n} a_{ik} b_{kj} = a_{i1} b_{1j} + a_{i2} b_{2j} + \cdots + a_{in} b_{nj}$$

for $i = 1, 2, \cdots, m$ and $j = 1, 2, \cdots, \ell$. □

Note that if $m \neq \ell$, then even though **AB** is possible, the product **BA** makes no sense and is undefined. Even for square matrices of the same order, the product **AB** is *not* necessarily equal to **BA**.

To illustrate, consider a row vector $\boldsymbol{v}$ with three components (i.e., a 1×3 matrix) multiplied with a 3×2 matrix **P**; namely,

$$\boldsymbol{\upsilon} = \boldsymbol{v}\mathbf{P} = [5\ 10\ 20] \begin{bmatrix} 1 & 12 \\ 3 & 13 \\ 7 & 17 \end{bmatrix} .$$

The resulting matrix is a row vector with two elements given by

$$\begin{aligned} \boldsymbol{\upsilon} &= [5 \times 1 + 10 \times 3 + 20 \times 7\ ,\ 5 \times 12 + 10 \times 13 + 20 \times 17] \\ &= [175\ ,\ 530]\ . \end{aligned}$$

However, the product of $\mathbf{P}\boldsymbol{v}$ is *not* possible because the number of columns of **P** is different from the number of rows of $\boldsymbol{v}$. As a further example, let the column vector **f** denote the first column of **P**. The product $\boldsymbol{v}\mathbf{f}$ is a 1×1 matrix, usually written as a scalar as

$$\boldsymbol{v}\mathbf{f} = 175\ ,$$

and $\mathbf{f}\boldsymbol{v}$ is a 3×3 matrix given by

$$\mathbf{f}\boldsymbol{v} = \begin{bmatrix} 5 & 10 & 20 \\ 15 & 30 & 60 \\ 35 & 70 & 140 \end{bmatrix} .$$

A.3 Determinants

The determinant is a scalar associated with every square matrix. Determinants play a role in obtaining the inverse of a matrix and in solving systems of linear equations. We give a theoretical definition of the determinant of a matrix in this section and give a more computationally oriented definition in the next section.

Before presenting the definition of determinants, it is important to understand permutations. Take, for example, the integers 1 through 6; their natural order is (1,2,3,4,5,6). Any rearrangement of this natural order is called a permutation; thus, (1,5,3,4,6,2) is a permutation. For the natural order, each number is smaller than any number to its right; however, in permutations, there will be some numbers that are larger than other numbers to their right. Whenever a number is larger than another one to its right, it is called an inversion. For example, in the permutation (1,5,3,4,6,2) there are 4 inversions; namely, (5,3), (5,4), (5,2), and (6,2). A permutation is called *even* if it has an even number of inversions; it is called *odd* if it has an odd number of inversions.

Definition A.6. The *determinant* of a square matrix **A** of order n is given by

$$|\mathbf{A}| = \sum(\pm)a_{1j_1}a_{2j_2}\cdots a_{nj_n} ,$$

where the summation is over all possible permutations $j_1 j_2 \cdots j_n$ of the set $\{1,2,\cdots,n\}$. The sign is positive (+) if the permutation $j_1 j_2 \cdots j_n$ is even and negative (−) if the permutation is odd. □

A permutation $j_1 j_2 \cdots j_n$ contains exactly one element from each row and each column. Thus, a square matrix of order n has $n!$ permutations. For example, the matrix

$$\mathbf{A} = \begin{bmatrix} a_{11} & a_{12} & a_{13} \\ a_{21} & a_{22} & a_{23} \\ a_{31} & a_{32} & a_{33} \end{bmatrix} .$$

has 3! or 6 permutations and the determinant is given by:

$$|\mathbf{A}| = a_{11}a_{22}a_{33} - a_{11}a_{23}a_{32} + a_{12}a_{23}a_{31} - a_{12}a_{21}a_{33} + a_{13}a_{21}a_{32} - a_{13}a_{22}a_{31}$$

where, for example, $a_{13}a_{22}a_{31}$ has a negative sign because the sequence of the second subscripts $(3,2,1)$ is a permutation with three inversions; namely, (3,2), (3,1), and (2,1). Similarly, the determinant of a square matrix **A** of order 2 is:

$$|\mathbf{A}| = \begin{vmatrix} a_{11} & a_{12} \\ a_{21} & a_{22} \end{vmatrix} = a_{11}a_{22} - a_{12}a_{21} .$$

(In the next section, this 2×2 determinant will form the basic building block on which the determinant for a general $n \times n$ matrix is obtained.)

The main properties of determinants can be summarized as follows:

1. If **A** is a square matrix, then $|\mathbf{A}| = |\mathbf{A}^T|$, where $\mathbf{A^T}$ is the transpose matrix of **A**. The transpose matrix $\mathbf{A^T}$ has elements $a_{ij}^T = a_{ji}$.
2. If any two rows (or columns) of a square matrix **A** are interchanged, resulting in a matrix **B**, then $|\mathbf{B}| = -|\mathbf{A}|$.
3. If any two rows or columns of a square matrix **A** are identical, then $|\mathbf{A}| = 0$.
4. If any row (column) of a square matrix **A** can be written as a linear combination of the other rows (columns), then $|\mathbf{A}| = 0$.
5. If any row or column of a square matrix **A** has all zeros, then $|\mathbf{A}| = 0$.
6. If any row or column of a square matrix **A** is multiplied by a constant k, resulting in a matrix **B**, then $|\mathbf{B}| = k|\mathbf{A}|$.
7. If matrix **B** results from replacing the uth row (column) of a square matrix **A** by that uth row (column) plus a constant times the vth $(v \neq u)$ row (column), then $|\mathbf{B}| = |\mathbf{A}|$.
8. If **A** and **B** are two $n \times n$ matrices, then $|\mathbf{AB}| = |\mathbf{A}||\mathbf{B}|$.
9. If a square matrix **A** of order n is upper or lower triangular, then its determinant is the product of its diagonal elements; in other words, $|\mathbf{A}| = a_{11}a_{22},\cdots,a_{nn}$.

A.4 Determinants by Cofactor Expansion

Evaluating the determinant using Definition A.6 is difficult for medium or large sized matrices. A systematic way of finding the determinant of a matrix is given by the cofactor expansion method. This method evaluates the determinant of a square matrix of order n by reducing the problem to the evaluation of determinants of square matrices of order $(n-1)$. The procedure is then repeated by reducing the square matrices of order $(n-1)$ to finding determinants of square matrices of order $(n-2)$. This continues until determinants for square matrices of order 2 are evaluated.

Definition A.7. The *minor* of a_{ij} (element of the square matrix **A** of order n) is the square matrix $\mathbf{M}_{ij}$ of order $n-1$ resulting from deleting the ith row and the jth column from matrix **A**. □

Definition A.8. The *cofactor* c_{ij} of a_{ij} is defined as $c_{ij} = (-1)^{i+j}|\mathbf{M}_{ij}|$. □

To illustrate, consider the following 3×3 matrix **A**

$$\mathbf{A} = \begin{bmatrix} 1 & 2 & 3 \\ 4 & 5 & 6 \\ 7 & 8 & 9 \end{bmatrix}.$$

The minor of a_{21} is given by

$$\mathbf{M}_{21} = \begin{bmatrix} 2 & 3 \\ 8 & 9 \end{bmatrix},$$

and the cofactor of a_{21} is given by

$$c_{21} = (-1)^{2+1}\,|\mathbf{M}_{21}| = -(2 \times 9 - 3 \times 8) = 6\,.$$

Property A.1. *Let* **A** *be a square matrix of order n with elements a_{ij}. Then the determinant of matrix* **A** *can be obtained by expanding the determinant about the ith row*

$$|\mathbf{A}| = \sum_{j=1}^{n} a_{ij} c_{ij}\,,$$

or by expanding the determinant about the jth column

$$|\mathbf{A}| = \sum_{i=1}^{n} a_{ij} c_{ij}\,.$$

For example, to evaluate the determinant of a square matrix of order 3, the following expansion of the first row could be considered:

$$
\begin{aligned}
|\mathbf{A}| = (-1)^{1+1} a_{11} \begin{vmatrix} a_{22} & a_{23} \\ a_{32} & a_{33} \end{vmatrix} &+ (-1)^{1+2} a_{12} \begin{vmatrix} a_{21} & a_{23} \\ a_{31} & a_{33} \end{vmatrix} \\
&+ (-1)^{1+3} a_{13} \begin{vmatrix} a_{21} & a_{22} \\ a_{31} & a_{32} \end{vmatrix} \qquad (A.1) \\
&= a_{11}(a_{22}a_{33} - a_{23}a_{32}) - a_{12}(a_{21}a_{33} - a_{23}a_{31}) \\
&\quad + a_{13}(a_{21}a_{32} - a_{22}a_{31}) \, .
\end{aligned}
$$

Notice that the determinant is the same if an expansion by any other row or column is used.

The cofactor expansion method can be used recursively so that the original $n \times n$ determinant problem is reduced to computing determinants of 2×2 matrices. For example, a square matrix of order 4 could be calculated by

$$
|\mathbf{A}| = a_{11}|\mathbf{M}_{11}| - a_{12}|\mathbf{M}_{12}| + a_{13}|\mathbf{M}_{13}| - a_{14}|\mathbf{M}_{14}| \, .
$$

Since each of the determinants above involves a 3×3 matrix, Eq. (A.1) can then be used to evaluate each $|\mathbf{M}_{ij}|$ using determinants of 2×2 matrices.

A.5 Nonsingular Matrices

Another important concept in the algebra of matrices is that of singularity. Square matrices can be classified as singular or nonsingular. Nonsingular matrices represent a set of linearly independent equations, while singular matrices represent a set where at least one row (column) can be expressed as a linear combination of other rows (columns).

Definition A.9. A square matrix $\mathbf{A}$ of order n that has a multiplicative inverse is called a *nonsingular matrix*. In other words, if $\mathbf{A}$ is nonsingular, there exists a square matrix, denoted by $\mathbf{A}^{-1}$, of order n such that $\mathbf{A}\mathbf{A}^{-1} = \mathbf{A}^{-1}\mathbf{A} = \mathbf{I}$. If no inverse exists, the matrix is called *singular*. For a nonsingular matrix, the matrix $\mathbf{A}^{-1}$ is called the *inverse* of $\mathbf{A}$. □

Property A.2. *If* $\mathbf{A}$ *is nonsingular, the matrix* $\mathbf{A}^{-1}$ *is unique, and the determinants of* $\mathbf{A}$ *and* $\mathbf{A}^{-1}$ *are nonzero. Furthermore,* $|\mathbf{A}^{-1}| = 1/|\mathbf{A}|$.

From the above property, it is easy to see that a matrix is nonsingular if and only if the determinant of the matrix is nonzero. For example, the following Markov matrix is singular

$$\mathbf{P} = \begin{bmatrix} 0.1 & 0.3 & 0.6 & 0.0 \\ 0.0 & 0.2 & 0.5 & 0.3 \\ 0.0 & 0.1 & 0.2 & 0.7 \\ 0.1 & 0.3 & 0.6 & 0.0 \end{bmatrix},$$

because the determinant of **P** is zero. (Use the fourth property of determinants listed on p. 384.)

A.6 Inversion-in-Place

When a square matrix is nonsingular and the elements of the matrix are all numeric, there are several algorithms that have been developed to find its inverse. The best known method for numerical inversion is the Gauss-Jordan elimination procedure. Conceptually, the Gauss-Jordan algorithm appends an identity matrix to the original matrix. Then, through a series of elementary row operations, the original matrix is transformed to an identity matrix. When the original matrix is transformed to an identity, the appended identity matrix is transformed to the inverse. This method has the advantage that it is intuitive, but has the disadvantage that the procedure is wasteful of computer space when it is implemented.

Methods that use memory more efficiently are those that use a procedure called inversion-in-place. The operations to perform inversion-in-place are identical to Gauss-Jordan elimination but savings in storage requirements are accomplished by not keeping an identity matrix in memory. The inversion-in-place requires a vector with the same dimension as the matrix to keep track of the row changes in the operations. This is necessary because the identity matrix is not saved in memory and row interchanges for the inversion procedure need to be reflected as column changes in the implicitly appended identity matrix.

We shall first give an overview of the mathematical transformation required for the inversion-in-place procedure and then give the specific steps of the algorithm. The algorithm uses a pivot element on the diagonal, denoted as a_{kk}, and transforms the original matrix using the elements on the pivot row and pivot column. Let a_{ij} denote an element in the original matrix, and let b_{ij} denote an element in the transformed matrix. The equations describing the transformation are:

pivot element

$$b_{kk} = 1/a_{kk}$$

pivot row

$$b_{kj} = a_{kj}/a_{kk} \qquad \text{for } j = 1,2,\cdots,k-1,k+1,\cdots,n$$

pivot column

$$b_{ik} = -a_{ik}/a_{kk} \qquad \text{for } i = 1,2,\cdots,k-1,k+1,\cdots,n$$

general transformation

$$b_{ij} = a_{ij} - a_{kj}a_{ik}/a_{kk} \text{ for } \begin{cases} i = 1,2,\cdots,k-1,k+1,\cdots,n \\ j = 1,2,\cdots,k-1,k+1,\cdots,n \end{cases}$$

The above equations are applied iteratively to the matrix for each $k = 1, \cdots, n$. After finishing the transformation using pivot element a_{kk}, the matrix (a_{ij}) is replaced by the matrix (b_{ij}) and the value of k is incremented by one. Notice, however, that the iteration could be done in place. That is, the equations presented above could be written with a_{ij}'s instead of b_{ij}'s, as long as the operations are performed in the correct sequence.

The only additional problem that may arise is when the pivot element, a_{kk}, is zero. In such a case, the equations cannot be applied since the pivot element is in the denominator. To solve this problem, the pivot row is interchanged with a row that has not yet been pivoted. (If a suitable row for the pivot cannot be found for an interchange, the matrix is singular.) The interchange must be recorded so that when the iterations are finished, the inverted matrix can be reordered.

To diminish round-off error, pivots are selected as large as possible. Thus, the inversion-in-place algorithm interchanges rows whenever the current candidate pivot is smaller in absolute value than other candidate pivots. In the description of the inversion-in-place algorithm, the following conventions will be followed: the matrix **A** is a square matrix of order n; the index k indicates the current row; the index *irow* indicates the row containing the maximum pivot element; the vector *newpos* is of dimension n, where $newpos(k)$ is the row used for the kth pivot; *piv* is the reciprocal of the pivot element; and *temp* is a temporarily held value. It should also be emphasized that because this is an inversion-in-place algorithm, as soon as a value of the matrix a_{ij} is changed, the next step in the algorithm will use the changed value.

Step 1. For each $k = 1, \cdots, n$ do **Steps 2** through **6**.

Step 2. Let *irow* be the index in the set $\{k, k+1, \cdots, n\}$ that gives the row that has the largest element, in absolute value, of column k; that is,

$$irow = \text{argmax}\{|a_{ik}| \ : \ i = k, \cdots, n\} .$$

(Remember, k is fixed.)

Step 3. If the largest element, in absolute value, on the kth column of rows $k, k+1, \cdots, n$ is zero, the matrix is singular, and the program should stop. In other words, if $|a_{irow,k}| = 0$, stop, no inverse exits.

Step 4. Update the vector that keeps track of row interchanges by setting $newpos(k) = irow$.

Step 5. If the pivot row, *irow*, is not equal to the current row, k, interchange rows k and *irow* of the matrix.

Step 6. Apply the operations for inversion-in-place using the following sequence:

Operation A. Set $piv = 1/a_{kk}$, and then set $a_{kk} = 1$.

Operation B. For each $j = 1, \cdots, n$ do

$$a_{kj} = a_{kj} \times piv$$

Operation C. For each $i = 1, 2, \cdots, k-1, k+1, \cdots, n$ do

Operation C.i. Set $temp = a_{ik}$, and then set $a_{ik} = 0$.

Operation C.ii. For $j = 1, \cdots, n$ do

$$a_{ij} = a_{ij} - a_{kj} \times temp \,.$$

Step 7. For each $j = n, n-1, \cdots, 1$ check to determine if $j \neq newpos(j)$. If $j \neq newpos(j)$, then interchange column j with column $newpos(j)$.

Step 3 of the algorithm is to determine if the original matrix is nonsingular or singular. When implementing the algorithm on a computer, some error tolerance needs to be specified. For example, set $\varepsilon = 0.0001$ and stop the algorithm if the absolute value of the pivot element is less than ε. (It is much better to check for a value that is "too" small than to check for exact equality to zero.) The inverse of the matrix is obtained after Steps 2 through 6 of the algorithm are completed for each k, but the columns may not be in the correct order. Step 7 is a reordering of the matrix using the information in the vector *newpos*. Notice that the reordering must be done in reverse order to "undo" the row changes made in Step 5.

If a matrix is upper or lower triangular, the inversion-in-place algorithm can be made much more efficient. (An upper triangular matrix is one in which all elements below the diagonal are zero; that is $a_{ij} = 0$ if $i > j$. A lower triangular matrix has zeros above the diagonal.) **For a triangular matrix**, make the following modifications to the inversion-in-place algorithm.

1. Check for any zeros on the diagonal. If there are any zeros on the diagonal, stop because the matrix is singular. If there are no zeros on the diagonal, the matrix is nonsingular (Use the ninth property of determinants listed on p. 384.)
2. The vector *newpos* is not needed and steps 2 through 5 can be skipped.
3. For an upper triangular matrix, the instruction for Operation B should be "For each $j = k, k+1, \cdots, n$ do". For a lower triangular matrix, the instruction for Operation B should be "For each $j = 1, \cdots, k$ do".
4. For an upper triangular matrix, the instruction for Operation C should be "For each $i = 1, \cdots, k-1$ do"; for $k = 1$, Operation C is skipped. For a lower triangular matrix, the instruction for Operation C should be "For each $i = k+1, \cdots, n$ do"; for $k = n$, Operation C is skipped.

Consider the following example where the inverse of a square matrix of order 3 is to be obtained using the inversion-in-place algorithm.

$$\begin{bmatrix} 2.0 & 4.0 & 4.0 \\ 1.0 & 2.0 & 3.0 \\ 2.0 & 3.0 & 2.0 \end{bmatrix}$$

We begin by looking at row 1. Because the first element of the first row is the largest element in the first column, there is no need to do row interchanges. (In other words, $newpos(1) = 1$.) The new matrix after performing Operations A through C is:

$$\begin{bmatrix} 0.5 & 2.0 & 2.0 \\ -0.5 & 0.0 & 1.0 \\ -1.0 & -1.0 & -2.0 \end{bmatrix}$$

We next look at the elements in column 2 starting at row 2, and we observe that the largest element, in absolute value, among the candidate values is the a_{32} element. (Note that $|a_{12}| > |a_{32}|$ but because a pivot operation has already been performed on row 1, the element in row 1 is not a candidate value.) Thus, $irow = 3$ and $newpos(2) = 3$ which indicates that row 2 and row 3 must be interchanged before any transformation can be performed. This interchange results in:

$$\begin{bmatrix} 0.5 & 2.0 & 2.0 \\ -1.0 & -1.0 & -2.0 \\ -0.5 & 0.0 & 1.0 \end{bmatrix} .$$

After performing Operations A through C in the matrix above we have:

$$\begin{bmatrix} -1.5 & 2.0 & -2.0 \\ 1.0 & -1.0 & 2.0 \\ -0.5 & 0.0 & 1.0 \end{bmatrix} .$$

The final pivot element is a_{33}. The resulting, matrix after performing Operations A through C, is:

$$\begin{bmatrix} -2.5 & 2.0 & 2.0 \\ 2.0 & -1.0 & -2.0 \\ -0.5 & 0.0 & 1.0 \end{bmatrix} .$$

The matrix has been inverted. However, the second and third rows were interchanged during the inversion process; thus the columns in the above matrix cannot be in the correct order. The inverse of the original matrix is thus obtained by "undoing" the row interchanges by changing the corresponding columns on the last matrix. (Or, to be more pedantic, we first observe that $newpos(3) = 3$ so no change is necessary to start with; $newpos(2) = 3$ so columns 2 and 3 must be interchanged; finally, $newpos(1) = 1$ so no change is necessary for the last step.) The inverse to the original matrix is thus given by:

$$\begin{bmatrix} -2.5 & 2.0 & 2.0 \\ 2.0 & -2.0 & -1.0 \\ -0.5 & 1.0 & 0.0 \end{bmatrix} .$$

A.7 Derivatives

Matrix equations will sometimes occur in optimization problems so that it may be necessary to take derivatives of expressions involving matrices. To illustrate, suppose that the elements of a matrix $\mathbf{Q}$ have been written as a function of some parameter, for example:

$$\mathbf{Q} = \begin{bmatrix} 1-\theta & \theta \\ 0 & 1-\theta^2 \end{bmatrix} . \tag{A.2}$$

The derivative of a matrix is taken to mean the term-by-term evaluation of the matrix:

$$\frac{d\mathbf{Q}}{d\theta} = \begin{bmatrix} -1 & 1 \\ 0 & -2\theta \end{bmatrix}.$$

The difficulty comes when the inverse of a matrix is part of an objective function. For example a common matrix expression when dealing with Markov chains is the matrix **R** whose (i, j) term gives the expected number of visits to State j starting from State i. Assume that for the matrix of Eq. (A.2), $\theta < 1$ and the matrix represents the transition probabilities associated with the transient states of a Markov chain. Then the matrix giving the expected number of visits to the two states is

$$\begin{aligned} \mathbf{R} &= \left(\begin{bmatrix} 1 & 0 \\ 0 & 1 \end{bmatrix} - \begin{bmatrix} 1-\theta & \theta \\ 0 & 1-\theta^2 \end{bmatrix} \right)^{-1} \\ &= \frac{1}{\theta^2} \begin{bmatrix} \theta & 1 \\ 0 & 1 \end{bmatrix}. \end{aligned}$$

As long as **R** can be written analytically, it is usually not difficult to give its derivative. In this case it would be

$$\frac{d\mathbf{R}}{d\theta} = -\frac{1}{\theta^3} \begin{bmatrix} \theta & 2 \\ 0 & 2 \end{bmatrix}.$$

However, when the matrix is larger than a 2×2, it may be impossible to express **R** analytically; therefore, we give the following property from [1] by which the derivative can be calculated for the more general case.

Property A.3. *Let* **Q** *be a matrix that is a function of the parameter* θ *and assume the potential of* **Q***, defined by*

$$\mathbf{R} = (\mathbf{I} - \mathbf{Q})^{-1},$$

exists. Then the derivative of **R** *with respect to the parameter is given by*

$$\frac{d\mathbf{R}}{d\theta} = \mathbf{R} \frac{d\mathbf{Q}}{d\theta} \mathbf{R}.$$

It is interesting to observe that the matrix result is similar to the scalar result: let $f(t) = (1 - x(t))^{-1}$, then $\frac{df}{dt} = f \frac{dx}{dt} f$.

References

1. See Hocking, R.R. (1985). *The Analysis of Linear Models*, Brooks/Cole Publishing Co., Monterey, CA.

Index